ADVANCE PRAISE FOR

"This book is as rich in scientific knowledge as it is passionate about scientific history and the complexity of plant life on earth. With the aim to merge various concepts in plant ecology, the thought-provoking framework it provides will certainly stimulate research in holistic plant functioning. Furthermore, it is simply an absolute pleasure to read!"

Joana Bergmann, PhD, Leibniz Centre for Agricultural Landscape Research (ZALF), Germany.

"A stout-hearted trek through the rugged landscape of plant ecological strategies, species traits, and how they translate into demographic success in some settings but not others. The writing is energetic and richly illustrated; Laughlin must be a lively teacher! An excellent read for research students and discussion groups."

Mark Westoby, Professor Emeritus, Department of Biological Sciences, Macquarie University, Sydney, Australia.

"This book is remarkable for the enthusiastic treatment of not only the critique of published ideas about plant strategies, but also the wide range of studies that underpin those ideas. The author seems equally at home as he reviews relevant findings (and gaps in understanding) in areas as different as plant morphology and physiology on the one hand and demography, evolution, and game theory on the other."

Peter Grubb, Professor Emeritus, Department of Plant Sciences, University of Cambridge, UK.

"The mission to predict the responses of plant communities to climate change is 'difficult, immense, and at times impossible', but Laughlin acts as a friendly and experienced guide to accompany you on that challenging journey. This accessible book is the first to explain how differences in the functional traits of plant species scale to different demographic patterns at the level of the community, which in turn has implications for ecosystem restoration and conservation. Step by step, the reader is patiently guided and encouraged into conceptual thinking in plant ecology, inevitably leading to new hypotheses and ways of testing them in plants at different ecological scales. After reading this book, you will never again consider plants as 'dull organisms', but instead as living entities capable of taking a multitude of strategic decisions, which in turn deserve our attention and care.

I specifically admire the tone of the book. When reading, I 'hear' the caring teacher sharing both the concepts and the personal stories beyond the ecology; the true excitement of the botanist discovering the secrets of plants; the experienced scientist leading us to the state-of-art of this field, and the connecting optimist believing that it is still possible to restore the plant diversity in our ecosystems. This is hardcore science, but in an encouraging and light hearted tone. This book is for the new (and old!) generations of students in plant science and ecology! I am deeply impressed."

Liesje Mommer, Professor of Plant Ecology and Nature Conservation, Wageningen University, The Netherlands.

PLANT STRATEGIES

Plant Strategies

The Demographic Consequences of Functional Traits in Changing Environments

Daniel C. Laughlin

University of Wyoming

OXFORD
UNIVERSITY PRESS

OXFORD
UNIVERSITY PRESS

Great Clarendon Street, Oxford, OX2 6DP,
United Kingdom

Oxford University Press is a department of the University of Oxford.
It furthers the University's objective of excellence in research, scholarship,
and education by publishing worldwide. Oxford is a registered trade mark of
Oxford University Press in the UK and in certain other countries

© Daniel C. Laughlin 2023

The moral rights of the author have been asserted

All rights reserved. No part of this publication may be reproduced, stored in
a retrieval system, or transmitted, in any form or by any means, without the
prior permission in writing of Oxford University Press, or as expressly permitted
by law, by licence or under terms agreed with the appropriate reprographics
rights organization. Enquiries concerning reproduction outside the scope of the
above should be sent to the Rights Department, Oxford University Press, at the
address above

You must not circulate this work in any other form
and you must impose this same condition on any acquirer

Published in the United States of America by Oxford University Press
198 Madison Avenue, New York, NY 10016, United States of America

British Library Cataloguing in Publication Data

Data available

Library of Congress Control Number: 2022951490

ISBN 978–0–19–286794–0
ISBN 978–0–19–286795–7 (pbk.)

DOI: 10.1093/oso/9780192867940.001.0001

Printed and bound by
CPI Group (UK) Ltd, Croydon, CR0 4YY

Cover image: "My Country" by
Ada Pula Beasley (Ampilatwatja Art Centre), with permission

Links to third-party websites are provided by Oxford in good faith and
for information only. Oxford disclaims any responsibility for the materials
contained in any third party website referenced in this work.

For mom and dad,
who passed on their propensity
to fill large wooden bookshelves

The journey is difficult, immense, at times impossible, yet that will not deter some of us from attempting it . . . we will travel as far as we can, but we cannot in one lifetime see all that we would like to see or learn all that we hunger to know.

—Loren Eiseley (1957, p. 12), *The Immense Journey*

Contents

Part III. Comparative Functional Ecology

Part V. The Effect of Traits on Demographic Rates

Preface

Whether anything is to be gained by trying to fit more complex strategy schemes to the
life history variation observed in plants remains to be seen. What does seem certain
is that plant ecologists will continue to search for them.

—Silvertown and Charlesworth (2001, p. 300), *Introduction to Plant Population Biology*

This is a book about how plants make a living. Some plants are gamblers, others are swindlers. Some plants are elaborate spenders while others are strugglers or miserly savers (During et al., 1985; Oldeman and Van Dijk, 1991). Plants have evolved a remarkable array of adaptive solutions to the existential problem of survival and reproduction in a world where disturbances can be deadly, resources are scarce, and competition is cutthroat.

Surviving on Earth is not a trivial accomplishment. The odds are very much stacked against plants. Consider that the chance a seedling successfully establishes from a seed can be less than one in a thousand, and the chance that this seedling reaches reproductive maturity can also be less than one in a thousand. This means that a seed has a one in a million chance of growing big enough to make more seeds.[1] Not many would bet on those odds, but I guarantee that anyone who bets against plants are sure to find themselves on the dole. Next time you wander through wild fields and ancient forests immersed in swards of grass, exaltations of wildflowers, and cathedrals of trees, remind yourself that every single individual wild plant succeeded despite everything that usually goes wrong 99.999% of the time. When you see a wild plant, you are witnessing something altogether improbable, yet despite these bafflingly low odds, it is rare to find even a square meter of land unoccupied by plants.

Against the odds, vegetation thrives because plants improve their odds by rolling loaded dice (Shipley, 2010). Plants have inherited phenotypic traits that increased their chance of success and these traits are indicators of strategies for establishment and survival. A plant strategy can be thought of as "how a species sustains a population" (Westoby, 1998, p. 214) because all successful strategies must have positive demographic outcomes in the habitats to which they are adapted. That is not to say that adaptation is perfection, but it is certainly better than average (Niklas, 1997).[2]

Few topics have both captured the imagination and furrowed the brows of ecologists than plant strategies. Yet no topic is more important for understanding the assembly

---

[1] Estimates of recruitment vary widely across species, but these numbers are within the range reported from studies in tropical and subtropical rain forests (Van Valen, 1975; Terborgh et al., 2014; Chang-Yang et al., 2021).

[2] A general definition of plant strategies is developed in Chapter 1.

of plant communities, predicting plant responses to global change, and enhancing the restoration of our degrading biosphere. Why, you may ask, am I perpetuating the use of the word "strategy" when it carries teleological baggage?[3] First, the word strategy inspires the imagination, much more than "plant functional type." It is also "old, and intuitive to every child" (Wilson et al., 2019, p. 241). Second, the term has been debated in the past, but after much malignment the word has simply stuck (Grubb, 1980). The word "strategy" was strongly criticized by John Harper in his essay *After Description* (Harper 1982), but by my own count, he used it at least ten times in his landmark monograph *Population Biology of Plants* (Harper, 1977).[4] Third, the word strategy implies that plants are playing an evolutionary game that they can either win or lose, where the prize for winning is the ability to keep playing and the cost of losing is extinction (Vincent and Brown, 2005). Strategies are integral to game theoretical approaches, which play a pivotal role in the development of modern plant strategy theory and our understanding of trait evolution in competitive milieus.

The vast array of plant strategy models now require synthesis. These models tend to emphasize either life history strategies based on demography, or functional strategies based on phenotypic traits (Garnier et al., 2016). The most recent monograph about functional strategies entirely ignored life history traits (Craine, 2009; Enquist, 2010). This lack of integration is nothing new. Indeed, the disciplinary divide between demography and functional ecology runs deep and the intellectual battles between these two camps have jumped off the pages of the literature over the last 40 years.

Harper criticized descriptive analyses of strategies that were not grounded in the demographic realities of life and death: "The detailed analysis of proximal ecological events is the only means by which we can reasonably hope to inform our guesses about the ultimate causes of the ways in which organisms behave" (Harper, 1982, p. 23). In response, Grime wrote "[What is] particularly harmful, in my view, is Harper's insistence upon the detailed study of proximal events in the field in contemporary populations as the only reliable way to gain a general understanding of vegetation processes. In golfing terms, this is equivalent to 'putting from the tee'" (Grime, 1984, p. 17), a sporting reference to the futility of explaining vegetation dynamics through reductionist methods that rely on endless counting of individuals and modules. Fortunately, this disciplinary divide has been breaking down (e.g., Bazzaz, 1996; Ackerly et al., 2000; Kimball et al., 2012; Moles, 2018; Rüger et al., 2018; Volaire, 2018; Kelly et al., 2021), but silos remain. Progress could be galvanized if demographers had more interest in physiology and if functional ecologists were more interested in demography.

[3] My own copy of the first edition of Grime (1979) has many hand-written notes in the margins from previous owners. One previous owner placed quotation marks around the word "strategies" in the heading of the first chapter—in red pen—clearly agitated by use of the word.
[4] You can find them yourself on pages 30, 82, 106, 107, 632, 651, 653, 690, 703, and 761. Or you can read his paper on reproductive strategies (Harper and Ogden, 1970). Harper disliked many other words that could potentially be ill-defined and used carelessly, especially "vegetative reproduction," "adaptation," and "stress," but all these words have remained in circulation. According to Turkington (2010), on Harper's fiftieth birthday, a few of his witty students posted an advertisement for a seminar titled "Vegetative Reproduction as an Adaptive Strategy in Stressful Environments," and the listed speaker was J. L. Harper!

In his writing, Grime often referred to the concept of "traits conferring fitness" (Grime, 2001, p. 65–6), yet surprisingly never proposed how to demonstrate the link between traits and demographic fitness. This singular phrase holds the key to unification. Functional ecologists measure traits, and demographers measure fitness as the per capita population growth rate. In this book I propose that it is only by screening traits and fitness across multiple species in contrasting environments can we conduct robust tests of plant strategy theory. After all, plant strategies should be "defined in reference to both demographic criteria and those features of life-history, physiology, and biochemistry that determine the responsiveness of plants to soils, land-use, and climatic factors" (Grime et al., 1997a, p. 123). This has long been the goal of many ecologists and evolutionary biologists. James White wrote that "the time has come to indicate that the separation between investigations of the demography and sociology of plants in natural vegetation is artificial, unnecessary and undesirable" (White, 1985b, p. 2). Jonathan Silvertown concurred, writing that "plant demography could benefit from a comparative approach, and any viable theory of life-history strategies must ultimately be founded on demography" (Silvertown et al., 1992, p. 135). Can we turn these visions into realities?

This book aims to articulate a coherent framework for studying plant strategies that unifies demography with functional ecology to advance prediction in plant ecology. Central to this framework are functional traits: the heritable morphological, physiological, and phenological attributes of plants that influence demography and therefore drive fitness differences among species. Global compilations of both demographic models and functional traits within the last decade are making this new synthesis possible (Díaz et al., 2016; Salguero-Gómez et al., 2016). Armed with a deeper understanding of the dimensionality of life history and functional traits, we are equipped to quantitatively link phenotypes to population growth rates across gradients of resource availability and disturbance regimes. I emphasize traits measured at the species level because they represent evolutionary experiments in phenotypic variation that we can use to empirically test for the existence of strategies. Within-species intraspecific variation is clearly important, otherwise natural selection would grind to a halt, but I have consciously chosen to focus on species-level variation because for most traits variation among species exceeds variation within species, and variation across 350,699 species is probably enough variation for one book.[5]

Part I introduces foundational concepts. In Chapter 1, I define the scope of plant strategy theory by emphasizing a contrast in perspectives and approaches that have been taken since the very conception of plant ecology as a discipline. While simple models have remained the most popular, more complicated models get us closer to explaining reality. Chapter 2 synthesizes the historical development of plant strategy models that have tended to focus on either life-history strategies based on demography or functional strategies based on phenotypic traits (Garnier et al., 2016). After admiring this menagerie of models, Chapter 3 articulates the dimensionality of plant strategy theory,

---

[5] This is the number of accepted species names according to theplantlist.org on October 5, 2022. Note that Harper (1977, p. 649) was a fierce critic of treating species populations as the level of study, whereas I agree with Grime that it is the most appropriate level of taxonomic organization.

which is the minimum number of dimensions required to explain the variation in plant form and function. The theory is defined by considering fitness, demographic rates, life history and functional traits, temperature, resource limitation, disturbance regimes, density dependence, and frequency dependence.

Part II explains the demographic perspective. There is no better place to begin other than plant population models, for all successful plant strategies must exhibit positive demographic outcomes. Chapter 4 describes the quantitative tools for modeling population dynamics, including matrix models, sensitivity analysis, and integral projection models. Chapter 5 explores plant life-history theory that provides a framework for understanding schedules of births and deaths and life-history trade-offs.

Part III explains the comparative approach to plant functional ecology. Chapter 6 explores the quantification of the multidimensional plant phenotype with a focus on roots, belowground storage organs, stems, leaves, and seeds. I propose five leading dimensions of plant strategies. In Chapter 7 I discuss their effects on fitness across each dimension of resource availability and disturbance regimes.

Part IV explains how to quantitatively link traits to fitness. Chapter 8 describes a modeling framework to quantify the net effect of traits on fitness under density-independent and density-dependent conditions. Chapter 9 describes how game theoretical approaches are used to predict the net effect of traits on fitness under frequency dependence, when the fitness of a strategy depends on the other strategies in the community. Chapter 10 explores how plant strategies can be applied to enhance conservation of declining species and restoration of degraded habitats.

Part V articulates a mechanistic framework to estimate fitness through trait-based demographic rates. Chapter 11 summarizes a large body of empirical and theoretical work that links traits to relative growth rates and survival. Chapter 12 illustrates how much more work needs to be done to understand the traits that promote recruitment. The reproduction kernel in population models is the most challenging demographic rate to measure and model, but recent advances have been made.

This book was written with graduate students of ecology and evolution in mind. Each chapter ends with a set of questions that may spark meaningful discussion in graduate seminars. One chapter can be read each week, but you might consider breaking up Chapters 2 and 6 into two-week sections, given their length and breadth. Predicting the fates of populations and communities using traits is famously viewed as the Holy Grail of ecology (Lavorel and Garnier, 2002; Lavorel et al., 2007; Funk et al., 2017), yet we sometimes lack clear evidence that functional traits live up to the hype. The time is right for a new synthesis. Shall we begin?

Prologue

You can imagine the topic of conversation when botanists from three countries share a meal together. Ruth and Aven Nelson, botanists at the University of Wyoming, were hosting two European guests at their cozy home in Laramie and this was their first evening together. The Nelsons were regaling their visitors with stories of excursions deep in the backcountry of the Rocky Mountains. Each was trying to outdo the other, tale after tale. Well into the evening Ruth decided it was time to tell the story about her wild day in the alpine when she was forced to use her wooden plant press as a shield against a violent hailstorm just below the Keyhole on Longs Peak in Colorado. This adventure was just about to win best of show when suddenly Aven pulled his best story out of the vault, recalling the day he and his team were hauling wagon loads of fresh herbarium specimens out of Yellowstone National Park.

"We were just about ready to go when a pair of male grizzlies wandered into camp," Aven told them. "The bears frightened my horses away and I couldn't catch them in time. All my 1,400 specimens were stranded in the wilderness! And to top that off, my field assistant stepped into scalding geothermic mud and severely burned his ankles." Shameful laughter mixed with disbelief erupted. After the ruckus calmed down and the conversation settled into gentle banter, Ruth heard the beckoning call of a vesper sparrow outside an open window, reminding her of their evening plans. Even though their guests probably could have used a quiet evening to recover from their long journeys by boat and train, they were much too eager to wander in the prairie, where the mountain flora mingles with the plants of the plains (Figure P.1).

The first to arrive was Leonty Grigoryevich Ramenskii, who travelled all the way from Voronezh State University in Russia. Leonty felt immediately at home in the grasslands of the high plains, which reminded him of the vast steppe of his mother country. The second to arrive on the very next train into town was Carl Georg Oscar Drude. Oscar was the Director of the Botanical Gardens at Dresden University of Technology in Germany and was also eager to explore the flora of the Rocky Mountain ecoregion. It was their first time to the American West.

"Shall we head off?" Ruth asked.

They scraped their plates, grabbed their notebooks, and set off toward Jacoby Ridge to the east. They walked up dry creek beds of mixed alluvium, along limestone ledges of the Casper Formation, and up and over ancient shale beds strewn with yucca (*Yucca glauca*). The guests made chivalric attempts to offer Ruth assistance when the terrain appeared too steep and rocky, but she waved them off, "Thank you, I can manage better on my own," she said politely, hiding her annoyance. "I can walk through this landscape blindfolded," Ruth declared, as she took the lead. She was a botanical leader in the region and had catalogued scores of plant species growing in remote regions along the Front

Figure P.1 *Mixed grass prairie on Jacoby Ridge in the Laramie Basin, Wyoming, with an early summer thunderstorm in the distance. Fuzzy beardtongue (*Penstemon eriantherus*), left, and two-lobed larkspur (*Delphinium nuttallianum) in bloom amidst the black sage (*Artemisia nova*) and cool-season bunchgrasses.*
Photo by the author.

Range of the Rockies. She was working on a new edition of her celebrated *Handbook of Rocky Mountain Plants* (Nelson, 1979), which kept her in the field on a regular basis. "Here we are, this is what we came here for," she said.

The bunchgrasses were the first to introduce themselves. They were surrounded by dense swards of needle-and-thread (*Hesperostipa comata*), a grass with impressively long awns, and many other tussock-forming graminoids appeared at their feet, including June grass (*Koeleria macrantha*) and Idaho fescue (*Festuca idahoensis*).

"*Stipa, Koeleria,* and *Festuca!*" proclaimed a contented Leonty. His familiarity with the Russian steppe gave him a taxonomic advantage in this mixed grass prairie. Aven Nelson, the founding Director of the Rocky Mountain Herbarium, pointed out one of his favorites, saying "*Oryzopsis hymenoides,* called Indian ricegrass, is prized by the Shoshone for its large grains that can be ground into flour to make a nutty-flavored mush. And these purple beauties are fuzzy beardtongues—*Penstemon eriantherus*—we look forward to these blooms every spring. Over here are the purple two-lobed larkspurs—*Delphinium nuttallianum.* They defend themselves from predators with a toxic alkaloid. In fact, the cattlemen avoid this patch, so their stock don't get sick."

"*Juniperus* and *Pinus* as well," Oscar muttered as he scribbled in his notebook. Rocky mountain juniper was growing in a dry arroyo, and a few scattered limber pine

(*Pinus flexilis*) and ponderosa pine (*Pinus ponderosa*) were arrayed linearly along an exposed limestone ridge. They continued to converse in Latin binomials for over an hour, exchanging names of genera and species as if they were nouns and verbs—a time-honored tradition for international gatherings of botanists in the field.

Aven cleared his throat. "My friends, I have a question for you. As you know, I am a systematist. I get no greater joy in my work than finding and naming a new species. But you are ecologists and I hear that you both have a knack for a different sort of classification. I've been reading your papers and have learned that you are classifying plants based on their relationship with climate and soil. Now that you have seen some of the plants in our prairie, I want to know: how many different kinds of plants do you see?"

Oscar was the first to respond as if he anticipated the question. "I have traveled the world and have cataloged 55 physiognomic life forms." His fame for devising such a thorough global list was exactly why Aven invited Oscar to this gathering. Aven had recently read *Okologie der Pflanzen*, Oscar's famous book (Drude, 1913), and was struck by his comprehensive approach to plant classification. Oscar continued, "But in this prairie I have seen only eleven life forms. There are both conifers and deciduous trees here. This hedgehog cactus belongs to my stem succulent group. I saw some forbs, and a few forbs that had the woody bases of the subshrub group. Other forbs have herbaceous basal rosettes. And that paintbrush is parasitic, no? There are tussock grasses and a few rhizomatous grasses, this flowering onion is clearly a geophyte, and I've seen annual and short-lived dicots along the trails. Yes, I believe that covers them all."

"Only eleven, Oscar? I expected more from you," Leonty responded with a twinkle in his eye. Oscar stared back, blank-faced. Leonty continued, "I agree with my colleague that if you are to base these groups on a jumbled combination of physical morphology and taxonomy, we can indeed list a great variety of plant life forms. But I only see three kinds of plants here: patients, violents, and explerents." Oscar shook his head, but Aven and Ruth were clearly intrigued. In his papers, Leonty proposed that three primary plant strategies are sufficient to explain the diversity of plant form and function. In Latin *violentum* means "tend to violence," *patiens* means "endurance," and *expleo* means "to fill up" (Ramenskii, 1938).

Leonty continued, "Violents grow fast, occupy territory and retain it. They suppress their neighbors through high resource uptake rates by roots and heavy shading by large leaves. Patients are successful by virtue of their tolerance of adverse environmental conditions. Explerents invade open sites quickly but are easily overtaken by violents." This simple model was both intuitive and memorable but was not well received among his contemporaries in Russia. "The patients are clearly dominant here," Leonty continued. "These include the bunchgrasses, shrubs, cacti, and junipers that persist and tolerate the cold and dry conditions in this wind-swept prairie. This *Grindelia* is an explerent, as it fills in the gaps but can be replaced by the faster growing violent forbs. I do not see many violent plants here, probably because the growing season is so short and the soil too thin. Yes, this prairie is dominated by patient plants."

Oscar retorted, "That is a very strange classification system, Leonty! Impractical if you ask me. How can you possibly place a species into one of those three categories with confidence? They sound more like animal behaviors than objective groups. I prefer a system that relies on objective characteristics that you can see with your eyes."

"My comrades in Voronezh say the same thing," Leonty lamented. "No offense, my friend, but I have grown weary of the endless lists of life forms that have been proposed in western Europe. An ecological classification should be based on how species acquire resources and respond to disturbances. Shouldn't that be our primary focus as ecologists?"

Aven was enjoying the lively debate. "It sounds like ecologists have as much disagreement about classification systems as taxonomists! I think your three-strategy system is very interesting, Leonty, but I agree with Oscar—I would find it difficult to place any species into one of those three groups. Have you made a dichotomous key?"

Ruth added, "I understand Oscar's approach much better, but I'd love to hear more about the three strategies, Leonty."

The sun was setting over the Snowy Range to the west, so the Nelsons suggested they turn around and walk back toward town. The walk home in the orange and pink light was contemplative, perhaps just a bit pensive between the two guests from afar, but the conversation back at the Nelsons' home regained its whimsy after Leonty brought out his bottle of Mariisnk. He kept their glasses full late into the chilly evening.[6]

---

[6] These characters are real people set in a real place, but these discussions are only what I imagine could have occurred if these botanists were to have converged in Wyoming 100 years ago. A tale of historical fiction was the most succinct way to capture the spirit of the last century of plant ecology to set the stage for the current debate.

Acknowledgments

This book would not have been possible without the help of many generous colleagues around the world. I am particularly grateful to Peter Grubb and Mark Westoby, who each read large portions of the book and provided insightful comments, constructive critique, alternative perspectives, and encouragement. I also thank Eric Garnier, Roberto Salguero-Gómez, Jennifer Funk, John Dwyer, Nadja Rüger, Peter Adler, Rachael Gallagher, Hendrik Poorter, Bill Shipley, Daniel Falster, and Dennis Knight for reading drafts of chapters and providing helpful feedback.

Many students and faculty attended my graduate seminar on the topic and provided important comments on early drafts, including Alice Stears, David Atkins, Hailey Mount, Sienna Wessel, Magda Garbowski, Andrew Siefert, Matthew Butrim, Bridger Huhn, Claudia Richbourg, Tim Terlizzi, David Tank, and Melissa DeSiervo. Many others generously provided use of their photographs and data, including Magda Garbowski, B. Adriaan Grobler, Gonzalo Navarro, Brandon Hays, Hudson Fontenele, Olga Demina, Diane Laughlin, Maxim Bobrovsky, Trevor Carter, Tessa Smith, Kaylan Hubbard, Bonnie Heidel, Paul Rothrock, Trevor Caughlin, Andrea Westerbrand, Alice Stears, Michael Remke, Joe Buck, Jacqueline Rose, Rolando Pérez, Richard Condit, Steven Paton, Jennifer Baltzer, Sabina Burrascano, and Max Licher. I also thank Lars Walker for providing tabular data on disturbance regimes that was used to make a figure in Chapter 3, and Ben Sullivan for keeping me up to date on the soil science literature. Kel Pendergrass from Studio Coe at the University of Wyoming was instrumental for training me to use image editing software.

I wrote this book while working at the University of Wyoming, which occupies the ancestral and traditional lands of the Cheyenne, Arapaho, Crow, and Shoshone Indigenous peoples, along with other Native tribes who call the Rocky Mountain, Great Plains, and Great Basin regions home. I recognize, support, and advocate alongside Indigenous individuals and communities who live here now, and with those forcibly removed from their Homelands.

Part I

Foundations

Part I establishes the foundations on which plant strategies have been built.

Chapter 1 provides a broad perspective on how ecologists have approached the study of plant strategies. It begins by grounding the book in the concept of convergent evolution, then it contrasts approaches that vary along a continuous versus categorical axis and along a simple versus complex axis. It defines plant strategies and highlights four major problems that modern plant strategy theory must strive to solve.

Chapter 2 provides a thorough overview of the development of plant strategy models over time. One of the main points of the chapter is that despite the overwhelming emphasis of Grime's model in the literature of plant strategies, a plurality of approaches has been proposed.

Chapter 3 articulates the dimensionality of plant strategy theory used throughout the rest of the book. Plant strategies have evolved in response to the complex interactions of abiotic selection pressures including temperature, disturbance regimes, and resource availability, and in response to density-dependent and frequency-dependent selection.

Part I

Foundations

1

Perspectives on Plant Strategies

Strong character convergence is noise to the systematist but signal to an ecologist seeking functional interpretations of diversity.

—William Bond (1997, p. 174), "Functional types for predicting changes in biodiversity: a case study of Cape fynbos"

Punctuationally speaking, wonder is a period at the end of a statement we've long taken for granted, suddenly looking up and seeing the sinuous curve of a tall black hat on its head, and realizing it was a question mark all along.

—David James Duncan (2002, p. 88), *My Story as Told by Water*

1.1 Convergent evolution provides inspiration to the student of plant strategies

In his Presidential Address to the British Ecological Society, Arthur Clapham (1956, p. 1) said "our primary concern as plant ecologists is to know why a plant of this species, and not of that, is growing in a given spot." Our primary concern remains unchanged. When out wandering I invariably ask "why"—Why do we only find the Queen of the Andes (*Puya raimondii*) growing above 3000 meters in Peru and Bolivia, yet it can grow and flower in a coastal botanical garden in San Francisco? Why does Parry's primrose (*Primula parryi*) only grow along trickling snow-fed streams in the Rocky Mountains but not along trout streams flowing through pine forest? Why does kauri (*Agathis australis*) regenerate on ridges more than gully bottoms throughout the warm temperate rain forests of Aotearoa New Zealand? These questions, rooted in wonder, propel plant ecology forward.

Since the late seventeenth century, brave botanists took to the high seas exploring the world to catalogue the diversity of plants on Earth (Egerton, 2018). They found repeated patterns in the growth forms of vegetation and surmised that climate controlled these patterns. The arduous journeys must have been both exhilarating and exhausting, but imagine the joy sprung in these weary travelers when they found plants in climates that reminded them of familiar country. Unrecognizable plant families passed as functionally identical forms. These early naturalists did not know that their discoveries

Plant Strategies. Daniel C. Laughlin, Oxford University Press. © Daniel C. Laughlin (2023). DOI: 10.1093/oso/9780192867940.003.0001

were some of the first examples of convergent evolution because the principles of natural selection were not yet fully articulated. They were so compelled by what they saw that they developed elaborate plant classification systems to make sense of the striking patterns (Barkman, 1988). The discovery that plant species can share similar traits because they grow in the same climate marks the dawn of the global search for plant strategies.

Karl Niklas (1997) suggests that the reappearance of convergent forms throughout the history of evolution is a consequence of physical and chemical processes that constrain the set of feasible strategies, saying "phenotypic correspondence among unrelated species provides strong circumstantial evidence for adaptive evolution because it shows that organisms differing in genetic and developmental capabilities can converge on comparable solutions to life's exigencies when confronted with the same or very similar selection pressures" (p. 316). Evidence of convergence in plants is widespread. According to McGhee (2011), the leaf did not arise once at the very beginning of plant evolution, but rather leaves evolved independently at least nine times. Plants have also converged on comparable solutions to resist herbivores: leaf spines have arisen independently seven times, stem spines arose nine times, toxic leaves arose 30 times, toxic seeds arose 17 times, and toxic roots arose seven times. The problem of dispersing seeds a suitable distance away from the mother was also solved convergently: tufted seeds arose independently five times, winged seeds arose 12 times, Velcro seeds arose seven times, exploding seeds arose eight times, and fleshy fruits arose 22 times (McGhee, 2011). It appears that some evolutionary innovations are inevitable.

Convergent evolution makes the classification of global biomes possible (Akin, 1991; Breckle, 2002). Those lucky enough to holiday in one of the five Mediterranean regions in the world will find evergreen shrubs with small, tough leaves. These sclerophylls arose across many different plant families in places as distant as California, Italy, Western Australia, South Africa, and Chile (Mooney and Dunn, 1970; Cody and Mooney, 1978). Mediterranean climate is defined by hot, dry summers and cool, wet winters, and the photographs and climate diagrams in Figure 1.1 illustrate how the same climate exerts strong selection for similar plant form and vegetation structure. Examples of the convergent sclerophyll growth form include the Cape sumach (*Colpoon compressum* in the family Santalacea) in South Africa and manzanita shrubs (*Arctostaphylos* spp. in the family Ericaceae) in California.

Another celebrated example of convergent evolution is the cylindrical stem succulent plant. This life form evolved independently in the Euphorbiaceae family in Africa and in the Cactaceae family in the Americas (Figure 1.2a, b).[1] This wildly successful strategy involves a suite of traits related to water conservation: storage of water in the photosynthetic stems, rapid water uptake by ephemeral roots in the short wet season, and high water-use efficiency as an outcome of crassulacean acid metabolism

---

[1] *Daviesia euphorbioides* is a pea from Western Australia that evolved the cylindrical succulent growth form with spines, but it does not use CAM photosynthesis.

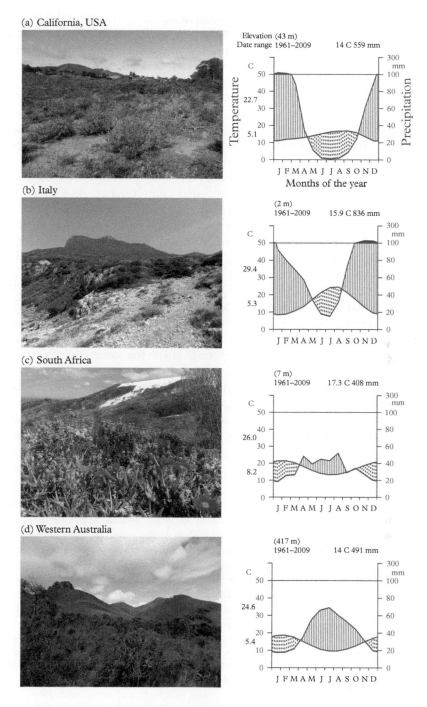

Figure 1.1 *Mediterranean vegetation from around the world (left column) and Walter–Leith climate diagrams for each site (right column). (a) Los Osos Oaks Reserve, San Luis Obispo County, California,*

continued

(CAM) (Heyduk et al., 2019). Given the complexity of the machinery, it is remarkable that CAM photosynthesis likely evolved independently over 60 times (Edwards and Ogburn, 2012)! The same is true for C_4 photosynthesis—the other carbon concentrating mechanism that enhances water-use efficiency and permits population expansion into hot, open habitats (Edwards and Ogburn, 2012). C_4 photosynthesis evolved in the Miocene independently 33 times (McGhee, 2011) and some have increased this estimate to 62 times (Sage et al., 2011)! C_4 grassy biomes are found across Africa, Asia, Australia, and the Americas (Scogings and Sankaran, 2019). These seasonally dry, fire-prone, and herbivore-rich savannas are prime examples of convergent vegetation types (Figure 1.2c, d), but even cold and dry grasslands appear similar from the Russian steppe to the mixed-grass prairies of Montana, USA (Figure 1.2e, f).

Forests are also full of examples of evolutionary convergence. As a child I was constantly drawn to trees to swing from their branches and climb into their canopies, but I was unaware at the time that the arborescent growth form evolved independently among at least nine separate clades, and water-conducting vessels evolved independently at least eight times (Niklas, 1997; McGhee, 2011). This convergence is on full display in tropical rain forests in Africa and Australia, where the verdant vegetation is so similar in appearance that each of these images (Figure 1.3a, b) could have been taken in either location and no one but an expert could tell the difference.

Temperate forests (Figure 1.3c, d) and boreal forests (Figure 1.3e, f) across the northern hemisphere also exhibit similar physiognomy, but these are not examples of evolutionary convergence because these forests share the same genera. For example, the circumpolar subalpine and boreal forests are dominated by different species of spruce (*Picea* spp.), so these vegetation types look similar because they are composed of related species. They do not meet the criteria of convergence, that is, unrelated species that evolved the same traits independently by a common agent of natural selection.[2] While unrelated convergent phenotypes provide prime examples of plant strategies,

---

Figure 1.1 (*continued*)
USA, in the California interior chaparral and woodlands ecoregion. Species include Acmispon glaber, Ceanothus cuneatus, Ericameria ericoides, Quercus agrifolia, *and* Salvia mellifera. *(b) Riparo Blanc—Promontorio del Circeo, Lazio, Italy, in the Italian sclerophyllous and semi-deciduous forests ecoregion. Species include* Chamaerops humilis, Helichrysum italicum *subsp*. pseudolitoreum, Juniperus oxycedrus *subsp*. macrocarpa, Myrtus communis, *and* Pistacia lentiscus. *(c) De Hoop Nature Reserve, Western Cape, South Africa, in the Fynbos shrubland ecoregion. Species include* Carpobrotus acinaciformis, Chironia sp., *and* Colpoon compressum. *(d) Stirling Range, Western Australia, in the Esperance mallee ecoregion. Species include* Andersonia axilliflora, Banksia montana, Darwinia collina, *and* Kingia australis.

(a): Photo taken by Jacqueline Rose and submitted to the Global Vegetation Project and reproduced here under the CC-BY-NC-SA 4.0 license. (b): Photo provided by Sabina Burrascano with permission. (c): Photo provided by Magda Garbowski with permission. (d): Photo by the author. Open-access photos available from the Global Vegetation Project (www.gveg.wyobiodiversity.org).

---

[2] Given that everything on the Tree of Life is related, phylogenetics defines "unrelated" or "distantly related" based on maximum likelihood models of common ancestry. Tips of a tree that share the same ancestral node would be too closely related to be an example of convergent evolution.

Convergent evolution of phenotypes: cylindrical stem succulents

(a) *Euphorbia polygona* in South Africa

(b) *Cereus forbesii* in Argentina

Savanna and the C4 grassy biome

(c) Savanna in Kenya

(d) Cerrado in Brazil

Cool-season grassland

(e) Steppe in Russia

(f) Grassland in Montana, USA

Figure 1.2 *(a)* Euphorbia polygona *(cylindrical stem succulent in the Euphorbiaceae family) in South Africa. (b)* Cereus forbesii *(foreground) and* Stetsonia coryne *(background) (two cylindrical stem succulents in the Cactaceae family) in the dry chaco ecoregion, Catamarca, Argentina. (c) Savanna in Tsavo West, Taita Taveta, Kenya. Species include* Acacia mellifera, Acacia nilotica, *and* Dobera glabdra. *(d) Parque Nacional de Brasília, Distrito Federal, Brasil, in the Cerrado ecoregion. Species include* Aristida riparia, Dalbergia miscolobium, Eriotheca pubescens, Kielmeyera coriacea, Paspalum stellatum, *and* Trachypogon spicatus. *(e) Temperate grassland in the Pontic steppe*

continued

related species often use the same strategy because heritable traits can be conserved through common descent. We revisit the relationship between phylogenies and strategies in Chapter 6.

1.2   A contrast in perspectives encapsulates the problem of plant strategy theory

The prologue recounts a tale of a postprandial walk in Wyoming taken by four botanists in the early twentieth century. This small dose of historical fiction ironically grounds this book in reality. The story of plant strategies begins with a contrast in perspectives. One view seeks the simplest explanation of the diversity of plant form and function. The other embraces a higher-dimensional view (Figure 1.4). The simplest explanations can sometimes become the most celebrated because they are perceived as being the most general, but when the simplest explanations leave us unsatisfied with our understanding of the natural world, then we must be willing to debate more complex arguments. The goal of this book is to articulate and operationalize a compromise perspective—one that seeks generality but embraces realistic complexity, and that is grounded in theory and can be tested empirically.

Oscar Drude was driven by his interest in the relationship between plant form and climate. His life form classification system, like many others in his day, included dozens of groups to be inclusive of the diversity of plant forms seen around the world. In the story, Drude epitomizes the complex view of reality. But it proved hard to be consistent with this complex perspective. His life form groups were a combination of taxonomic (monocot vs. dicot), morphological (woody vs. herbaceous), and life history (annual vs. perennial) characteristics. Inconsistencies in classification systems were even criticized by Drude himself, yet he retained these inconsistencies in his own systems (Barkman, 1988).

Is there a more general, and perhaps simpler approach to categorizing species into phenotypic groups that accounts for differences in physiology and life history? Leonty Ramenskii took a radically different approach and based his simple three-strategy model on plant resource use and demographic behavior. In fact, his approach foreshadowed one of the most celebrated plant strategy models today (Grime, 2001). Reducing the number of plant species from ~350,000 to a few functional types is no small task, but "classical taxonomy will have to give way to functional classifications" (Heal and Grime, 1991, p. 15). Functional classifications that transcend taxonomy are essential if we are to

---

Figure 1.2 (*continued*)
ecoregion, Lysogorka, Rostov, Russia, dominated by Stipa pulcherrima. *(f) Temperate grassland in the mixed-grass prairie, Ft. Keough, Miles City, Montana, USA, dominated by* Hesperostipa *(*Stipa*)* comata, Bouteloua gracilis, Pascopyrum smithii, *and* Poa secunda.

(a): Photo provided by B. Adriaan Grobler with permission. (b): Photo provided by Diego E. Gurvich with permission. (c): Photo provided by Kimberly Medley with permission. (d): Photo provided by Hudson Fontenele with permission. (e): Photo provided by Olga Demina with permission. (f): Photo by the author. Open-access photos available from the Global Vegetation Project (www.gveg.wyobiodiversity.org).

Tropical rain forest

(a) Ivindo National Park, Gabon

(b) Daintree, Queensland, Australia

Temperate deciduous forest

(c) Cluny Hill, Forres, Scotland

(d) Fenelton, Pennsylvania, USA

Temperate coniferous forest

(e) Pechora-Ilychsky Reserve, Russia

(f) Snowy Range, Wyoming, USA

Figure 1.3 *(a) Ivindo National Park, Ogooué-Ivindo, Gabon, in the Northwest Congolian lowland forests ecoregion. Species include* Baillonella toxisperma, Coula edulis, Dacryodes buettneri, Lophira alata, Mammea africana, *and* Panda oleosa. *(b) Bush Tucker trail, Daintree Discovery Center, Australia, in the Queensland tropical rain forests ecoregion. Species include* Angiopteris evecta, Archontophoenix alexandrae, Cerbera floribunda, Cyathea rebeccae, *and* Licuala ramsayi. *(c) Beech forest on Cluny Hill near Nelson's Tower, Forres, Scotland, UK, in the Celtic broadleaf forests ecoregion. Three beeches (*Fagus sylvatica*) stand at this site where three witches*

continued

make general predictions at global scales (Duarte et al., 1995). If a model is too complex it likely can't predict outcomes in new places, but if it is too simple it likely can't make good predictions. Is there an optimal middle ground?

1.3   Plant strategies can be categories or arranged along continua

I started this discussion by treating plant strategies as categories, which is natural because we enjoy lumping entities into groups to bring order to our universe. It is also how plant ecology began. The author of the first textbook of plant ecology wrote "just as species are the units in systematic botany, so are growth-forms the units in oecological botany" (Warming, 1909, p. 5). A great number and variety of terms for categorical plant strategies have been proposed over the last century, including "guilds," "functional groups," "character syndromes," "leagues," "cliques," and "plant functional types," to name just a few (Hawkins and MacMahon, 1989, Gitay and Noble, 1997). The distinctions between these terms are somewhat nuanced. Here, I follow Phil Grime and colleagues (1997a) and Westoby and Leishman (1997), who consider them to be the same pursuit. Emphasizing their subtle distinctions are a distraction so I treat the attempts to group species into strategies as a broad field of enquiry where contributions have been made by scholars from all corners of plant ecology.

The second contrast in perspectives is the preference to base plant strategy models on categorical differences versus basing models on continuous variation (Figure 1.4). Rather than asking "how many different groups?" some ecologists ask, "how many different dimensions?" This alternative approach seeks to arrange strategies along continuous dimensions based on quantitative traits that measure differences in plant function (Westoby et al., 2002; Reich et al., 2003). The challenge then becomes one

---

Figure 1.3 (*continued*)
predicted a kingly future for Macbeth and Banquo responded, "If you can look into the seeds of time and say which grain will grow and which will not, speak then to me." (d) Oak–hickory forest near Fenelton, Pennsylvania, USA, in the Appalachian mixed mesophytic forests ecoregion. Species include Acer rubrum, Quercus velutina, *and* Sassafras albidum. *(e) Pechora–Ilychsky Nature Reserve, Komi Republic, Russia, in the Urals montane forest and taiga ecoregion. Species include* Abies sibirica, Aconitum septentrionale, Diplazium sibiricum, Paeonia anomala, *and* Picea obovata. *(f) Snowy Range, Wyoming, USA, in the Colorado Rockies forests ecoregion. Species include* Abies lasiocarpa, Arnica cordifolia, Carex rossii, Pedicularis racemosa, Picea engelmannii, Ribes montigenum, *and* Vaccinium scoparium.

(a): Photo provided by Brandon Hays with permission. (b): Photo by the author. (c): Photo by the author. (d): Photo provided by Diane Laughlin, the author's mother. (e): Photo provided by Maxim Bobrovsky with permission. (f): Photo provided by Trevor Carter with permission. Open-access photos available from the Global Vegetation Project (www.gveg.wyobiodiversity.org).

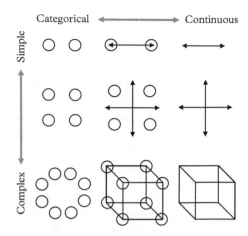

Figure 1.4 *Two gradients in perspectives on plant strategy theories. Models vary from simple models with a few strategies to complex models with dozens of strategies. Models also vary from categorical to continuous descriptions of strategies. Circles represent categorical strategies and arrows represent continuous axes of trait variation. For example, the top left illustrates a simple model with two categorical strategies, the top middle illustrates a model that emphasizes two end points as categorical strategies along a single continuous gradient, and the top right illustrates a purely continuous model with one axis of trait variation. Intermediate models are grounded in continuous trait variation yet still aim to classify species into groups. The distance between categorical strategies is arbitrary.*

of assessing the number of independent dimensions rather than binning species into groups. Southwood (1977) wondered whether the search for the leading dimensions of ecological strategies would be analogous to the chemists' painstaking yet successful search for the periodic table of the elements. Pianka and other animal ecologists continue to pursue this tantalizing idea (Pianka et al., 2017). The complexities of ecological systems cast doubt on the usefulness of this particular tabular approach to plant strategy theory (Bahr, 1982; Steffen, 1996).

Psychological models of human personalities may provide a more useful analogy (Westoby et al., 2002). A brief digression is warranted here, given that theories of human personality are more mature than theories of plant strategies. Just like the fast-growing field of trait ecology,[3] the field of trait psychology grew rapidly in the 1980s. Multidimensional theories of personality traits flourished. Trait psychologists define personality traits as "dimensions of individual differences in tendencies to show consistent patterns of thought, feelings, and actions" (McCrae and Costa, 2003, p. 18). Much like ecologists, psychologists struggled with the desire to classify personality types versus recognizing that personality traits vary continuously. Evolutionary biologists have the same

---

[3] When asked about his role in the field of trait-based ecology, Mark Westoby replied, "I cannot conceive of an ecology that is anything other than trait-based" (paraphrase of personal communication).

quandary: can we justifiably name species, genera, and families given the continuous genetic variation that exists among individuals?

You may have heard of the Myers–Briggs personality model (Myers and Myers, 1980). This popular model has four main binary traits: extraversion/introversion, intuition/sensing, thinking/feeling, and judging/perceiving. The model recognizes 16 different personality types based on all possible combinations of these four binary traits. I took the test at the following website: https://www.16personalities.com. I was scored to be 64% Introverted, 62% Intuitive, 60% Feeling, and 75% Judging. A fifth category about identity was 61% Assertive. They call me an Advocate.

Personality traits vary continuously. I am not 100% introverted, I am 64% introverted. I enjoy a large and festive gathering of friends and colleagues as long as I can slip away unnoticed to go take a walk in the woods with my pup. Other personality models emphasize this continuous variability rather than treating traits as binary. The five-factor model of human personality includes basically the same four factors as the Myers–Briggs model—extraversion, openness, agreeableness, and conscientiousness—but includes a fifth continuous factor called neuroticism. This theory proposes that human personalities vary continuously throughout a five-dimensional space of personality traits. There is better support for the five-factor model (McCrae and Costa, 2003), but the Myers–Briggs is more popular, likely because no one wants to be told how neurotic they are.[4]

It is not possible to measure "introversion" directly, but we can measure indicators of introversion, by scoring answers to questions, such as "Do you enjoy large parties at the end of a stressful week?" These continuous indicators are imperfect measures of introversion, but multiple indicators of the same personality trait are correlated with each other, and hopefully are independent of other traits. Plant ecologists use similar statistical tools to test the dimensionality of plant traits (Laughlin, 2014a; Díaz et al., 2016), a concept discussed in more depth in Chapter 3. Much as psychologists have searched for the correct number of personality traits and have debated over whether personalities are categorical or continuous, plant ecologists have searched for the true number of strategies and have struggled over classifying strategies as categorical or continuous. Unsurprisingly, this variety of different approaches has become a bit muddled in the literature. If we fail to clarify exactly which approach we are talking about, then we run the risk of exacerbating this confusion. This confusion is palpable at conferences, in working groups, in conversations with colleagues, and in reviews of manuscripts and grant proposals. These different perspectives are important because they imply different quantitative approaches for evaluating the evidence for plant strategies. They also imply that there are different definitions of plant strategies, so it is time to be explicit about how plant strategies are defined.

---

[4] The HEXACO model is also gaining support among psychologists, and it includes a sixth trait: honesty–humility.

1.4   A general definition of plant strategies

Plant strategies can be thought of as "ways of making an ecological living" (Westoby and Leishman, 1997) and "how a species sustains a population" (Westoby, 1998). Grime defines strategies as "a grouping of similar or analogous genetic characteristics which re-occurs widely amongst species or populations and causes them to exhibit similar ecology" (Grime et al., 1988, p. 3), that is, species that respond to the environment in the same way share the same set of genetically determined traits. Craine (2009, p. 6) provides a more recent definition of a plant strategy as "a set of interlinked adaptations that arose as a consequence of natural selection and that promotes growth and successful reproduction in a given environment." This definition is a bit better because it incorporates demography; however, it neglects survival, unless survival is implicitly subsumed in the word growth. Survival, growth, and reproduction are the vital demographic rates that determine the per capita growth rate—a measure of fitness—in a given environment. These definitions use the words "group" and "set," implying a categorical perspective on strategies, yet they both ground their theories in continuous variation in traits. The key point is that a strategy axis is composed of multiple correlated phenotypic traits. Scholes and colleagues (1997), bless them, defined strategies as "attribute package deals," in reference to the fact that if you exhibit one attribute you likely exhibit several others because correlated traits come in sets. Westoby and colleagues (2002, p. 126) defined an ecological strategy as "the manner in which species secure carbon profit during vegetative growth and ensure gene transmission into the future," which links resource use to long-term fitness. This definition makes no judgment about whether strategies are categorical or continuous, likely because Westoby is a champion of the continuum approach. Evolutionary game theory defines strategies as "heritable phenotypes that have consequences for the players' payoffs" and may be a fixed or variable trait (Vincent and Brown, 2005, p. 73). John Maynard Smith (1982, p. 10) suggested that "the word strategy could be replaced by the word phenotype; for example, a strategy could be the growth form of a plant." To synthesize these various perspectives, I use the following definition of plant strategies in this book:

> Plant strategies are phenotypes resulting from natural selection that enable a population to persist in a given environment.

Strategies are quantified by phenotypic traits. These traits have been under selection. Natural selection can be viewed as an optimization of fitness in a given environment to enable population persistence, and the environment includes both the abiotic and biotic context. Plant strategies are therefore intimately intertwined with the concept of adaptations (Ackerly, 2003; Ackerly and Monson, 2003). Plant adaptations have been defined as "any heritable feature (or set of associated features) whose presence increases the probability that the individual bearing it will survive or successfully reproduce under a given set of environmental conditions" (Niklas, 1997; p. 8). Taken together, these statements make it clear that plant strategies are adaptive traits that carry demographic consequences.

Phenotypes can be considered either as continuously varying traits in one-to-many dimensions or categories of correlated traits that span one-to-many dimensions. Most of the work over the twentieth century focused on categorical strategies, but more recent empirical approaches emphasize the continuous variation in traits and strategies. Grime's approach to plant strategies, which we explore at depth in Chapter 2, can be considered a sort of hybrid approach—his model is often called a "three-strategy model" because he described it as such (Grime, 2001), but Grime's underlying arguments and empirical tests are firmly grounded in continuous variation in traits that represent fundamental trade-offs. For example, slow-growing species tend to have leaves with low nitrogen concentration, but fast-growing species tend to have leaves with high nitrogen concentration. Grime and his colleagues even developed methods for using continuous trait values to bin species into categorical strategies (Hodgson et al., 1999; Pierce et al., 2013). Grime's model falls in the middle of the gradient between categorical and continuous but falls firmly toward the simple side of things (Figure 1.4). It is based on continuous variation in traits but ultimately relies on a classification system into three strategies (actually, seven if we include the intermediate types).

The limitations of the English language are partly to blame for these complications. Mathematics can easily handle continuous variation in traits and multiple dimensions of independent traits, but as soon as we start to use words to describe these continuous traits, we instantly fall back into categorical descriptions. In the previous paragraph I described high or low nitrogen concentration when describing the broad variation in this continuous trait that spans an order of magnitude. And recall that my personality type defined me as introverted, even though my introversion score was 64%, not 100%. Adequately expressing the subtle differences that exist between low, medium-low, medium, medium-high, and high is a serious challenge for describing continuous theories. It is also dry and repetitive. Our repertoire of words to describe continuous traits is fair to middling. Thankfully, mathematical models thrive on continuous variation.

1.5   Problems addressed by plant strategy theory

The tension between developing simple versus complex models is intertwined with the tension between developing categorical versus continuous plant strategy models. These perspectives color the various approaches to how ecologists have addressed the central problems for plant strategy theory.

What do I mean by "plant strategy theory?" In this book, plant strategy theory is a general framework to explain which traits optimize fitness in a given environment. In this book I propose an analytical framework for testing the existence of plant strategies, but I make no claim to be the originator of a new theory. Rather, in order to shine a light on the most fruitful path forward, I synthesize the large body of work that has contributed toward building this framework. We already know that traits matter, so this theory is not something that can be rejected outright; however, we still lack the contextual details. With sufficient data I think that we will be able to make more specific claims about when and where traits matter and for which reasons. There are four problems that a modern plant strategy theory must solve to advance our understanding of plant ecology.

First, most ecologists would agree that disturbance and site productivity (the inverse of "stress") are the most general drivers of plant strategy evolution (Huston, 1994; Grime, 2001). Mark Westoby (1998, p. 216) wrote, "In my view it is widely agreed that exploiting opportunities for fast versus slow growth, and coping with disturbance, are among the most important forces shaping the ecologies of plants within landscapes." However, plant strategy theory must explicitly acknowledge the multiple dimensions of disturbance and productivity (Figure 1.5). Plant growth is limited by temperature, and the availability of nutrients, water, and light. Note that all of these drivers of productivity can vary over time within a single site depending on the species present and the stage of vegetation development (Grubb, 1980). Moreover, disturbance is another catch-all concept that can be continuous or periodic, it can include different types (grazing vs. fire), and can vary in frequency, severity, and extent. For plant strategy theory to mature into a general predictive theory, it must decompose these two "habitat templets" (Southwood, 1977) into multiple dimensions. Doing so increases the complexity of the theory, yet not doing so oversimplifies the problem. Determining the dimensionality of plant strategies and the causal structure of their drivers is a major research priority.

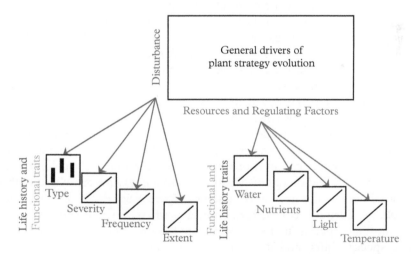

Figure 1.5 *Plant strategies are phenotypes that evolved under selection by disturbance regimes and factors that affect site productivity, including resource availability and regulating factors. Plant strategy theory must decompose disturbance regimes and gradients of productivity into their component parts and consider how both life history and functional traits jointly respond to all these factors.*

Second, plant strategy theory must articulate the causal relationships between functional traits and life history traits. Functional traits are heritable morphological, physiological, or phenological attributes that indirectly influence fitness by affecting growth, survival, and reproduction. Life history traits are emergent properties of population models computed from demographic rates, that is, growth, survival, and reproduction. Chapter 2 describes how plant strategy models fall into two loosely defined camps: those that emphasize demographic rates and those that emphasize functional traits (Garnier et al., 2016). The preface described the deep disciplinary divide between plant demography and ecophysiology (Harper, 1982; Grime, 1984), and recently, major steps have been taken to bridge this divide. Analyzing the dimensionality of traits is the first step, but this step merely defines what is phenotypically and demographically possible. The second, arguably more critical step, is to develop predictive models of the fitness consequences of phenotypes in specific abiotic and biotic contexts. Phenotypes that exhibit poor demographic performance will be selected against and go extinct. Interestingly, Chapter 2 shows that plant strategy models that emphasize demographic rates tend to explain temporal dynamics in response to disturbances, whereas those that emphasize functional traits tend to explain responses to resource limitation. Integrating these perspectives into a coherent framework for predicting fitness along all of these gradients is a major research goal.

Third, plant strategy theory must distinguish between the fundamental and realized niches of phenotypes—an important goal in ecology for decades, yet true progress has been limited (Austin, 1990). The limits to abiotic tolerance in the absence of competition define the fundamental niche. These physiological limits may contract (or perhaps expand) in the presence of neighbors, and these new limits define the realized niche of the phenotype. In other words, the fitness of a phenotype depends not only on its own traits in a given environment, but also on the traits of others. Consider the hypothetical example in Figure 1.6. The first example (left column) represents fitness landscapes in the absence of invading competitors. The blue resident species achieves peak fitness in the absence of competition in the environment indicated by the grey region (Figure 1.6a). However, the fitness landscape when expressed as a function of traits within this grey environment suggests that another strategy with lower trait values could potentially invade this grey environment (Figure 1.6b). There are two possible outcomes if the gold invader species with this lower trait value does invade: stable coexistence (middle column) or competitive exclusion (right column). The second example in the middle column illustrates stable coexistence, where both species achieve peak fitness in the grey environment (Figure 1.6c, d). The third example in the right column illustrates competitive exclusion, where the gold species could invade and drive the blue species to local extinction (Figure 1.6e, f). The realized niche of the blue species shifts, and it is forced to occupy the lower portion of the environmental gradient. This book describes approaches to predicting fitness from traits using both empirical approaches and game theoretical models upon which this figure is based. From my reading of the literature, the empirical approaches have too often neglected game theory. Ideally, empirical approaches can integrate the lessons learned from game theoretical models that explicitly account for species interactions. Quantifying the fitness of a phenotype when it depends on both the environment and the traits of other individuals in the community is a major challenge for plant strategy theory.

Fourth, plant strategy theory seeks to understand trait adaptations to spatial gradients in climate and soil as well as to temporal gradients in disturbance and small-scale heterogeneity within a local habitat. Evolutionary convergence across continents inspires plant strategy theory, but large local variation of traits within a local habitat challenges our understanding of trait adaptations. At first, this seems like a simple hierarchical problem: trait variation across habitats should be large and explain physiological tolerances across climatic gradients, whereas other traits will explain the responses to local disturbances and small-scale heterogeneity within habitats. It is common to hear ecologists describe species in a dryland habitat as "drought tolerant" or species within a tropical rain forest as "productive," but there can be nearly as much variation in traits within a habitat as there is across habitats. According to Reich and colleagues (2003, p. S159), "selection

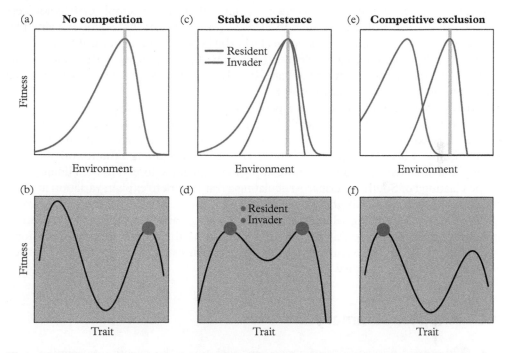

Figure 1.6 *The fitness of a species is a function of the local environment, its traits, and the traits of others. The top row illustrates fitness landscapes as a function of the environment. The bottom row represents fitness landscapes as a function of traits in the environment indicated by the sliver of grey in the top row. (a) The fundamental niche of the blue species in the absence of competition achieves peak fitness in the grey environment. (b) However, within this grey environment it is possible that a species with a smaller trait value could invade because its fitness would be higher. There are two possible outcomes if the brown species does invade. First, as illustrated in the middle column (c and d) the two species could stably coexist in this environment. Second, as illustrated in the right column (e and f) the brown species could invade and drive the blue species to local extinction, forcing the realized niche of the blue species downward along the environmental gradient.*

for functional traits is roughly equally as strong across strategies within a site as across local, regional, or continental environmental gradients." Consider an example: specific leaf area (SLA), the leaf area per unit mass, is an important leaf trait that varies across global climate gradients (Figure 1.7a), yet it also varies strongly within a community (Figure 1.7b). Disturbance-driven gap dynamics and patchy distribution of resources are the most likely culprits that drive within-site variation of SLA (Falster et al., 2017). Disturbances create vacant patches within a habitat that are colonized by fast-growing, light-demanding early colonizers that have high SLA. These can be replaced by slower-growing, shade-tolerant species with low SLA after decades of vegetation development (Figure 1.7b). But the problem gets more complicated still. Open ecosystems are not always dominated by fast-growing phenotypes with high SLA. In fact, large regions of the world are dominated by slow-growing, shade-intolerant plants that sport adaptations to chronic consumption by herbivory and fire; they are not simply early colonizers (Bond, 2019). Some of the variation in SLA across the planet that is not explained by climate or soil could possibly be explained by herbivores and fire regimes, although other traits are likely under far greater selection (spines, bark thickness, etc.). Reconciling how traits have been selected among habitats globally as well as within habitats locally is a major problem for plant strategy theory.

These fundamental challenges for plant strategy theory are also many of the core questions facing ecologists in general. That is no coincidence. If we can solve the problems addressed by plant strategy theory, then we will make significant advances for the larger discipline of ecology as a whole. If we succeed in developing a framework for plant strategies in an African savanna, can we apply it to subalpine forests in the Himalaya, or to the Caatinga of South America? Articulating a framework to explain variation in plant form and function that is both sufficient to account for local diversity and generalizable to global scales is a major goal in plant ecology. Linking traits to demography is the most promising path to achieve this goal. Before articulating the minimum ingredients for a general theory, let us first synthesize the diversity of plant strategy models that have been proposed throughout history.

1.6   Chapter summary

- Widespread observations of convergent evolution—the selection for similar traits in similar environments among unrelated taxa—continue to motivate interest in plant strategies.

- Some ecologists prefer simple models with as few strategies as possible whereas others propose dozens of strategies to explain the diversity of plant form and function.

- Some models describe strategies as distinct categories but continuous variation in plant traits suggests that strategies align along continuous dimensions.

- Language makes it difficult for us to discuss continuous strategies. When we describe variation in a continuous trait, we naturally emphasize the categorical extremes. Thankfully, mathematical models thrive on continuous variation.

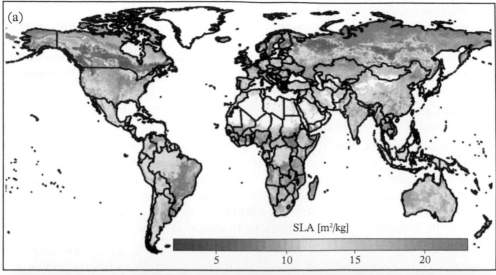

Figure 1.7 *Traits vary both across global climate gradients and within ecological communities.*
(a) Map of modelled average specific leaf area (SLA), the leaf area per unit mass, at the global scale.
(b) Community dynamics over temporal gradients of time since disturbance. Early colonists exhibit
more acquisitive trait values (higher SLA is indicated by the brighter green leaf color) than the
shade-tolerant trees that are better competitors for light in the long-term (lower SLA is indicated by the
darker green leaf color). Note that physiological differences among plants drive gap dynamics within
many forested ecosystems, but the classic concept of a predictable and linear successional dynamic is
clearly too simplistic and should not be applied everywhere (Binkley 2021). In open ecosystems, the
shade-intolerant plants that dominate in full sun may actually exhibit conservative traits, grow slowly,
and require adaptations to fire and herbivory (Bond 2019).
(a): Reproduced from Butler et al. (2017) with permission.

- Psychological models of human behavior provide an analogous framework for understanding plant strategies. Psychologists have also debated over the merits of simple versus complex and categorical versus continuous models of personalities. The five-factor model proposes that humans exhibit continuous variation along five distinct traits. Plant strategies may also be best described by continuous variation in independent traits.

- Plant strategies are phenotypes resulting from natural selection that enable a population to persist in a given environment.

- To make progress, plant strategy theory must (1) unpack the multiple dimensions of productivity and disturbance gradients into their component parts; (2) articulate the causal relationships between functional traits and life history traits; (3) be able to distinguish between the fundamental and realized niches of a phenotype; and (4) differentiate between trait adaptations to large-scale spatial gradients in climate and soil from trait adaptations to temporal gradients in disturbance and small-scale heterogeneity within local habitats.

1.7   Questions

1. Why do traits that have evolved independently multiple times throughout evolutionary history represent prime examples of plant strategies?

2. Why does Oscar Drude epitomize the complex perspective, and why does Leonty Grigoryevich Ramenskii represent the simple perspective? Recall the Prologue if needed. Which model is foreshadowed by Ramenskii's model?

3. Are human personalities a good analogy for plant strategies? What potential problems would arise if we classified plant strategies based on percentiles of continuous trait scores using the same approach used by the Myers–Briggs personality model?

4. Do other languages more clearly articulate multidimensional variation in continuous plant traits?

5. Which traits influence response to disturbance and which traits influence response to resource limitation?

6. Are there searches for animal strategies that parallel the search for plant strategies?

7. Can strategies evolve through a mechanism of evolution other than natural selection, such as drift?

2

A Menagerie of Plant Strategy Models

When you look at a piece of delicately spun glass you think of two things: how beautiful it is and how easily it can be broken.

—Tennessee Williams (1945, p. 7), *The Glass Menagerie*

2.1 Categorical life forms partially explain responses to global environmental gradients

The goal of plant strategy theory is to predict the demographic consequences of plant traits in any given abiotic and biotic context. A host of plant strategy models has been proposed by plant ecologists throughout history to generate such predictions. It is well known that "all models are wrong . . . but . . . may be useful nonetheless" (Box, 1976).[1] Like the glass figurines in Tennessee Williams's play, many of these models are elegant representations of reality, but because realities are so complex, the predictions of these models are easily broken. In eight parts, this chapter reviews the conceptual origins of plant strategies and provides a balanced synthesis of the contributions to this broad field of study.

First, we begin with the categorical classifications of plant life forms that started thousands of years ago and continues to this day (Raunkiaer, 1934; Box, 1981). Second, we explore models that emphasized variation in life history traits that evolved in response to disturbance regimes, including the vital attributes model (Noble and Slatyer, 1980), forest gap dynamics (Shugart, 1984), shrubland fire dynamics (Bond, 1997), and the evolution of the seeder–resprouter continuum (Pausas and Keeley, 2014). Third, we transition to models that emphasized variation in functional traits that evolved in response to resource limitation, including responses to nutrient limitation (Chapin, 1980), water limitation (Volaire, 2018), and light limitation (Purves et al., 2008). Fourth, we evaluate models that consider how strategies have evolved along multiple gradients in resource availability and disturbance (Tilman, 1988; Shipley et al., 1989; Smith and

---

[1] Note that the common quote attributed to George Box "all models are wrong, but some are useful" is actually a paraphrase, rather than a direct quote from his paper "*Science and Statistics*" (Box, 1976, p. 792). This compound quote uses ellipses to make the same point, but without paraphrasing.

Plant Strategies. Daniel C. Laughlin, Oxford University Press. © Daniel C. Laughlin (2023). DOI: 10.1093/oso/9780192867940.003.0002

Huston, 1989). Fifth, we consider Grime's (1977) CSR model, which continues to inspire discussions of plant strategies today. Sixth, we look at how other triangular models attempted to understand demographic variation (Silvertown et al., 1992) and defense syndromes (Agrawal and Fishbein, 2006). Seventh, plant ecologists expanded Grime's stress-tolerant strategy into multiple strategies (Grubb, 1998a; Craine, 2009). Lastly, we look at strategies proposed to be measured using continuous variation in functional traits (Westoby et al., 2002), an approach taken by evolutionary game theory (Vincent and Brown, 2005). Some of these models were formulated to explain strategies within a single habitat, whereas others were proposed to be far reaching and globally generalizable across all habitats. This menagerie of models provides a foundation on which to construct a general framework for plant strategy theory.

This search for order amid the bewildering global diversity of plants can be traced at least back to Ancient Greece. Theophrastus of Eresos (371–287 BCE) was Aristotle's close colleague and successor at the Lyceum. One of his most enduring works was his *Historia Plantarum* (*Enquiry into Plants*), in which he used a plant life form system that is still in use today: trees, shrubs, sub-shrubs, and herbs (i.e., literally "grasses") (Theophrastus, 287 BCE; Morton, 1981; Barkman, 1988; Weiher et al., 1999). The fundamental differences between woody and herbaceous plants have been appreciated since antiquity, and these Theophrastean forms conform to our general definition of plant strategies as distinct phenotypes—trees are taller, display their meristems aboveground, and exhibit secondary thickening of the xylem; grasses are shorter, display their meristems at the soil subsurface, and generate no secondary thickening of the xylem.

Natural philosophers including Albertus Magnus, Andrea Cesalpino, Joachim Jung, Johann Wolfgang von Goethe, and Augustin Pyramus de Candolle, debated different theories about the nature of plant form (Arber, 1950), but the modern discussion about plant form began when the first phytogeographers developed classification systems of plants. Confusingly, the terms "growth form" and "life form" are often confounded or considered synonymous (Braun-Blanquet, 1964; Barkman, 1988), but growth form is typically reserved to differentiate plant architecture free from any hypothesis about adaptation. Growth form is strictly about morphological distinctions (e.g., tree, shrub, forb, graminoid). Life forms, on the other hand, emphasize morphological and physiological adaptations to environmental conditions (Barkman, 1988), so for our purposes, life forms are of most interest.

Alexander von Humboldt (1806) developed one of the earliest classification systems and recognized 19 plant life forms. Most were named after a typical genus or family, such as palms, cactus, or grasses. Grisebach developed another system that included 60 physiognomic types. These were grouped into seven main categories that were meant to reflect climatic differences and are familiar to most biologists today: woody plants, succulents, climbing plants, epiphytes, forbs, grasses, and cryptograms (Grisebach, 1872). Recall Oscar Drude from the prologue, who criticized these approaches as muddled mixtures of morphologies and adaptive characters and proposed his own list of 55 life forms (Drude, 1886, 1913), which were also muddled mixtures. This muddling of phylogenetic clades, life histories, and morphological variation continues to plague the general application of plant functional types even today. In 1909, Eugenius Warming criticized these approaches: "A purely physiognomic system is devoid of scientific significance, which is introduced only when physiognomy is founded upon physiological and oecological facts" (Warming, 1909, p. 4). Warming developed a functional classification based on lifespan, flowering periods, vegetative expansion, rhizomatous branching, and mode of hibernation (Warming, 1884, 1895) and was primarily interested in understanding adaptations to environmental conditions.

Christen Raunkiaer developed one of the most enduring plant life form classifications, which divided plants into groups based on the height of their meristems aboveground (Figure 2.1). Meristems, also called hibernating buds, are the basic units by which plants grow. A focus on how plants protect meristems to overwinter, tolerate periods of drought, or permit resprouting after disturbance, or to use a more general term, perennate, is a simple and sensible way to categorize plants because the location of perennating meristems is connected to climate (Raunkiaer, 1934). He proposed five general categories. Phanerophytes are trees or shrubs that produce their perennating meristems high above ground. Chamaephytes are also woody but their perennating meristems are produced close to the ground and can often be buried by winter snow. Hemicryptophytes produce their meristems at the soil surface, such as grasses and sedges. Cryptophytes

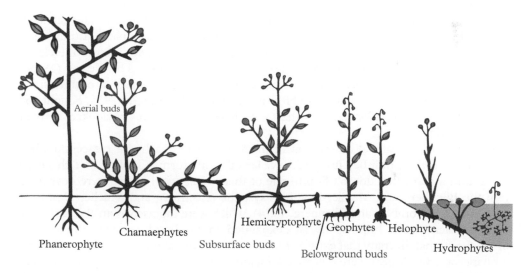

Figure 2.1 *Plants can be divided into life forms based upon the location of their dormant meristems. Bud exposure decreases from left to right. Phanerophytes (trees) expose the meristems aboveground on the tips of branches, while Hemicryptophytes maintain buds just below the soil surface. Cryptophytes (including Geophytes, Helophytes, and Hydrophytes) protect the meristems beneath the soil. Brown represents perennial organs, green represents ephemeral organs that die back in the unfavorable season, and orange represents sexual reproductive organs.*

Redrawn from Goldsmith and Harrison (1976) with permission, which was reproduced from Raunkiaer (1934).

produce the meristems beneath the soil surface or under water. Finally, therophytes are annual plants that survive unfavorable time periods as seeds. These five groups can be expanded to ten by including different kinds of phanerophytes, cryptophytes, succulents, and epiphytes. Raunkiaer demonstrated that the distribution of these life forms is not random across biomes, and the system is still being used today (Salguero-Gómez et al., 2016). However, "Although [Raunkiaer's system is] a useful classification, it does not have significant resolving power to differentiate between plants with the same structural properties but occurring in very different ecological situations" (Woodward and Kelly, 1997, p. 47).

The development of growth form and life form classification systems became a sort of cottage industry within plant ecology. For more than a century, plant ecologists all over the world worked feverishly to bring order to the vast diversity of plant form and function (Clements, 1920; Du Rietz, 1931; Dansereau, 1951; Ellenberg and Mueller-Dombois, 1967; Mueller-Dombois and Ellenberg, 1974; Hallé et al., 1978, Gillison, 1981). The development and relationships among the many classification systems have a rather complex history and I refer the interested reader to the expert reviews by Cain (1950), Barkman (1988), and Duckworth and colleagues (2000). One of the inherent problems with categorical classification systems is that they have too many degrees of freedom: it is not clear when to stop adding more categories because it is so easy to constantly add more groups.

Before we settle into the latter half of the twentieth century, I here briefly acknowledge that a group of ecologists in the North American prairies pursued a trait-based approach to understand vegetation patterns (Steiger, 1930). In addition to the classic measurements and drawings of rooting depth among prairie species by Weaver (1919), growth forms and functional traits were being used through the 1960s to understand vegetation dynamics in the prairies and forests of Wisconsin (Knight, 1965; Knight and Loucks, 1969). While pioneering at the time, these studies were overlooked by their contemporaries.

The search for a global description of plant life forms reached a crescendo when Elgene Box (1981) published his impressive list of 77 life forms (Table 2.1). I challenge you to find a plant that doesn't fit into one of these categories! These were developed from the Ellenberg and Mueller-Dombois systems (Ellenberg and Mueller-Dombois, 1967; Mueller-Dombois and Ellenberg, 1974), which were themselves an elaboration of the Raunkiaer system. Importantly, Box expanded on the Raunkiaer system to account for the functional diversity that exists within each type. For example, more than 30 types of phanerophytes (trees) were delineated, including tropical rain forest[2] trees (such as trees in the Lauraceae), temperate broad-evergreen trees (such as *Quercus* spp.), rain-green broad-leaved trees (such as *Tectona* spp.), and summergreen broad-leaved trees (such as *Acer* spp.). His system grouped species into life forms based on structural types (e.g., tree, shrub, graminoid), plant size, leaf type (i.e., broad, narrow, graminoid,

---

[2] I retained the original names from Box (1981) in the table, but in this book I aim to follow the precedent set by Whitmore (1987, p. 24) to refer to tropical and temperate "rain forest," as opposed to "rainforest," given that we never write "dryforest, moistforest, thornforest," etc.

Table 2.1 *Plant life forms of the Box system and their relationship with the Raunkiaer (1934) life form system.*

Plant form	Examples	Raunkiaer life form
Trees (Broad-leaved)		Phanerophytes
Evergreen		
1. Tropical Rainforest Trees	Lauraceae, Rubiaceae, Dipterocarpaceae, Lecythidaceae, and Myristicaceae	
2. Tropical Evergreen Microphyll Trees	Fabaceae, Meliaceae, Simaroubaceae	
3. Tropical Evergreen Sclerophyll Trees	Eucalyptus	
4. Temperate Broad-evergreen Trees		
a. Warm-temperate	*Quercus virginiana, Magnolia, Persea, Prunus*	
b. Mediterranean	*Quercus ilex, Arbutus, Olea europaea*	
c. Temperate Rainforest	Magnoliaceae, Lauraceae	
Deciduous		
5. Raingreen Broad-leaved Trees		
a. Monsoon mesomorphic	*Tectona*, Dipterocarpaceae	
b. Woodland xeromorphic	*Acacia, Adansonia,* Caesalpinaceae	
6. Summergreen Broad-leaved		
a. Typical-temperate mesophyllous	*Quercus, Acer, Fagus*	
b. Cool-summer microphyllous	*Betula, Populus, Nothofagus*	
Trees (Narrow and Needle-leaved)		Phanerophytes
Evergreen		
7. Tropical Linear-leaved Trees	*Podocarpus, Agathis*	
8. Tropical Xeric Needle-trees	*Juniperus procera, Widdringtonia*	
9. Temperate Rainforest Needle-trees	*Tsuga, Thuja, Sequoia*	

continued

Table 2.1 *continued*

Plant form	Examples	Raunkiaer life form
10. Temperate Needle-leaved Trees		
a. Heliophilic Large-needled	*Pinus taeda, Pinus caribbea*	
b. Mediterranean	*Cedrus, Cupressus, Pinus pinea*	
c. Typical Temperate	*Pinus strobus, Pinus ponderosa*	
11. Boreal/Montane Needle-trees	*Picea, Abies*	
Summergreen		
12. Hydrophilic Summergreen Needle-trees	*Taxodium, Metasequoia*	
13. Boreal Summergreen Needle-trees	*Larix, Pseudolarix*	
Small and Dwarf-trees		Phanerophytes
14. Tropical Broad-evergreen Small Trees	Rainforest understory, Fabaceae, Melastomataceae, Rubiaceae	
15. Tropical Broad-evergreen Dwarf-trees	Campo cerrado treelets	
16. Cloud-Forest Small Trees	*Podocarpus*, Ericaceae	
17. Temperate Broad-evergreen Small Trees	*Ilex, Nothofagus, Berberis*	
18. Broad-raingreen Small Trees	Fabaceae	
19. Broad-summergreen Small Trees	*Prunus, Nothofagus, Betula tortuosa*	
20. Needle-leaved Small Trees	*Juniperus, Actinostrobus*	
Rosette-trees		Phanerophytes
21. Palmiform Tuft-trees	Palms, Caricaceae	
Rosette-treelets		Phanerophytes
22. Palmiform Tuft-treelets	Understory palms, cycads	
23. Tree Ferns	Cyatheaceae, Dicksoniaceae	
24. Tropical Alpine Tuft-treelets	*Senecio, Espeletia*	
25. Xeric Tuft-treelets	*Yucca, Dracaena, Xanthorrhoea*	

Arborescents   Phanerophytes

26. Evergreen Arborescents   Mallee eucalyptus

27. Raingreen Thorn-scrub   *Acacia, Commiphora*

28. Summergreen Arborescents   *Prosopis, Salix*

29. Leafless Arborescents   *Haloxylon, Calligonum*

Krummholz   Phanerophytes

30. Needle-leaved Treeline   *Picea, Abies, Juniperus*
    Krummholz

Shrubs   Phanerophytes

31. Tropical Broad-evergreen   *Coffea*, Rubiaceae, Ericaceae
    Shrubs

32. Temperate Broad-evergreen
    Shrubs

   a. Mediterranean   Proteaceae, *Quercus dumosa,*
                                    Rhamnus

   b. Typical Temperate   *Ilex, Ligustrum, Daphne*

   c. Broad-ericoid   *Rhododendron*

33. Hot-desert Evergreen Shrubs   Zygophyllaceae, *Acacia*
                                    aneura (mulga)

34. Leaf-succulent Evergreen   *Crassula argentea*
    Shrubs/Treelets

35. Cold-winter Xeromorphic   *Artemisia*
    Shrubs

36. Summergreen Broad-leaved
    Shrubs

   a. Mesomorphic   *Rosa, Vaccinium*

   b. Xeromorphic   Deciduous chaparral, šibljak,
                                    Protea, Grevillea

37. Needle-leaved Evergreen   *Juniperus communis*
    Shrubs

Dwarf-shrubs   Chamaephytes

38. Mediterranean Dwarf-shrubs   *Thymus, Salvia, Eriogonum*

39. Temperate Evergreen Dwarf-   Heath and arctic/alpine
    shrubs   Ericaceae

continued

Table 2.1 *continued*

Plant form	Examples	Raunkiaer life form
40. Summergreen Tundra Dwarf-shrubs	*Betula nana, Salix reptans*	
41. Xeric Dwarf-shrubs	*Ephedra, Anabasis, Retama*	
Cushion-shrubs		Chamaephytes
42. Perhumid Evergreen Cushion-shrubs	*Azorella selago*	
43. Xeric Cushion-shrubs	Puna/ Patagonian hard cushions	
Rosette-shrubs		Phanerophytes
44. Mesic Rosette-shrubs	Understory and ground palms	
45. Xeric Rosette-shrubs	*Agave, Yucca, Aloë*	
Stem-succulents		Phanerophytes
46. Arborescent Stem-succulents	*Carnegiea gigantea, Euphorbia candelabrum*	
47. Typical Stem-succulents	Unbranched barrel cacti, *Mammillaria*	
48. Bush Stem-succulents	Branched *Opuntia*	
Graminoids		Hemicryptophytes (mostly)
49. Arborescent Grasses	bamboos	
50. Tall Cane-grasses	*Imperata, Arundinaria*	
51. Typical Tall Grasses	*Andropogon, Festuca*, prairie grasses	
52. Short Sward-grasses	*Cynodon dactylon, Bouteloua gracilis*	
53. Short Bunch-grasses	*Festuca, Stipa, Agropyron*	
54. Tall Tussock-grasses	Pampas and Patagonian grasses (e.g., *Stipa*)	
55. Short Tussock-grasses	Puna grasses, *Festuca novae-selandiae*	
56. Sclerophyllous Grasses	"spinifex" (*Triodia*), *Scleropoa*	
57. Desert Grasses	*Aristida* (wire grass)	

Forbs   Hemicryptophytes

  58. Tropical Evergreen Forbs   Cannaceae, *Begonia*,
                                       Zingiberaceae

  59. Temperate Evergreen Forbs   *Gaultheria, Chimaphila,*
                                       Hexastylis

  60. Raingreen Forbs   Fabaceae, Asteraceae

  61. Summergreen Forbs   Forest dicots, geophytes,
                                       Asteraceae, Campanulaceae,
                                       Apiaceae, Lamiaceae

  62. Succulent Forbs   *Portulacca, Sedum, Semper-*
                                       vivum

Undifferentiated small herbs   Hemicryptophytes
                                                                   (mostly)

  63. Xeric Cushion-herbs   *Saxifraga, Dryas, Draba*

  64. Ephemeral Dry-desert Herbs   Annuals, dwarf-geophytes,   Therophytes
                                       graminoids

  65. Summergreen Cold-desert   Dwarf-geophytes, graminoids
      Herbs

  66. Raingreen Cold-desert Herbs   Geophytes, graminoids

Vines and lianas   Lianas (not recognized
                                                                   originally)

  67. Tropical Broad-evergreen   *Ficus, Calamus*, stranglers
      Lianas

  68. Broad-evergreen Vines   *Philodendron, Lonicera,*
                                       Smilax

  69. Broad-raingreen Vines   Fabaceae, *Ipomoea*

  70. Broad-summergreen Vines   *Vitis, Parthenocissus, Rhus*
                                       radicans

Ferns   Hemicryptophytes

  71. Evergreen Ferns   Rainforest ferns (e.g.,
                                       Polypodium)

  72. Summergreen Ferns   Temperate ferns (e.g.,
                                       Aspidiaceae)

Epiphytes   Epiphytes (vascular)

  73. Tropical Broad-evergreen   Bromeliads, orchids, aroids,
      Epiphytes   cacti

  74. Narrow-leaved Epiphytes   Ferns, mosses, *Tilandsia*

  75. Broad-wintergreen Epiphytes   Mistletoes (*Loranthaceae*)

continued

Table 2.1 *continued*

Plant form	Examples	Raunkiaer life form
Thallophytes		Thallophytes (non-vascular cryptograms)
76. Mat-forming Thallophytes	Forest and tundra mosses, folious lichens	
77. Xeric Thallophytes	Crustose lichen	

NB: This list is actively being updated by Elgene Box and a new version with about 130 life forms will soon be available. A few of the example taxa in this table have been updated in consultation with Peter Grubb, but the list of plant forms itself remains unchanged from the original publication.
Adapted from Box 1981), with permission.

absent), leaf size (i.e., macrophyll, mesophyll, microphyll, nanophyll), leaf structure (i.e., herbaceous, coriaceous, sclerophyll, succulent, ligneous, pubescent), and photosynthetic habit (e.g., evergreen, summer-green, raingreen).[3] What made Box's system more powerful than others was that he modeled each type based on climatic drivers (such as temperature and precipitation) in one of the first global models of vegetation distribution. Note that some genera (e.g., *Festuca, Betula, Picea*) are provided as examples across multiple groups, indicating that adaptive radiations have occurred within clades. Box is actively updating this list and will soon publish a revision of about 130 plant life forms, with the new inclusion of a laurophyll leaf type (pers. comm.).

At about the same time Box (1981) published his monograph, the first generation of Dynamic Global Vegetation Models (DGVMs) were being built and the relevance of a list of plant functional types (PFTs) became paramount. The primary goal of these models was to use a short list of PFTs to model terrestrial carbon dynamics based on how vegetation responds to climate and CO_2, and on how vegetation feeds back to affect climate. These DGVMs used only a few functional types, in spite of the fact that Box published a list of 77 plant life forms. For example, the Sheffield-DGVM includes broadleaf evergreen, needleleaf evergreen, broadleaf deciduous, and needleleaf deciduous as the four tree PFTs to distinguish how leaf senescence and tissue quality impacts albedo and CO_2 fixation over time (Sitch et al., 2008). The Ecosystem Demography model (ED v3.0) predicts the global distribution of seven PFTs (Ma et al., 2022).

Plant functional types, for all intents and purposes, are life forms. In the 1980s, it was felt that a better system of life forms for DGVMs could probably be devised with just a bit of concerted effort. The length of such a list would comfortably fit within the realm of 20–50 PFTs. As our understanding of global climate processes and vegetation feedbacks improved, the International Geosphere-Biosphere Programme (IGBP) was launched to address climate change—one of the most pressing crises of our time. As a result of an IGBP workshop, Smith and colleagues (1997) published a series of papers from a variety

---

[3] There are a lot of terms in this sentence! But let's not get bogged down in terminology right now—see Box's original monograph for details.

of perspectives (some more optimistic than others[4]) about the possibility of defining this set of PFTs that could be used in DGVMs. However, many years have gone by and the list of PFTs that are applied in DGVMs has basically remained the same—you can count them on your fingers (Lavorel et al., 2007). To put it bluntly, the global community of plant ecologists has failed to answer the question that was set forth by the IGBP. There is still no consensus list of PFTs produced by plant ecologists for climate modelers (Parmesan and Hanley, 2015). Perhaps the main distraction from solving this problem was the warranted fascination with variation in continuous traits among plants (Reich et al., 1997), and the recognition of the high variation in traits within a given life form (Lavorel et al., 2007). An alternative approach to modeling vegetation–climate interactions is to simulate continuous traits, rather than use categorical types, which requires a thorough reparameterization of the models and is an active area of research (Pavlick et al., 2013; Meng et al., 2015). Eco-evolutionary optimization approaches leverage known trade-offs along continuous axes of plant traits to parameterize DGVMs to predict large scale vegetation change (Franklin et al., 2020; Harrison et al., 2021). Others have argued that functional types should be based on evolutionary lineages given that traits are phylogenetically conserved (Anderegg et al., 2022), but it is unclear how this approach could be globally generalized from one region to another if the regions are dominated by distinct lineages (e.g., Pinaceae forests in the northern hemisphere versus Podocarpaceae forests in the southern hemisphere).

2.2   Life history strategies evolved in response to disturbance dynamics

Since the 1970s, the recognition that disturbance drove much of the dynamics in natural systems inspired a new generation of ecologists to discover the mechanisms that determined these dynamics (Pickett and White, 1985). Disturbances are events that remove biomass in an ecosystem and often increase the availability of space and resources. Many models were developed to understand vegetation dynamics following disturbances and have zeroed in on the variety of ways plants evolved under various disturbance regimes. These models have a strong focus on life history traits, as opposed to functional traits, because they describe temporal dynamics of populations.

One of the earliest models of plant response to disturbance originated in rangeland science, where ecologists classified species according to their response to livestock grazing (Dyksterhuis, 1949). Highly palatable species were called "decreasers" because they continually declined with increasing grazing pressure (Figure 2.2). Other native species that were less palatable were called "increasers" because they increased in abundance under light to moderate grazing pressure. They increased because they fill the gaps left behind after the decreasers were reduced in abundance. However, even the "increasers" eventually decrease under extreme grazing pressure because the livestock eat anything

---

[4] Lauenroth and colleagues (1997) suggest that a global classification scheme would not be applicable at local scales. They discuss how two *Bouteloua* grasses exhibit quite different life histories, yet because they belong to the same genus and are morphologically similar, they would be classified in the same group.

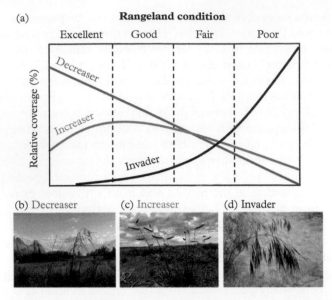

Figure 2.2 *(a) A model of three potential responses to grazing pressure, which increases from left to right along a gradient in rangeland condition. (b) Needle-and-thread grass (*Hesperostipa comata*) is an example of a decreaser. (c) Blue grama (*Bouteloua gracilis*) is an example of an increaser. (d) Cheatgrass (*Bromus tectorum*) is an example of an invader.*

(a): Adapted from Dyksterhuis (1949) with permission. (b): Photo taken in Canyonlands National Park, Utah by Magda Garbowski, reproduced with permission. (c): Photo taken by the author in Laramie, Wyoming. (d): Photo taken in Canyonlands National Park, Utah by Magda Garbowski, reproduced with permission.

they can find. Finally, species that are not native to the rangeland that increase with increasing grazing pressure are called "invaders." These classifications are still used in rangelands around the world (Vesk and Westoby, 2001; Del-Val and Crawley, 2005; Carmona et al., 2013; Morris, 2016).

The vital attributes model (Noble and Slatyer 1980), one of the most highly influential models of temporal dynamics in plant communities, used life history traits of species to predict their performance after a disturbance. The term "vital attributes" conjures the concept of vital rates, which are synonymous with demographic rates: growth, survival, and reproduction.[5] Noble and Slatyer (1980) identified three key attributes: the method of persistence during and after a disturbance; the ability to establish and grow to maturity; and the time it takes to reach critical stages of life history. By combining multiple categories of each attribute together, they identified 30 different species types, but reduced it down to 15 distinct patterns of behavior following disturbances. They also called attention to the importance of understanding how species disperse, establish, and persist in communities that are periodically disturbed. Figure 2.3 illustrates how they

---

[5] Nurses measure vital rates such as blood pressure and heart rate, which provide information about the health of a human being. Vital rates in population models provide information about the pulse of the population: whether it is healthy and growing, or ill and shrinking.

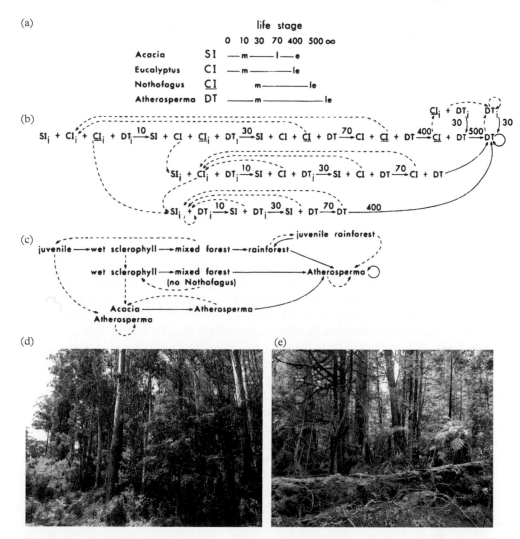

Figure 2.3 *The vital attributes model of vegetation dynamics over time-since-fire applied to wet sclerophyll and rain forest vegetation in Tasmania, Australia. (a) The age at reproductive maturity (m), the longevity of the population (l), and the longevity of its propagule pool (e) for four indigenous species:* Acacia dealbata, Eucalyptus regnans, Nothofagus cunninghamii, *and* Atherosperma moschatum. *(b) SI, CI, CI, and DT are different strategy types from their model, and each species has a different code from (a). Solid arrows show transitions in the absence of disturbance and the yearly intervals between transitions are shown above each solid arrow. Dashed arrows represent transitions after a disturbance. (c) Simplified vegetation transition model representing the model in (b). (d) Early-stage wet sclerophyll vegetation on Myers Creek Road, Toolangi, Victoria, including* Eucalyptus regnans, Acacia dealbata, Pomaderris apetala, *and* Acacia melanoxylon. *It is likely regrowth from the 1939 Black Friday bushfires. (e) Late-stage cool temperate rain forest with* Nothofagus cunninghamii, Atherosperma moschatum, *and* Dicksonia antarctica *on the Overland Track, Tasmania.*

(a–c): Reproduced from Noble and Slatyer (1980) with permission. (d, e): Photos by Tessa Smith, reproduced with permission.

applied their model to summarize the temporal dynamics of wet sclerophyll and rain forest vegetation in Tasmania, Australia. Bushfires have the potential to reset the clock in these forests. Silver wattle (*Acacia dealbata*) and mountain ash (*Eucalyptus regnans*) are some of the first to establish a wet sclerophyll canopy after fire. In the absence of subsequent fires, these species are then replaced by the slightly longer-lived rain forest species myrtle beech (*Nothofagus cunninghamii*) and southern sassafras (*Atherosperma moschatum*). Similar dynamics occur in subtropical and tropical regions of Australia, where the same genera play similar roles as they do in temperate Tasmania (John Dwyer, pers. comm.).

Models of life history attributes were soon applied to very different habitats. Arnold van der Valk developed a similar model to describe vegetation dynamics in prairie wetlands using three key life history traits: lifespan, seed or propagule longevity, and seed germination or propagule establishment requirements (van der Valk and Davis, 1978; van der Valk, 1981, 1985). Rather than using continuous traits to represent these life history attributes, he used three categories of lifespan (annual, short-lived perennial, long-lived perennial), two categories of propagule longevity (seed bank vs. dispersal-dependent), and two categories of establishment requirements (submerged vs not submerged). This three-way scheme created 12 possible life history categories, which van der Valk and Davis (1978) used to describe the general dynamics of vegetation following flooding and muskrat disturbance in prairie wetlands (Figure 2.4).

Research into the life history variation in response to disturbance continued in gap dynamics models of forests and shrublands. Shugart (1984) developed a model of forest dynamics, with four strategies that could be arranged in a 2 × 2 table. The strategies were defined by whether a tree creates a gap in the canopy when it dies or not, and whether it requires a gap for regeneration or not; large trees produce gaps when they die, small trees do not, and light-demanding trees require gaps for regeneration, shade-tolerant trees do not. For example, *Liriodendron tulipifera* (tulip tree) is an example of a tree that both produces a gap when it dies, and it requires a gap for regeneration because it is light demanding. *Fagus grandifolia* (American beech) is an example of a species that produces a gap when it dies, but it doesn't require a gap for regeneration because it is shade tolerant. Australia's red ash tree (*Alphitonia excelsa*) is a species that doesn't produce a gap when it dies, but it requires a gap for regeneration. Finally, the rainforest tree brush bloodwood (*Baloghia lucida*) in eastern Australia is an example of a tree that neither produces a gap when it dies, nor requires a gap for regeneration. These combinations lead to an interesting set of rules of replacement where some strategies are self-reinforcing: the large tree that requires gaps will replace itself when it dies, whereas the small tree that does not require gaps will simply regenerate beneath its own shade.

William Bond (1997) modified Shugart's forest model and applied it to the fire-prone fynbos of the Cape Floristic region in South Africa. This hyper-diverse landscape contains a bewildering number of shrubs and simplifying them down to four main strategies was no small task. Most of these species would be classified as "stress-tolerant," but Bond further classified species into four types that could also be arranged in a 2 × 2 table. The

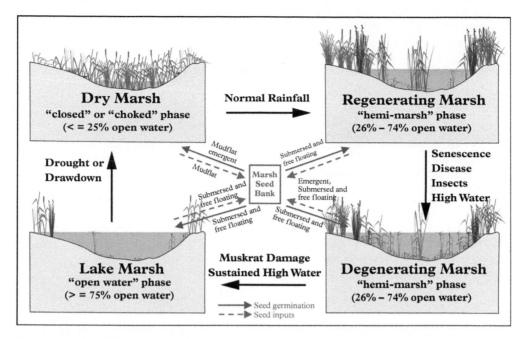

Figure 2.4 *A model of vegetation dynamics in a prairie glacial marsh in Iowa, USA, driven by seed bank emergence in relation to flooding and muskrat damage. Twelve life history strategies were identified that occupy the niches in this dynamic ecosystem.*

Adapted from van der Valk and Davis (1978) with permission. This adaptation incorporated the improvements by Johnson et al. (2016)

strategies were defined by whether they are killed by fire or not, and whether they require fire for regeneration or not. Species that survive fire as adults by resprouting from buds and recruit following fire exhibit episodic regeneration and overlapping generations that persist on the landscape. Species that are killed by fire as adults and recruit following fire from soil seed banks exhibit episodic recruitment and non-overlapping generations. Broad-leaved trees and shrubs with an affinity to forest and thickets such as *Rhus* spp. and *Diospyros* spp. survive fire by resprouting but do not regenerate following fire, and thereby exhibit continuous recruitment and overlapping generations. Finally, the rarest group of species are killed by fire, do not regenerate following fire, and therefore exhibit relatively continuous recruitment.

Interest in evolutionary responses to fire regimes continues today as megafires burn at unprecedented scales in a warming climate (Attiwill and Adams, 2013; Abella and Fornwalt, 2015). Two plant strategies in fire-prone ecosystems are "seeders" and "resprouters." Pausas and Keeley (2014) developed a life history model based on the adult-to-offspring survival ratio, where vegetative survival and regeneration would be important if this ratio was high and where regeneration by seed would be favored if this

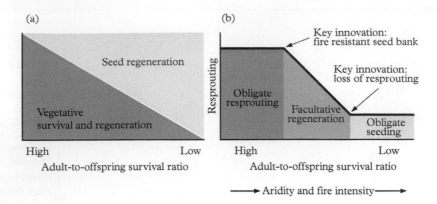

Figure 2.5 *(a) A model of dominant modes of regeneration along a gradient of the adult-to-offspring survival ratio. (b) Hypothesized changes in the probability of resprouting along an adult-to-offspring survival ratio, which varies concomitantly along an aridity and fire intensity gradient.*
Adapted from Pausas and Keeley (2014) with permission.

ratio was low (Figure 2.5a). From this model, they hypothesized that, since resprouting is an ancient, widespread trait that is not limited to fire-prone vegetation, obligate resprouters are the ancestral strategy. The development of a fire-resistant seed bank was the first innovation that led to the evolution of facultative seeders, and the loss of resprouting was the key innovation that led to the evolution of obligate seeders (Figure 2.5b). The aboveground shoots in resprouters are top-killed in a fire, but dormant buds can resprout and the plant recovers relatively rapidly. Some seeders exhibit enhanced germination rates following fire. However, if fires are too frequent, even resprouters can decline due to the exhaustion of their carbohydrate reserves.

Memorable names of different strategies emerged in the ecological lexicon. Bormann and Likens (1994) distinguished between "exploitative" and "conservative" species. Oldeman and Van Dijk (1991) developed a similar scheme for tropical rain forests, where they distinguished between the "gamblers" and the "strugglers." Gamblers were fast-growing, shade-intolerant trees, sometimes called gap species, and strugglers were slow-growing, shade-tolerant trees, sometime called non-gap species (Hartshorn, 1978). During and colleagues (1985) distinguished forbs in chalk grasslands as either "spenders" or "savers." Spenders are short, shallow-rooted forbs that produce copious amounts of seed with short residence times in the soil seed bank, and they spend their capital quickly when there are large openings in which to establish and reproduce vigorously. On the other hand, savers are taller and deeper-rooted forbs that produce few seeds that persist for long periods in the soil seed bank, and they save their capital and generally prefer denser stands of vegetation.

While some ecologists were focused on life history traits, others integrated plant morphology and physiology into our understanding of vegetation dynamics. For example, Fakhri Al-Bazzaz (1979) summarized the physiological traits that distinguished early successional species growing in open fields from late successional species in broad-leaved deciduous forest. Plants adapted to high light in recently disturbed sites exhibit high light-compensation points, fast rates of photosynthesis, respiration, and transpiration. Table 2.2 lists many more traits that distinguish between high- and low-light environments driven by time-since-disturbance. What is the relationship between life history traits and physiological traits? Do the physiological traits determine the life history variation? These questions are addressed throughout the remainder of the book. We now turn to a series of models developed to describe the evolution of phenotypes in response to resource limitation, as opposed to disturbance. We first examine models developed to understand strategies for single resource gradients, then later explore models for multiple environmental gradients.

Table 2.2 *Physiological traits of early and late successional plants.*

Attribute	Early successional plants	Late successional plants
Seeds		
Dispersal in time	Long	Short
Induced dormancy	Common	Uncommon?
Seed germination		
Enhanced by		
Light	Yes	No
Fluctuating temperatures	Yes	No
High NO_3^- concentrations	Yes	No?
Inhibited by		
Far-red light	Yes	No
High CO_2 concentrations	Yes	No?
Light saturation intensity	High	Low
Light compensation point	High	Low
Efficiency at low light	Low	High
Photosynthetic rates	High	Low
Respiration rates	High	Low
Transpiration rates	High	Low
Stomatal and mesophyll resistances	Low	High
Resistance to water potential	Low	High

continued

Table 2.2 *continued*

Attribute	Early successional plants	Late successional plants
Acclimation potential	High	Low
Recovery from resource limitation	Fast	Slow
Ability to compress environmental extremes	High	Low?
Physiological response breadth	Broad	Narrow
Resource acquisition rates	Fast	Slow?
Material allocation flexibility	High	Low?

Reproduced from Bazzaz (1979) with permission.

2.3   Functional strategies evolved in response to resource gradients

Ecophysiologists have been particularly interested in determining how trait variation explains responses to individual resources. Let's start with nutrient limitation. Terry Chapin and Rien Aerts (Chapin, 1980; Aerts and Chapin, 1999) distinguished between "competitive/ruderal" and "nutrient-stress tolerant" strategies, where relative growth rates played a central role in distinguishing the strategies (Figure 2.6). Fast growth was easy to explain in productive sites: the fastest growing phenotypes were able to dominate. However, the advantages of slow growth in poor sites proved more difficult to understand, and models were developed to explain why slow growing perennials with

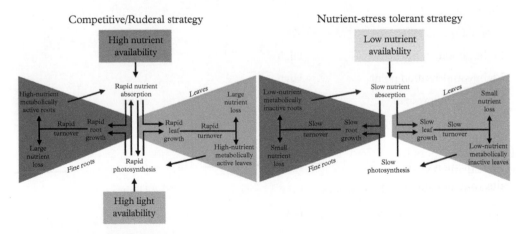

Figure 2.6 *A model of phenotypic traits associated with high and low nutrient availability.*
Adapted from Chapin (1980) with permission.

nutrient-conserving strategies dominate nutrient-poor habitats (Aerts and van der Peijl, 1993). Low leaf nutrient concentrations were originally thought to have been selected for in slow-growing species as a consequence of low nutrient uptake rates and high allocation to defense and storage. Later on, slow growth was associated with slow turnover of tissues, low growth respiration rates, and allocation of carbon to secondary metabolites (Chapin et al., 1993). These low and high nutrient strategies broadly overlap with Grime's stress-tolerant and competitive/ruderal strategies (Grime, 1979), respectively, which are discussed later in the chapter.

Water supply is one of the most powerful filters on plant communities globally (Keddy and Laughlin, 2022). In fact, we are currently witnessing one of the greatest and most widespread tree die-off events in recorded history (Allen et al., 2010; Hartmann et al., 2022) driven by hotter droughts in a changing climate (Breshears et al., 2005). Given all the mortality we are seeing in forests and woodlands, scientists have sharpened their focus on identifying the causes of death. Manion (1991) proposed that an individual's progression toward mortality could be conceptualized as a "death spiral" where factors either predispose, incite, or directly contribute to plant death (Griffin-Nolan et al., 2021). One particularly fascinating debate is whether trees are dying because of hydraulic failure caused by embolisms in xylem that prevent water flow, or because of carbon starvation caused by reduced photosynthesis. Tree biologists have proposed that species can be sorted along a continuum from isohydric to anisohydric (Klein, 2014). Isohydric species exhibit strong control of leaf water potential by closing their stomates in response to water deficit. In contrast, anisohydric species lack strong stomatal regulation and continue to transpire across a wide range of leaf water potentials. Isohydric species may be at greater risk of carbon starvation because they stop photosynthesizing early in drought (McDowell et al., 2008), whereas anisohydric species may be at higher risk of hydraulic failure (Choat, 2013). This debate continues, although evidence seems to be pointing toward a widespread role of hydraulic failure in large-scale tree die-off events (Adams et al., 2017).

One of the earliest drought strategy schemes was a model that included three-strategies:

1. drought escape;
2. drought avoidance (endurance with high internal water content); and
3. drought tolerance (endurance with low internal water content) (May and Milthorpe 1962, Fischer and Maurer 1978, Levitt 1980).

Drought escapers include mostly annuals and perhaps a few short-lived perennials that grow during seasons of ample water supply (Kooyers, 2015). Drought avoiders have traits that permit maximum water uptake or minimize water loss. Drought tolerators have tissues that are less sensitive to dehydration.

Florence Volaire (2018) proposed a revised terminology and defined six strategies arrayed along a gradient of decreasing soil water potential and increasing water deficit (Figure 2.7). The dehydration escape strategy is comprised of mostly desert annuals that rely on timing recruitment to seasons where soil moisture is greatest. The classification

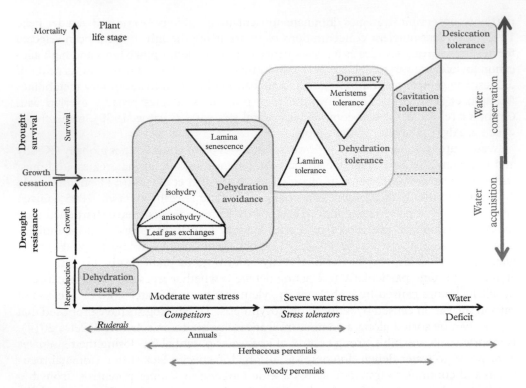

Figure 2.7 *A model of six proposed plant strategies related to drought: dehydration escape, dehydration avoidance, dehydration tolerance, dormancy, cavitation tolerance, and desiccation tolerance. These vary continuously along a gradient of increasing water deficit.*
Reproduced from Volaire (2018) with permission.

of these desert annuals as "ruderal" phenotypes (Grime, 1979) has been challenged and can alternatively be viewed as specialized stress-tolerators (Madon and Médail, 1997). The dehydration avoidance strategy includes the gradient from anisohydry to isohydry, phenotypes that can access water deep in the water table (called "phreatophytes" or "water spenders"), as well as those species that exhibit drought-deciduous leaves to avoid water loss (Baquedano and Castillo, 2006; Volaire, 2018; Bueno et al., 2019). Volaire links these strategies to demographic rates by indicating that these last two strategies have life histories with high importance of reproduction and growth compared to survival. The dehydration tolerance strategy includes "water saver" phenotypes that can continue to photosynthesize under severe water stress and phenotypes that remain alive because dormant meristems persist. Volaire adds a cavitation tolerance strategy, although it is unclear why the previous two strategies do not include embolism resistance as relevant traits. Lastly, desiccation tolerance is the most extreme strategy and includes a few species that can completely dry out for long periods of time and resurrect themselves after rewetting. These last three strategies have life histories with high importance of survival compared to growth and reproduction.

Plant strategy models have also been developed to explain responses to light limitation. One recent model called the perfect-plasticity approximation (PPA) focused on height structured asymmetric competition for light. The model does away with all the complexities of other forest gap models and was shown to produce reliable predictions in temperate forests (Adams et al., 2007; Purves et al., 2008). This approach showed that two continuous species-level traits distinguish early from late successional species: H_{20} and $Z^\star$. H_{20} is the height of a 20-year-old tree when grown in the open and is therefore a measure of the early-successional niche. $Z^\star$ is the height of a tree when grown in monoculture with itself and is therefore predictive of the species that will eventually win the slow race to canopy dominance. While the traits are continuous, they do tend to only pick out two tree strategies: early and late successional species.

Can plants escape, avoid, and tolerate shade in a similar way that plants can escape, avoid, or tolerate low water supply? Shade-tolerant species can regenerate and persist in deep shade for decades and have long been of great interest to silviculturists and forest ecologists (Valladares and Niinemets, 2008), but Gommers and colleagues (2013) identified the existence of both shade-tolerant and shade-avoidance strategies. Shade avoiders are the species with fast growth rates that overtop their neighbors. So how would plants "escape" the shade? There is a whole group of herbaceous understory species that escape light limitation by timing their phenology to grow during a time when the overstory canopy has not yet leafed out. The beloved spring ephemerals, which greet winter-weary wanderers each spring, are excellent examples of species that escape the shade, even though they persist in it for most of the year (Schemske et al., 1978).

Circling back to nutrients, can plants escape, avoid, and tolerate soil infertility? Chapin's (1980) low nutrient strategy can be thought of as an infertility tolerator. For this three-strategy model to extend to soil fertility, infertility avoiders would need to somehow tap into patches of resource availability in a manner similar to the "drought avoiders" that tap into deep water tables (the phreatophytes). Plants that associate with nitrogen fixing bacteria fit the bill. We call them "nitrogen-fixing plants" even though the bacteria do all the work in exchange for habitat in root nodules and sugars produced by photosynthesis. Nitrogen-fixing plants themselves are abundantly supplied with nitrogen and therefore avoid soil infertility by producing the nitrogen themselves. They are often indicators of infertile soil, especially in scoured deglaciated terrain or recent volcanic eruptions with little biological legacy following the disturbance (Walker and del Moral, 2003; Prach and Walker, 2020). What about infertility escapers? Infertility escapers would need to time their phenology to grow during the season when nutrient limitation is at its lowest. Given that nutrient availability is strongly tied to water supply (Grime, 1994), perhaps infertility and drought escapers are overlapping strategies.

Incidentally, similar language is used for wetland plants dealing with oxygen limitation in submerged conditions. Two strategies have been proposed to handle low oxygen levels in submerged habitats (Bailey-Serres and Voesenek, 2008; Akman et al., 2012). The "low oxygen escape" strategy rapidly elongates the shoot to restore contact with the atmosphere. These species often have large amounts of aerenchyma in belowground tissues for gas transport. The "low oxygen quiescence" strategy is typified by slower growth and conservation of energy reserves; this strategy is best if floods are transient because stems would break if water levels dropped. These strategies have primarily been studied within species on a small set of species like rice, *Rumex* spp., *Rorippa* spp., and *Solanum dulcamara,* and should be evaluated on larger sets of species simultaneously to evaluate its generality in wetlands around the world.

2.4   Functional strategies evolved in response to multiple environmental gradients

Section 2.3 described models of strategies that evolved in response to variation in a single resource, but plants must adapt to multiple environmental gradients simultaneously. Not surprisingly, there has been a long-standing interest in ranking species performance along multiple environmental gradients. Perhaps the most impressive rankings are the Ellenberg indicator values of European plant species (Ellenberg, 1988). Heinz Ellenberg proposed an ordinal ranking of species based on their performance along abiotic gradients, including light, temperature, moisture, continentality, soil pH, nitrogen, and salinity. For light, the numbers range from 1 (deep shade) to 9 (full sunlight). For nitrogen, the numbers range from 1 (least amount) to 9 (excessive supply). These rankings were based on the realized niche of the species, that is, where the species are found in nature, not based on experimental physiological results. No other continent has attempted such a feat for an entire flora. The closest analog in North America would be the Coefficient of Conservatisms, which rank species by their preference for most disturbed (zero) to least disturbed (10) habitats, and can be used to calculate floristic quality assessments of nature preserves (Matthews et al., 2015).

Niinemets and Valladares (2006) ranked northern hemisphere tree species based on their tolerances to shade, drought, and waterlogging. These types of rankings are not functional traits per se because they are not measurable phenotypic properties measured without reference to the environment (Garnier et al., 2016). Rather, these rankings reflect the species observed distributions in nature—their realized niches. Nevertheless, they have proven themselves to be powerful tools for comparing species and their global distributions.

David Tilman (1988) developed his plant strategy theory based on species responses to the relative abundances of soil nutrients and light (Figure 2.8). He developed his theory from his mathematical model of resource ratios, where species that can persist at the lowest resource level ($R\star$) wins in competition (Tilman, 1982). His model is therefore grounded in a demographic model of population dynamics in a community of competitors and directly challenges Grime's model. While we haven't evaluated Grime's model yet, let us briefly contrast the two here. Grime viewed the best competitor as the one that acquires the resource first, whereas Tilman viewed the best competitor as the one that could survive on the lowest availability of the resource. Grime's best competitor is the fastest-growing species in pairwise competition experiments, whereas Tilman's best competitor is the one that outlasts the short-lived species in successional sequences. Tilman criticized short-term competition experiments because they are limited to describing "transient dynamics" following disturbances, not long-term outcomes

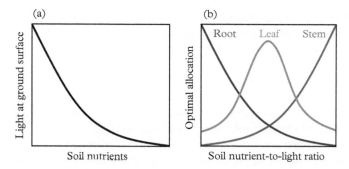

Figure 2.8 *(a) Increasing soil resources inevitably leads to light limitation at the ground surface because the vegetation becomes more productive, density increases, individual plant crowns overlap, and the canopy closes. (b) Tilman's model proposed that plant strategies separate along a gradient defined by the ratio of soil nutrients-to-light and that the fractions of roots, leaves, and stems would vary accordingly. Phenotypes with high root-to-shoot ratios are predicted to dominate in sites with low nutrients and high light, whereas phenotypes with low root-to-shoot ratios are predicted to dominate in sites with high nutrients and low light.*
Adapted from text in Tilman (1988).

of competition after settling into an equilibrium. The slower-growing species that persists at low resource levels will be the ultimate winner in competition. One of Grime's most controversial ideas was that competition is unimportant in stressful sites. Unlike Grime, Tilman viewed competition as ubiquitous everywhere and not just limited to productive environments (Tilman, 1988; Aerts et al., 1991) because there is clear evidence that density dependence is an important driver of population dynamics everywhere. Tilman's views on competition for limited resources are his most important contribution to our current understanding of competitive plant strategies (Craine, 2009). I refer the interested reader to Grace (1990) and Goldberg (1990) for the most insightful discussions of the differences between Grime and Tilman's views of competition.

In Tilman's theory, light and soil nutrients are inversely correlated. Light is limited in sites with fertile soil because a productive closed canopy develops. On sites with poor soil, canopies cannot close, and light does not become limiting. Therefore, Tilman treats the soil nutrient-to-light ratio as a single environmental gradient and proposed that species that grow slower and invest more carbon in roots will be more competitive for nutrients, and species that grow fast and invest more carbon into shoots will be more competitive for light. However, after many experimental tests, these predictions are not well supported by empirical data (Craine, 2009). In fact, Tilman's root-to-shoot ratio model never really caught on outside of the US because coping with low nutrients was not about having large root mass (Berendse and Elberse, 1990; Aerts and van der Peijl, 1993) and coping with shade was not about having large leaf mass (Kitajima, 1994). The merits of using root-to-shoot ratio as a predictor of relative growth rates is discussed in greater detail in Chapter 11.

At about the same time Tilman was developing his model, Smith and Huston (1989) developed a model of plant strategies based on tolerances to shade and drought (Figure 2.9). According to their model, species with the fastest growth rates are adapted to high light and well-watered habitats, but a trade-off exists between species that can tolerate shade versus those that can tolerate drought. They argued that shade-tolerance selects for thin leaves and low root mass fraction—traits that are both detrimental in dry environments. While it may be true that evolutionary constraints prevent a species from being both shade and drought tolerant, Smith and Huston's original premise is probably incorrect. The constraint is more likely driven by the need to retain leaves in shade, the need to maintain metabolic processes from declining at low water potentials and prevent death of young tissues, and the need to make seeds large enough to produce enough roots to survive a dry season (Grubb, 2016). Regardless, Smith and Huston (1989) use the model to show that successional dynamics can be interpreted as temporal shifts in species dominance in response to shade tolerance, and zonation can be interpreted as a spatial shift in species dominance.

Zonation in wetlands provided a tractable environmental gradient for studying plant adaptations to flooding. Shipley and colleagues (1989) measured 20 traits on 25 species of marsh plants to classify them into categorical plant strategies that occur along gradients of soil fertility, frequency of gap formation, and size of individual gaps. Traits measured on juvenile plants included seed size, germination requirements, and maximum relative growth rate. Traits measured on adult plants included morphological attributes such as the distance between successive shoots, the height of the plant, and the width of the canopy. Traits of juveniles were decoupled from traits of adults, but species could be grouped into four strategies that occupied different environmental conditions (Figure 2.10). "Stress-tolerant" plants (Type IV) exhibited large seeds, slow growth rates, and slow seed germination rates and were found in undisturbed sites with infertile

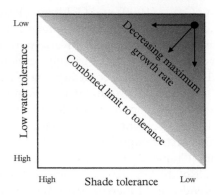

Figure 2.9 *A model of woody plant strategies under light and water limitation. The highest growth rate is in the upper right and growth rates decrease with increasing tolerance to low light and water.*
Reproduced from Smith and Huston (1989) with permission.

(a)

(b)

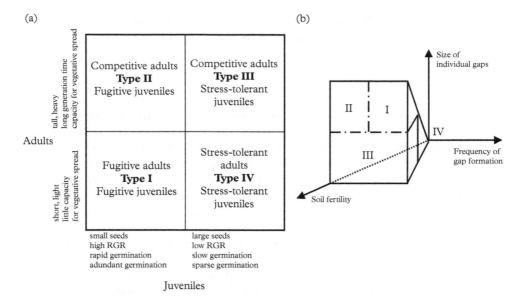

Figure 2.10 *(a) A model of four categorical plant strategies based on continuous variation in 20 plant traits. (b) Hypothesized relationships between the four strategies and environmental characteristics.* Reproduced from Shipley et al. (1989) with permission.

soil. In contrast, "fugitive" plants (Type I) exhibited small seeds, fast growth rates, and rapid germination and were found in frequently disturbed sites with fertile soil.

This leads us to the next model of plant strategies proposed as a globally generalizable explanation of plant responses to resource limitation and disturbance. It has been difficult to wait to describe this model until now because it is the benchmark against which all other plant strategy models have been and will be compared. It is worth noting that all these other approaches discussed here deserve mention by their own right. However, given the prominence of this next model, it deserves its own section and a thorough appraisal. You might use this break in the chapter to fire up the kettle, brew a cup of tea, and admire the nearest plant before diving back in to read about Grime's preeminent model.

2.5 Grime's CSR model: plants evolved in response to stress and disturbance

Grime's (1974, 1977, 1979, 2001) CSR plant strategy model is undoubtedly the most widely known and cited, as well as the most intuitive yet controversial. Life history and functional traits evolved in a complex array of abiotic and biotic factors, and one of the most important advances in ecology was Grime's expansion of the dimensionality of abiotic factors proposed to drive the evolution of plants. When most people think of plant strategies they are thinking of Grime's model. It is celebrated for its simplicity and proposed generality. Recent books on trait-based ecology include it as one of the most important plant strategy models (Garnier et al., 2016; de Bello et al., 2021).

Grime was working during a time when r-K selection theory was wildly popular. r-K selection was originally proposed as a trade-off between population growth rate and resistance to density-dependent mortality (MacArthur and Wilson, 1967; Boyce, 1984), but quickly became associated with a contrast between early and late successional species (Pianka, 1970), where early-successional species were considered r-selected, and vice versa. This misapplication of the original theory as described by Grubb (1987) was associated with the proposal that disturbance rate and the relative density of competitors drove r-K selection. Grime (1979) referred to this as the "duration of the habitat." Accordingly, r-selected species exhibited fast rates of reproduction and were well adapted to frequently disturbed environments. Short-lived thistles (e.g., *Cirsium* and *Carduus*) come to mind as examples of r-selected plant species. K-selected species exhibited slow rates of reproduction and were well adapted to stable environments. The long-lived bristlecone pine (*Pinus longaeva*) comes to mind as an example of a K-selected plant species. Frequently disturbed habitats could be rapidly colonized by fast growing populations of r-strategists, whereas stable habitats could become dominated by large and long-lived K-strategists (MacArthur and Wilson, 1967; Pianka, 1970).

Grime (1979) argued that the K-strategists are not all the same—there are different types of long-lived species based on whether they are adapted to productive or resource-limited conditions. Figure 2.11a illustrates this historic expansion of K-strategists into "competitors" and "stress-tolerant" strategies.[6]

"Competitors" (denoted by the letter C) fill the undisturbed and productive habitats by rapidly acquiring resources in crowded vegetation and do so with rapid production of surface area of both leaves and roots. They are known for rapid proliferation of leaves into the canopy and roots into the soil and can adjust their allocation to leaves and roots as needed to maximize resource uptake (Table 2.3).[7]

"Stress-tolerators" (denoted by the letter S) fill the undisturbed and unproductive habitats by nature of their persistent and long-lived organs and conservation of resources, which enables them to persist and endure in chronically stressful sites. They are known for their slow growth rates, evergreen habit, long-lived organs, slow turnover of carbon, nutrients, and water, infrequent flowering, and being well defended against herbivory.

The r-strategists were unchanged in Grime's model and were called the "ruderal" strategy, even though this name was primarily used to describe species of "waste places," like parking lots and sidewalks, not disturbed wild landscapes (Rabotnov, 1985). "Ruderals" (denoted by the letter R) fill the disturbed and productive habitats by exhibiting short lifespans and investing their rapidly acquired capital into reproduction, which enables their populations to persist in the face of frequent disturbance.

---

[6] Grime also claimed that the frequency of ruderals, competitors, and stress-tolerators spanned the r-K continuum in that order—see Figure 11 in Grime (1979).

[7] Many ecologists dislike the name "competitor" because it implies that ruderals and stress-tolerant plants don't compete for resources.

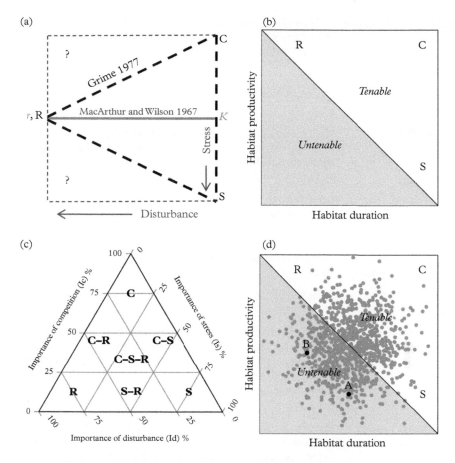

Figure 2.11 *(a) Grime's CSR model expanded the K end of the r-K selection gradient to include long-lived organisms in productive and stressful sites. (b) Grime proposed that strategies evolved along gradients of habitat productivity and duration and argued that no strategy could occupy both highly disturbed and highly stressed environments (the "untenable" lower triangle). (c) Grime converted the upper right triangle in (b) into an equilateral triangle and proposed seven categorical strategies as representatives of a space defined by the importance of competition ("Ic"), stress ("Is"), and disturbance ("Id"). (d) Assuming a random, normally distributed sample of environments (represented as dots in the figure) that span variation in habitat productivity and duration, half of the possibilities are ignored in Grime's model because he throws out the entire lower triangle. Points A and B represent examples of strategies that are likely to exist but fall just beyond the edge of triangle. (b): Reproduced from Grime (2001) with permission. (c): Reproduced from Grime (1977) with permission.*

They are characterized by annual or short-lived perennial life histories, high rates of dry matter production, copious seed production, and fast seed ripening. However, the *r*-selected ruderals were not expanded along a productivity gradient into two strategies of short-lived plants, one that thrives in productive conditions and one that thrives in resource-limited conditions. Why is this?

Table 2.3 *Traits associated with each of the three primary strategies in Grime's CSR model.*

Trait	Competitors (C)	Stress-tolerators (S)	Ruderal (R)
(i) Morphology			
1. Life forms	Herbs, shrubs, and trees	Lichens, herbs, shrubs, and trees	Herbs
2. Morphology of shoot	High dense canopy of leaves; extensive lateral spread above and below ground	Extremely wide range of growth forms	Small stature, limited lateral spread
3. Leaf form	Robust, often mesomorphic	Often small or leathery, or needlelike	Various, often mesomorphic
(ii) Life history			
4. Longevity of established phase	Long or relatively short	Long–very long	Very short
5. Longevity of leaves and roots	Relatively short	Long	Short
6. Leaf phenology	Well-defined peaks of leaf production coinciding with period(s) of maximum potential productivity	Evergreens, with various patterns of leaf production	Short phase of leaf production in period of high potential productivity
7. Phenology of flowering	Flowers produced after (or, more rarely, before) periods of maximum potential productivity	No general relationship between time of flowering and season	Flowers produced early in the life history
8. Frequency of flowering	Established plants usually flower each year	Intermittent flowering over a long life history	High frequency of flowering
9. Proportion of annual production devoted to seeds	Small	Small	Large
10. Perennation	Dormant buds and seeds	Stress-tolerant leaves and roots	Dormant seeds
11. Regenerative strategies*	V, S, W, B_s	V, B_{sd}, W	S, W, B_s

(iii) Physiology

12. Maximum potential relative growth rate	Rapid	Slow	Rapid
13. Response to stress	Rapid morphogenetic responses (root–shoot ratio, leaf area, root surface area) maximizing vegetative growth	Morphogenetic responses slow and small in magnitude	Rapid curtailment of vegetative growth, diversion of resources into flowering
14. Photosynthesis and uptake of mineral nutrients	Strongly seasonal, coinciding with long continuous period of vegetative growth	Opportunistic, often uncoupled from vegetative growth	Opportunistic, coinciding with vegetative growth
15. Acclimation of photosynthesis, mineral nutrition, and tissue hardiness to seasonal change in temperature, light, and moisture supply	Weakly developed	Strongly developed	Weakly developed
16. Storage of photosynthate and mineral nutrients	Most photosynthate and mineral nutrients are rapidly incorporated into vegetative structure, but a proportion is stored and forms the capital for expansion of growth in the following spring	Storage systems in leaves, stems, and roots	Confined to seeds

(iv) Miscellaneous

17. Litter	Copious, not usually persistent	Sparse, often persistent	Sparse, not usually persistent
18. Palatability to unspecialized herbivores	Often high	Low	Usually high

---

[*] V = vegetative expansion, S = seasonal regeneration in vegetation gaps, W = numerous small wind-dispersed seeds or spores, Bs = persistent seed bank, Bsd = persistent seedling bank. Reproduced from Grime (2001), with permission.

Grime was adamant that it is impossible for plants to be adapted to high levels of stress and disturbance simultaneously. It is, in his words, an "untenable" strategy. This decision was not justified by much data, just some conceptual logic. But logic can be flawed. And new data can be brought to light. Other ecologists proposed that the *r-K* selection gradient could be expanded into two dimensions by adding a gradient of "impoverishment" selection to create four primary strategies (Taylor et al., 1990), but unfortunately this approach never caught on. Let's explore this decision to deny the existence of this fourth strategy in some detail.

Grime proposed that the adaptive responses of plants are explained by the importance of habitat productivity and habitat duration. Stress limits habitat productivity and disturbance limits habitat duration. Agents of stress, such as nutrient, water, and light limitation, and sub-optimal temperatures limit biomass production. On the other hand, disturbance destroys biomass through physical forces such as herbivory, fire, or windstorms. Grime (1979) focused on the corners of the two-dimensional plane in his description of the primary plant strategies. By focusing on the corners, four extreme possibilities emerge. Note that by emphasizing the extremes Grime uses a categorical description of a concept that is inherently based on continua.

According to Grime, only three corners are viable: no plant can be adapted to highly disturbed and unproductive habitats, leaving the fourth corner "untenable" (Figure 2.11b). Grime described secondary strategies that are intermediary between the three extremes: stress-tolerant ruderals (SR), competitive ruderals (CR), stress-tolerant competitors (SC), and the stress-tolerant-competitive ruderal (CSR). These intermediate categories are helpful because they emphasize that the model is grounded in continuous variation in traits.

This is where things get strange. Grime cut the two-dimensional plane in half into two right triangles and declared that no viable strategies can be found within one of the triangles (Figure 2.11b). He maintained this position even though his research group observed species that could cope in the lowest-nutrient and most-trampled conditions in an experiment where nutrients and trampling varied orthogonally (Campbell et al., 1991). The "untenable" triangle is drawn with too sharp of a pen. There are so many more possibilities in this simple hypothetical figure. If we relax the assumption that nothing is tenable below the line and instead consider that plants have evolved to adapt to, say, average rates of disturbance and slightly above average stress (point A in Figure 2.11d), or average stress levels and slightly higher than average rates of disturbance (point B in Figure 2.11d), then other possibilities emerge. Possibilities outside the boundary of the right-triangle seem plausible, but Grime eliminated their possibility by throwing out the entire lower triangle! Such rigidity is difficult to justify in a world where continuous variation in life history traits, functional traits, and environmental gradients exist.

It is unsurprising that many ecologists have already challenged Grime's untenable triangle. Told that something does not exist, some scientists will spend the rest of their careers searching for it, for example, dark matter, or the Higgs boson. Likewise, several ecologists searched high and low for plants that persist in simultaneously stressful and frequently disturbed sites, and they were not that hard to find. For example, there is a remarkably rare species of beardtongue called blowout penstemon (*Penstemon haydenii*) that grows on shifting sands in Wyoming and Nebraska. One population grows on the Ferris Dunes at the downwind end of the Great Basin Divide in Wyoming (Figure 2.12) (Heidel et al., 2014; Tilini et al., 2017). This environment is both unstable due to frequent wind, and climatically stressful—it only receives about 300 mm of precipitation annually, the mean annual temperature is around 5 °C, and the sand is infertile. Is this species an anomaly? Peter Grubb (1985) does not think so, and argued that plant species that persist on continuously disturbed sand dunes in regions that also experience drought are excellent examples of plants that have solved the problem that Grime thought untenable. Grubb (1980) also cited *Xanthorrhoea australis* and *Helichrysum obtusifolium* as species that are benefitted by fire yet grow in the dry and nutrient-deficient heaths of South Australia. Plants that grow on frequently disturbed scree fields in cold alpine habitats also might fit the bill.

Herben and colleagues (2018) used Ellenberg indicator values to test this idea. Recall that Ellenberg values are neither traits nor environments, they are ordinal values based on the typical environment in which a species is found across their range. These authors plotted Ellenberg nutrient values against disturbance intensity values and found that this space was completely filled, indicating that many species were found on sites with poor soils that were frequently disturbed (Figure 2.12). These authors also plotted nutrient values against disturbance severity values and saw that very few plants actually grew on sites with poor soils that were severely disturbed, suggesting that no plants can tolerate extreme infertility and extreme disturbance severity. Perhaps no strategy is tenable under extreme stress and disturbance, but the bivariate data in Figure 2.12 refute the rigid boundary line illustrated in Figure 2.11b.

Another problem that defies the logic of algebra and geometry arises when we examine the inherent assumptions of a quantitative analysis that projects the "tenable triangle" in Figure 2.11b into a ternary space illustrated in Figure 2.11c. Grime started with a two-dimensional figure defined by habitat duration and productivity, then cut that plane in half to create a right-angle triangle, then transformed it into an equilateral triangle defined by three dimensions: the importance of competition (Ic), the importance of stress (Is), and the importance of disturbance (Id). In such a ternary space, all three variables must sum to a constant (often 100%), which means that these three variables are not independent. In other words, if $Ic + Is + Id = 100\%$ and we know the values of Ic and Is, then we automatically know the value of Id (i.e., $Id = 100\% - Ic - Is$). All three values cannot vary independently from one another, so there are only two degrees of freedom in this triangular space. These work well in situations like soil texture triangles that are defined by three fractions of particle size that sum to 100%. However, the translation of the "tenable upper triangle" into a ternary space assumes that we can quantify Ic, Is, and Id independently. The problem is that they are theoretical constructs, bereft of empirical quantification. Loehle (1988b, p. 286) identified the geometric distortions that are generated by such triangular models and concluded that "the triangular data representation scheme distorts data, results in loss of information, and generates overly restrictive assumptions about strategic or environmental trade-offs." In other words, a triangle cannot adequately represent a tesseract.

Sand dunes are constantly shifting, low-productivity habitats

(a)   (b)

Scatterplot of Ellenberg values across species with the CSR triangle superimposed

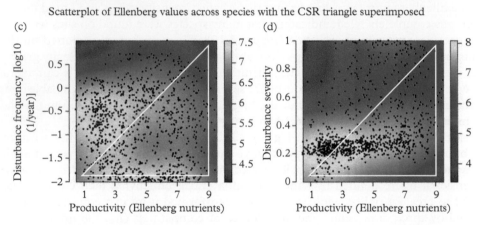

(c)   (d)

Figure 2.12 *(a) Sand dunes are an example of a habitat that is unstable because of frequent wind and are low in productivity due to infertile sands and a cold–dry climate. The Ferris Dunes are at the downwind end of the Great Basin Divide in Wyoming, USA. (b) Blowout penstemon (*Penstemon haydenii*) lives in these sand dunes and appears to be adapted to both constantly shifting sands and infertile soil. (c) Plot of species-level Ellenberg values (each dot represents a species) arrayed along gradients in productivity and disturbance frequency, and (d) productivity and disturbance severity. Grime's CSR triangle is superimposed in white onto each of the scatterplots to indicate that the triangular model does not recognize many species that grow in low productivity and high disturbance environments.*

(a): Photo by Kaylan Hubbard, used with permission. (b): Photo by Bonnie Heidel, used with permission. (c, d): Reproduced from Herben et al. (2018) with permission.

Ecologists continue to search for traits that serve as proxies for the importance of competition, stress, and disturbance. Methods have been developed to take measurements of plant traits and project these into a ternary space to quantify the CSR strategies in both herbaceous and woody plant species. Hodgson and colleagues (1999) used six functional traits (canopy height, leaf dry matter content (LDMC), flowering period, lateral spread, leaf dry weight, specific leaf area (SLA), and flowering start date for forbs) to classify species into the ternary space of plant strategies. This approach relies on the assumption that combinations of these traits are good proxies for the three non-independent variables Ic, Is, and Id. Pierce and colleagues (2013) simplified things even further and proposed that all we need to know is the SLA, LDMC, and the leaf area to determine the location of species in the triangular space. SLA was viewed as a proxy for Id, where higher SLA was associated with the ruderal strategy. Leaf area was a proxy for Ic, where higher leaf area was associated with the competitor strategy. LDMC was a proxy for Is, where higher LDMC was associated with the stress-tolerator strategy. Curiously, it was deemed sufficient to use three mathematically dependent traits measured on a single plant organ (leaves) to classify plants into three distinct strategies. They are mathematically dependent because SLA uses leaf area in the numerator and dry leaf mass in the numerator and LDMC uses dry leaf mass in the numerator. This approach implies that root traits, stem traits, flowering traits, or vegetative spread tell us little about plant strategies. I am heartened by these attempts to operationalize the classification of Grime's CSR model using continuous functional traits, but I remain puzzled that leaf traits are used to represent the concepts Ic, Is, and Id, rather than independent variables that actually quantify competition, stress, and disturbance.

Why don't we measure disturbance and productivity directly? Grime (2001, p. 105) wrote that "there are no easy and reliable methods of measuring productivity and disturbance in the field." Basing plant strategy theory on quantities that cannot be measured sounds more like string theory than ecology. The lack of independent units on the axes is perhaps one of the most problematic features of the triangular CSR model—robust ecological theories must be operational. By this I mean that the theory should be testable from the essential defining features of the theory. Habitat duration is an essential driver, yet how is this measured? And what about habitat productivity—do we measure this in terms of gross primary production? Without measurable data, everything else is hand waving.

Let us say for a moment that we can quantify them. Let us simulate global rates of net primary production and disturbance and assume that these variables are independent and identically distributed. Figure 2.11d shows a scatterplot of these simulated environments, which fill the plane as any multivariate normal data would do. These simulated points certainly do not look like a triangle, and the "untenable triangle" denies the possible existence of all the possible points that are located in the lower triangle. It is clear that Grime dismisses the very real possibility that strategies could evolve in approximately double the number of situations defined in his model. In other words, CSR exhibits a heavier reliance on hand-drawn axes than on the principles of geometry, algebra, and statistical probability.

Grime's most enduring legacy is his leadership in demonstrating the importance of screening plant traits systematically (Keddy and Laughlin, 2022). Grime and his colleagues used linear multivariate analysis to test this ternary model on one of the most impressive screenings of plant traits under standardized conditions ever to have been undertaken (Grime et al., 1997b). The ordination that was produced suggested that at least three-dimensions were needed to explain variation in traits among these British species. The authors concluded that the first dimension represented an axis of acquisition versus conservation (which we now call the leaf economics spectrum), the second axis represented a phylogenetic contrast between dicots and monocots, and the third axis represented variation in life history traits. If we conveniently ignore the second axis and plot the first and third axes, then a two-dimensional ordination appears, that is, a plane of points.

The authors then did something rather curious. They fit a triangular peg into a square hole by imposing the CSR triangle over the two-dimensional plane as evidence for the existence of the CSR model. According to Wilson and colleagues (2019, p. 245), "the fit of a scatter to a triangle could be questioned: a triangle can be placed across any swarm of points, and in fact a trapezoid fits better!" This superimposition projected a three-dimensional ternary space onto a two-dimensional plane, and the result was less than convincing.

Grime knew he was not the first to propose a three-strategy model, and the similarities between his model, first published in 1974, and that proposed by Ramenskii (1938) are uncanny (recall the Prologue). Rabotnov (1985) lamented how the writings of Ramenskii have made little impact on the non-Russian speaking botanists—it makes me wonder how many powerful ideas have been lost due of the diversity of languages and the barriers to communication that exist.[8] It is also worth noting that the proposed fundamental drivers of Grime's (1974) three-strategy model (productivity and disturbance) were on the minds of his contemporaries as well, and three-strategy models appeared in a variety of other papers (Table 2.4). Van Valen (1971) proposed the existence of three species types: colonizing species were good dispersers, equilibrium species were good competitors, and resistant species could tolerate adverse environmental conditions. Stress and disturbance are the same axes in Southwood's (1977) habitat template where species are adapted to gradients in habitat "favorableness" and "predictability." Whittaker and Goodman's (1979) demographic models generated support for a third "adversity" strategy—they described three demographic strategies as exploitation selected, saturation selected, and adversity selected. Greenslade (1983) also proposed a strategy that was orthogonal to the r-K selection gradient called the "adversity gradient." All in all, as is true for all good ideas in biology (recall the controversy over whether it was Darwin or Wallace who originated the theory of evolution by natural selection), many of the same concepts and ideas were independently brewing in the minds of ecologists around the world in the middle of the twentieth century. More recently, Grime (2001) was very pleased to see the spatial modeling results of Bolker and Pacala (1999, p. 591) who showed that "colonization, rapid exploitation, and tolerance . . . emerge naturally from a simple model of spatial competition and local dispersal."

---

[8] Relevant examples in the opposite direction also exist: In his presidential address, Harper (1967) emphasized Yoda's self-thinning rule, which may have been relegated to a lesser-read Japanese publication (Yoda, 1963) if Harper hadn't popularized it.

Table 2.4 *Parallel concepts among three-strategy models for limiting resources, general three-strategy models, and a three-strategy demographic model, and how they relate to avoidance, tolerance, and escape strategies.*

Models	Three Strategies		
	Acquisitive	Conservative	Proliferous
Strategies for specific resources			
Drought (May and Milthorpe, 1962, Fischer and Maurer, 1978, Levitt, 1980)	Avoid	Tolerate	Escape
Shade (Schemske et al., 1978, Gommers et al., 2013)	Avoid	Tolerate	Escape
Soil infertility	Avoid	Tolerate	Same as drought escapers?
Low oxygen (Akman et al., 2012)	NA	Quiescence	Escape
General strategy models			
Ramenskii (1938)	Violents	Patients	Explerents
Grime (1979)	Competitor	Stress-tolerator	Ruderal
Van Valen (1971)	Equilibrium	Resistant	Colonizing
Greenslade (1983)	K-selected	Adversity-selected	r-selected
Whittaker and Goodman (1979)	Saturation-selected	Adversity-selected	Exploitation-selected
Bolker and Pacala (1999)	Exploitation	Tolerance	Colonization
Demographic models			
Silvertown's demographic triangle (Silvertown et al., 1992)	Growth	Survival	Reproduction

2.6   Three-strategy models tend to mirror three demographic rates

Grime proposed his triangular CSR model to explain plant strategies globally. Grime himself discussed arid and cold habitats as examples of stressful sites dominated by stress-tolerant species and he discussed riparian zones or temperate meadows as productive sites dominated by competitors. But other ecologists emphasized Grime's model as a way to interpret strategies within a habitat, where different strategies exploit the spatial and temporal dynamics generated by disturbance and resource heterogeneity. Shugart (1997, p. 34) wrote that CSR "is a set of rules that can be used to predict a pattern of strategies (and associated life-forms) expected under a particular environmental regime." Mark Westoby similarly emphasized that strategies should explain patch dynamics within a habitat as well as environmental differences across habitats (Westoby and Leishman, 1997). We saw earlier that life history strategies were developed most often to explain responses to patch dynamics within a habitat, which suggests that demographic rates that drive life history variation are keys to understanding plant strategies.

A compelling pattern has emerged from the summary of different three-strategy models that begs the question about how these strategy models relate to plant demography (Table 2.4). The avoid–tolerate–escape strategies can be viewed as responses to limited water, light, and nutrients. The avoidance strategy is most closely aligned with growth and acquisition, the tolerance strategy is aligned with survival and being conservative, and the escape strategy is aligned with reproduction and being a proliferator. This perspective contrasts with Grime's view that only the stress-tolerant strategy is adapted to limited resources, and the similarity of the avoid–tolerate–escape strategies with Grime's CSR is undeniable (Volaire, 2018). This suggests that Grime's model may be helpful as a conceptual description of different strategies for dealing with resource limitation through different demographic responses. Indeed, natural selection can produce alternative strategies for the same environmental conditions (Marks and Lechowicz, 2006).

Jonathan Silvertown and Miguel Franco compared the importance of growth, survival, and reproduction across a range of plant species and plotted the results in a ternary space that they called the "demographic triangle"(Silvertown and Franco, 1993; Silvertown et al., 1993; Franco and Silvertown, 2004). Each corner of the triangle represents a demographic rate. Species can be plotted within the triangle based on the relative importance of growth, survival, and reproduction to their population dynamics. Chapter 4 evaluates in more detail the demographic methods used to generate these triangles, but it is important to mention here because Silvertown compares the demographic triangle with Grime's triangle directly (you could glance at Figure 4.6 if you wish). Silvertown showed that annuals cluster together where reproduction is the most important demographic rate. Semelparous perennials (plants that flower once then die) cluster where growth and reproduction are important. Iteroparous herbaceous perennials (plants that flower continuously throughout their life) of open habitats tend toward the middle of the triangle, whereas iteroparous herbaceous perennials in forests tend to cluster together because they exhibit lower fecundity. Long-lived shrubs and trees tend to cluster where survival is most important.

Grime (1997a, p. 123) recognized the importance of this demographic approach by saying that it was a "very welcome development." Silvertown proposed that his demographic triangle could provide a demographic interpretation of Grime's CSR triangle. He hypothesized that fecundity would be indicative of ruderalness, growth would be indicative of competitiveness, and that survival would be indicative of stress tolerance.

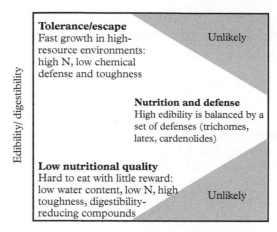

Figure 2.13 *A three-way model of plant defense syndromes.*
Adapted from Agrawal and Fishbein (2006) with permission.

He directly compared the demographic traits of 18 herbaceous plants to their locations in the CSR triangle, but he found no correlation between his demographic triangle and Grime's triangle (Silvertown et al., 1992). This negative result does not mean that demographic and functional approaches are incompatible. Rather, it simply means that we haven't figured out how to synthesize them yet. Grime's model combined life history and ecophysiological traits (Table 2.4), yet it never made the link between traits and fitness clear. Chapter 3 shows how we can take a demographic approach to comparative functional ecology to advance plant strategy theory.

Another example of a three-way model emerged from proposals to generalize our understanding of plant response to herbivory and the evolution of plant defense strategies (Stamp, 2003; Agrawal, 2007). Any attempt to develop of general theory of plant strategies that ignores the powerful role that herbivores have played in the evolution of plants is doomed to fail. Indeed, plants would be far more efficient if they did not have to mount defenses against constant attacks. Grime viewed his acquisitive–retentive axis as a response to selection for preventing leaves from being eaten in resource-limited environments (Table 2.3). Fast-growing palatable species often have higher protein content and less lignified structural tissue in their leaves than slow-growing unpalatable species. Agrawal and Fishbein (2006) proposed a three-way plant defense syndrome triangle defined by variation in edibility/digestibility, which aligned with Grime's acquisitive–retentive axis, and toxicity/barriers to feeding (Figure 2.13). The presence of toxic chemicals and trichomes is most likely to occur in plants that exhibit intermediate edibility because plants with large investment in toxins should not have the resources to produce leaves with high nutrition and plants with leaves that are low in nutrients do not need to invest in toxins.

2.7 Grubb and Craine proposed additional stress-tolerant strategies

In 1877, Leo Tolstoy began *Anna Karenina* with the following observation: "Happy families are all alike; every unhappy family is unhappy in its own way" (Tolstoy, 2014, p. 3, translated from Russian by Rosamund Bartlett). This "Karenina" principle can be applied to wild habitats: there is one way for a habitat to be productive, that is, no resource is strongly limiting; but a deficiency in any one resource can lead to multiple ways of being unproductive. Similar ideas were proposed in the centrifugal model of plant communities, where there is one way to be a favorable habitat, but multiple ways for a habitat to be unfavorable, including flooding, infertility, shallow soil, etc. (Keddy and MacLellan, 1990; Wisheu and Keddy, 1992). Grime, however, lumped all species adapted to sub-optimal conditions into the same stress-tolerator strategy, regardless of which resource was limiting, and claimed that they all grow slowly and conserved their tissues for extended periods, and nothing seemed to irk Grubb more.

Grubb thought that Grime's simple model, while generally useful, was hopelessly simple and could never explain the great variation of strategies seen in nature. Grubb promoted a "positive distrust in simplicity" (Grubb, 1992, p. 586). Grubb argued, in a variety of compelling papers and book chapters (Grubb, 1985, 1992, 1998a), that there are more than three primary strategies considering the simple fact that plants can be consumed by fire, buried by floods, and eaten by plagues of locust, and their growth is limited by no less than low water availability, infertile soils, shade, toxins, frost, and heat. In his words, "Simple generalizing schemes as reference points for thought and communication must be evolved, but . . . the simplest schemes cannot serve for adequate thought and communication because they fail to capture many of the major axes of variation found in nature" (Grubb, 1998a, p. 4).

Grubb objected to many things about the CSR model (Grubb, 1976, 1985), but three things stand out. First, he thought the idea of an "untenable" strategy was not logical and argued that many species tolerate frequent disturbance in unproductive environments. Second, he disliked that disturbance was treated as a single gradient, given that disturbances can be both persistently chronic and catastrophically abrupt, and given the multiple dimensions of disturbance regimes including frequency, intensity, timing, and extent. Third, he objected to a single stress-tolerant strategy (Grubb 1998a),[9] on which I elaborate here.

Grubb was averse to simple models. He predicted that the end point in plant strategy theory would be an "oligo-dimensional, somewhat untidy schema" (Grubb, 1998a,

---

[9] Grubb, like John Harper, hated the word "stress" and argued that Grime's stress axis was more accurately called a "productivity" axis (Grubb, 1985). Thom Kuyper was sympathetic to this argument, and once told me that "as long as we cannot buy a stress meter, we cannot quantify it, so its usefulness in ecology is extremely questionable, to put it mildly" (pers. comm.).

p. 25). But to his credit he also provided rational arguments for how to expand on Grime's ideas—not just abandon them altogether. Notably, he proposed to expand the stress-tolerant strategy into three distinct strategies: low-flexibility, switching, or gearing-down. The names for these have not caught on in the literature, so I briefly review them here. It is perhaps somewhat surprising that these ideas have not been thoroughly tested given that interest in intraspecific trait variation and plasticity has experienced a revival in the last decade (Siefert et al., 2015). Like all good, well-defined strategies, these three come with a set of coordinated traits.

The low-flexibility strategy exhibits long-lived leaves, slow growth rates, and inflexible form and gas exchange rates (photosynthesis and respiration) even when supplied with the limiting resource. This is perhaps the closest analog to Grime's stress-tolerator, but Grubb saw two other types of species that survived in resource poor conditions. The switching strategy includes plants that are like Grime's stress-tolerator when immature, but which can exhibit fast growth rates in adulthood. The gearing-down strategy was proposed for plants that can drop their respiration rates when confronted with limited resources. Grubb provided detailed examples of each strategy with respect to mineral nutrients, water, and light. Interestingly, Grubb emphasized the ability to survive under resource scarcity, rather than the growth rates under those conditions which is emphasized by Chapin's (1980) and Grime's (1979) models. Both are important. Grubb (1977) also emphasized the importance of the regeneration niche. He proposed that each species has a unique set of optimal circumstances for recruitment, and this helps to explain coexistence in species-rich communities. This emphasis on regeneration was novel given that many approaches emphasized growth rates and survival. Integrating the regeneration niche into plant strategy models remains an important challenge.

Joe Craine (2009) was also dissatisfied with the way that Grime's model lumps all the adaptations to limiting resources into a single stress-tolerant category. Craine's *Resource Strategies of Wild Plants* focuses entirely on unpacking stress tolerance into distinct strategies, but he took a different approach from Grubb. Craine focused on the traits that enhance growth under each limiting resource and had no interest in evaluating plasticity. Craine deconstructed one of Grime's (2001) most provocative claims in his revised treatise. Recall that Grime lumped all the stress-tolerant species into a single category, regardless of their adaptation to low nutrients, low water, or low light, because they all grow slowly and conserve their acquired nutrients and tissues. Grime (2001, p. 65–66) suggested that the similarity in traits is due not to the convergence of these traits under different limiting resources, but that they are all caused by the same underlying factor: limited nutrients. Craine called these the "convergence" and "commonality" hypotheses, respectively. The commonality hypothesis proposes that shade-tolerant and drought-tolerant species exhibit stress-tolerant traits because their most limiting resource is mineral nutrients, not light or water.

Despite the fact that Craine ignored life history traits in a monograph on plant strategies (Enquist, 2010), he carefully articulated the sets of traits that represent strategies for low nutrients, low light, low water, and low CO_2 while simultaneously integrating herbivore defense mechanisms (Figure 2.14). He geometrically expanded Grime's triangle

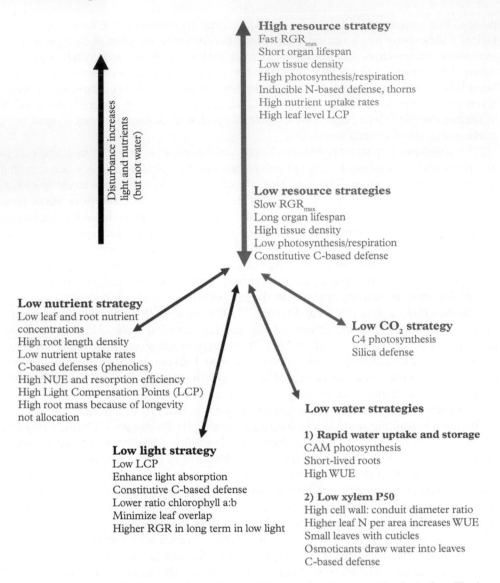

High resource strategy
Fast RGR_{max}
Short organ lifespan
Low tissue density
High photosynthesis/respiration
Inducible N-based defense, thorns
High nutrient uptake rates
High leaf level LCP

Disturbance increases
light and nutrients
(but not water)

Low resource strategies
Slow RGR_{max}
Long organ lifespan
High tissue density
Low photosynthesis/respiration
Constitutive C-based defense

Low nutrient strategy
Low leaf and root nutrient
concentrations
High root length density
Low nutrient uptake rates
C-based defenses (phenolics)
High NUE and resorption efficiency
High Light Compensation Points (LCP)
High root mass because of longevity
not allocation

Low CO_2 strategy
C4 photosynthesis
Silica defense

Low water strategies

1) Rapid water uptake and storage
CAM photosynthesis
Short-lived roots
High WUE

2) Low xylem P50
High cell wall: conduit diameter ratio
Higher leaf N per area increases WUE
Small leaves with cuticles
Osmoticants draw water into leaves
C-based defense

Low light strategy
Low LCP
Enhance light absorption
Constitutive C-based defense
Lower ratio chlorophyll a:b
Minimize leaf overlap
Higher RGR in long term in low light

Figure 2.14 *Craine's (2009) pyramid model of plant strategies, as interpreted by the author. The four low resource strategies all share a set of common traits listed under Low resource strategies. The traits that are unique are listed beneath each individual strategy separately.*

into a pyramid. While this may be conceptually attractive, it lacks a quantitative justification because it implies that only one end of the main axis radiates into four orthogonal axes. The principal gradient in Craine's model is a contrast between the high-resource strategy and all other low-resource strategies (Craine et al., 2012a).

The high-resource strategy includes plants with fast growth rates, short organ lifespans (Craine et al., 1999), low tissue density (Craine et al., 2001), high rates of photosynthesis and respiration (Tjoelker et al., 2005), high light compensation points (Lessmann et al., 2001), and high nutrient uptake rates (Chapin, 1980). In high-resource environments, plant defense mechanisms include inducible nitrogen-based chemicals because nitrogen is not limiting (Ohnmeiss and Baldwin, 1994). These species are most common in ecosystems where light and nutrients are continually replenished through disturbance. There is no difference between species adapted to high light versus high nutrients or high water.

In contrast, low-resource strategies are separated into low-nutrient, low-light, low-water, and low-CO_2 strategies. All of these strategies share a common set of traits: slow RGR (Grime and Hunt, 1975), long-lived organs (Craine et al., 2002), high tissue density (Craine et al., 2001), low rates of photosynthesis and respiration (Poorter et al., 1991; Wright et al., 2004; Tjoelker et al., 2005), and constitutive carbon-based defenses (Craine, 2009). These are the traits that exhibit "convergence" across different resource limitations. However, each low-resource strategy exhibits unique sets of traits that evolved in response to the unique features of each resource, which Craine used as arguments against Grime's commonality hypothesis.

The low nutrient strategy exhibits low leaf and root nutrient concentrations (Grime et al., 1997b), high root length density (Craine et al., 2001), low nutrient uptake rates (Chapin, 1988), high nutrient-use efficiency and resorption efficiency (Berendse et al., 1992; Aerts, 1995; Aerts and Chapin, 1999; Vitousek, 2018), high light compensation points, and high root mass because of longevity not allocation. Plant defenses are carbon-based because nitrogen is too precious to allocate to defense.

The low light strategy exhibits low light compensation points (Craine and Reich, 2005; Lusk and Jorgensen, 2013), a higher number of mesophyll cells to enhance light absorption (DeLucia et al., 1996), a lower ratio of chlorophyll a:b (Niinemets and Tenhunen, 1997), minimized leaf overlap, and high RGR in the long term in low light (Sack and Grubb, 2001). Plant defenses tend to be constitutive carbon-based defenses when light is limiting.

The low water strategy includes two types. The first type has evolved in response to water pulses and is exemplified by succulents that exhibit rapid water uptake and storage, short-lived roots (Dubrovsky et al., 1998), and CAM photosynthesis that gives them high water-use efficiency (Nobel, 1991). The second type evolved in response to chronically low water availability and exhibit small leaves (Fonseca et al., 2000), high cell wall-to-conduit diameter ratios to reduce stem vulnerabilities to embolism formation (Hacke et al., 2001; Choat et al., 2012), higher leaf nitrogen per area to increase water-use efficiency (Wright et al., 2001), leaf cuticles, and osmoticants that draw water into leaves (Adams et al., 1981). Plant defenses tend to be carbon based when water is limiting.

Lastly, the low CO_2 strategy includes C_4 photosynthesis (Sage, 2004) and silica-based defenses (Hodson et al., 2005). Craine argued that silica-rich C_4 grasses evolved when atmospheric concentrations of CO_2 were historically low. Using silica as a defense to slow herbivore ingestion allows this strategy to allocate carbon to growth and respiration.

The four low-resource strategies form the base of the pyramid, and the high-resource strategy forms the point at the top. This is yet another triangular-type model without defined axes and a conspicuous lack of empirical data to support this geometry, but the model is a useful conceptual description of different ways to tolerate different limiting resources. Indeed, it follows the Karenina principle rather well. A recent analysis of ordinal rankings of species tolerance to drought, shade, and waterlogging indicated that there are indeed multiple dimensions to stress tolerance (Puglielli et al., 2021). Species that are drought tolerant are intolerant of water logging, and shade tolerance was independent of both drought- and water-logging tolerance. It would be useful to conduct a similar analysis using continuous traits to test the geometric assumptions of Craine's model. And with that, we have reached the most recent turning point in the development of plant strategy models—the modern emphasis on multidimensional models of continuous plant traits.

2.8 Westoby's continuous three-dimensional plant strategy scheme

Mark Westoby was fascinated with the idea of comparing species globally based on a core set of continuous traits. But after spending some time in Sheffield, he concluded he could not fully agree with Grime's model and became frustrated with the difficulties of classifying species into CSR types (Westoby, 1998). Westoby sought a solution that did not rely on ill-defined concepts mired in unresolvable controversy such as the "untenable corner," "stress," and "competition," and proposed a plant strategy scheme[10] where the strategy of any vascular plant species can be computed as the location in a three-dimensional space defined by SLA, canopy height at maturity,[11] and seed mass (Figure 2.15) (Westoby, 1998, 2007). This Leaf–Height–Seed (LHS) strategy scheme was proposed to operationalize species comparisons at a global scale. All these traits influence dispersal, establishment, and persistence (Weiher et al., 1999), and he chose these traits carefully to represent the key fundamental trade-offs faced by plants. In his words:

---

[10] I have learned throughout my travels abroad that one word can have different meanings in the English language. In virtually all English-speaking countries except the United States, the word "scheme" is used innocently as a plan or a program. The teller at my bank in New Zealand always tried to get me to sign up to a new "scheme" they were offering—I invariably gazed back at them with suspicion. In the United States, the word "scheme" often suggests a devious plot, the action of a swindler, a plan not to be trusted. Funnily enough, a synonym for scheme is "strategy," so the subheading for this section could be "plant strategy strategy." Because of my North American bias, I refrain from using the phrase "plant strategy scheme" unless it is in direct reference to Westoby's model.

[11] The meaning of "height at maturity" evolved over time. Westoby (1998) thought of height at maturity as the asymptotic height, that is, "the canopy height that species have been designed by natural selection to achieve." Westoby and colleagues (2002) also called this "potential canopy height." This is the data that is available at global scales in the TRY database. It wasn't until Falster and colleagues (2017) when they brought in H_{mat}, which is the height at which allocation switches from height gain to seed production.

It is ineluctable that a species cannot both deploy a large light-capturing area per gram and also build strongly reinforced leaves that may have long lives; cannot support leaves high above the ground without incurring the expense of a tall stem; cannot produce large, heavily-provisioned seed without producing fewer of them per gram of reproductive effort (Westoby, 1998, p. 218).

Grime's CSR strategies and Craine's pyramid are composed of suites of correlated traits, but the LHS scheme is defined by axes of single continuous traits. This is a non-trivial seismic shift in how strategies are quantified and a departure from perspectives on plant strategies that emphasize spectrums of correlated traits. It does not deny the existence of trait correlations, rather it aims to utilize easily measured traits to represent spectrums of correlated traits. For example, SLA represents variation along a leaf economics spectrum, which is composed of multiple correlated traits (Wright et al., 2004). This distinction between the use of a single trait to represent a trade-off versus combining multiple correlated traits into a reduced dimensional axis through data reduction seems to be a major sticking point among some ecologists with whom I have worked.

Grime supported Westoby's initiative because the traits would be easy to collect and one could rapidly score plants at a global scale, but he cast doubt on the scheme because the LHS traits "do not reflect many of the vegetation processes and ecosystem properties of greatest interest to ecologists and land managers" (Grime, 2001, p. xx). Grime continued, "More traits will be needed, but not the prohibitively large number recommended by Grubb (1998a)!" Grime especially doubted that SLA could be a useful

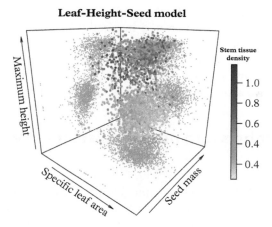

Figure 2.15 *Illustration of the Leaf–Height–Seed plant strategy scheme (Westoby 1998) using the open-access dataset of Carmona et al. (2021). The three-dimensional cloud of points is projected onto the flat surfaces of each pairwise plain as smaller grey points. The points are colored according to variation in stem tissue density. Note that taller species generally have larger seeds and exhibit denser stems. Species form two distinct clouds generated by the fundamental distinction between herbaceous and woody plants.*

predictor given its high intraspecific variability (Grime, 2001), which is curious, since SLA was a strong contributor to Grime's C-S axis (Grime et al., 1997b).

Westoby (1998) thought LHS would succeed if a sufficient proportion of future publications reported these three traits to enable global comparisons. Given that these traits formed the basis of the largest plant trait comparative study ever to have been conducted in the history of plant ecology (Díaz et al., 2016) it appears that his proposal was wildly successful! I found empirical support for these three axes in my own research in the ponderosa pine forest of northern Arizona (Laughlin et al., 2010). I measured these three traits, as well as several others, including fine root traits, on 133 understory plant species. In this paper I asked whether the LHS axes were independent axes of trait variation. Remarkably, SLA, height, and seed mass each loaded strongly on separate principal component axes, demonstrating that these traits represented important and independent axes of functional variation among the dominant species in that ecosystem. However, these three traits are not orthogonal at larger scales (Díaz et al., 2016): evidence shows that seed mass and maximum height are correlated with stem tissue density, reflecting the fundamental contrast between herbaceous and woody plants (Figure 2.15). Many applications have substituted maximum (asymptotic) height (H_{max}) for height at maturity (H_{mat}) because far more data is available for H_{max} and the two variables are positively correlated (Thomas, 2011). Westoby and his colleagues quickly expanded the scheme to include leaf area and twig size (Westoby et al., 2002), as well as stem tissue density and root traits (Westoby and Wright, 2006), but the true number of plant trait dimensions is still an open question (Laughlin, 2014a).

The LHS traits were also included in a core set of plant traits proposed to reflect the key problems facing plants, which brings us back to where this chapter began— in the Lyceum of Ancient Greece. Weiher and colleagues (1999) posed a challenge to Theophrastus's simple categorization and inspired a brave new search for traits that could allow us to differentiate plants according to dispersal, establishment, and persistence (Table 2.5). Seed mass should reflect variation in dispersal and establishment, SLA should reflect variation in establishment and persistence, and height should reflect variation in competitive ability. Other traits may be just as important.

2.9   Demographers and physiologists must work together

We have covered a lot of ground. This chapter has provided a broad overview of the plant strategy models proposed throughout history. I hope that you can now fully appreciate that Grime's model is just one of dozens of ways of approaching plant strategy theory, but that these multiple approaches require synthesis. These models originated from local attempts to explain strategies along single resource gradients to models designed to explain global gradients in plant form and function in response to multiple environmental gradients. Along the way we found that there is a curious fascination with

Table 2.5 *Three aspects of life history can be quantified using continuous variation in plant traits (hard vs. soft traits), an approach that was posed as a challenge to Theophrastus's classical plant classification.*

Challenge	Hard trait	Easy trait
1. Dispersal		
Dispersal in space	Dispersal distance	Seed mass, dispersal mode
Dispersal in time	Propagule longevity	Seed mass, seed shape
2. Establishment		
Seedling growth	Seed mass	Seed mass
	Relative growth rate	Specific leaf area (SLA), Leaf water content (LWC)
3. Persistence		
Seed production	Fecundity	Seed mass, above-ground biomass
Competitive ability	Competitive effect and response	Height, above-ground biomass
Plasticity	Reaction norm	SLA, LWC
Holding space/ longevity	Lifespan	Life history, stem density
Acquiring space	Vegetative spread	Clonality
Response to disturbance	Resprouting ability	Resprouting ability
Stress and disturbance avoidance	Phenology, palatability	Onset of flowering, SLA, LWC

Reproduced from Weiher et al. (1999), with permission.

strategies in sets of three (Table 2.4), yet recent models have expanded to capture the higher dimensionality imposed by resource co-limitation and disturbance regimes.

Garnier and colleagues (2016) noted that there are two types of plant strategy models: "life-history" models that emphasize demography and "ecological" models that emphasize functional adaptations. As described in the preface, the disciplinary division between demography and ecophysiology runs deep in the history of plant ecology (Harper, 1982; Grime, 1984). This age-old debate over whether understanding plants will require a reductionist demographic approach (Harper, 1977) or whether it will require an ecophysiological approach (Grime, 2001) has never fully been resolved, partly because plant strategy models have not adequately integrated demography and physiology into an empirically tractable framework. Despite recent interest in merging these disciplines (Kimball et al., 2012; Moles, 2018; Kelly et al., 2021), divisions remain. Most ecologists would consider both to be essential to a coherent plant strategy theory, but ecologists differ in the traits that they emphasize in their models.

To summarize all these models, I have placed them into a coordinate space (named by their primary authors) defined by first, an emphasis on continuous versus categorical strategies, and second, an emphasis on life history traits versus functional traits (Figure 2.16). The models are color coded by their complexity. This figure will no doubt invoke consternation among authors that disagree with where they have been placed, but after reading this chapter I hope you will agree that some models emphasize life history strategies, others emphasize functional strategies based on ecophysiology, and that the models range from simple to complex and from categorical to continuous. Demographic differences have been especially useful for explaining vegetation dynamics in response to disturbance within a habitat, and models based on functional differences are useful for explaining responses to resource limitation and climate.

The goal of this book is to synthesize these perspectives by articulating a demographic approach to functional ecology. Most empirical tests of plant strategy models are based on observing species with particular traits growing in particular environments. However, mere observations of a phenotype in a habitat cannot constitute a rigorous test of plant strategies because they cannot determine how suitable or unsuitable the habitat is if there is no information on the dynamics of the population. Chapter 3 describes all the necessary ingredients for developing a coherent plant strategy theory to unify demographic and functional approaches by linking traits to fitness.

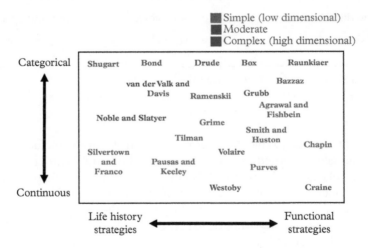

Figure 2.16 *Plant strategy models span three gradients: (1) a gradient in emphasis on life history traits versus functional traits, (2) a gradient from categorical to continuous, and (3) a gradient from simple to complex (colors). This imperfect classification is not meant to be exact because it results from the subjective perspective of the author. Efforts to synthesize these perspectives have emerged, but a full synthesis is still lacking.*

2.10   Chapter summary

- Categorical life forms only partially explain species distributions along climatic gradients. Raunkiaer's life form system is still used today, but it does not differentiate between plants with the same meristem structure that occur in different environments. Box's life form system is the most comprehensive to-date and is currently going through another revision. Plant ecologists have failed to produce a consensus-based short-list of plant functional types for use in dynamic global vegetation models for modeling vegetation-climate feedback.

- Plant strategy models that emphasize demographic life history traits were designed to explain temporal vegetation dynamics in response to disturbances such as grazing, fire, floods, and tree fall.

- Plant strategy models that emphasize morphological, physiological, and phenotypic traits (i.e., functional traits) were developed to explain responses to gradients in resource availability. Some models were designed for single resources, whereas others included more than one resource.

- Strategies of escape, avoidance, and tolerance have been proposed as responses to water, light, and nutrient limitation. The avoidance strategy is most closely aligned with rapid growth, the tolerance strategy is aligned with survival, and the escape strategy is aligned with reproduction.

- Tilman's model assumes that the soil nutrient-to-light ratio is sufficient to explain variation in light and soil resource availability. He proposed that species that grow slower and invest more carbon in roots will be more competitive for nutrients, and species that grow fast and invest more carbon into shoots will be more competitive for light.

- Grime's CSR model is the most widely recognized and cited plant strategy model. The CSR model assumes that habitat productivity and duration are the primary agents of selection and that three main strategies evolved in response to them: competitors, stress-tolerators, and ruderals. Grime expanded the K end of the r-K selection gradient into long-lived competitors and stress-tolerators. He declared that no plant could simultaneously adapt to high stress and high disturbance. The strategies were arranged in an equilateral triangle. Given the variety of trade-offs and environmental gradients that exist in nature, it is unlikely that a triangle can adequately represent a tesseract.

- Grubb and Craine each expanded the stress-tolerant strategy into additional groups. Grubb expanded the stress-tolerant strategy into low-flexibility, switching, or gearing-down strategies for each limiting resource. Craine argued that distinct strategies exist in low-nutrient, low-light, low-water, and low-CO_2 environments.

- Westoby proposed a Leaf-Height-Seed model where strategies can be quantified by the location of a species in a trait space defined by SLA, height at maturity, and seed mass. These traits were chosen because they represent fundamental trade-offs faced by plants. This signaled a significant switch to a fully continuous strategy model where single traits represented spectrums of correlated traits to stimulate global comparisons. Additional trait axes have been added over time, but the true number of axes remains an open question.

- Some ecologists emphasize demographic differences among species and others emphasize functional differences. The goal of this book is to synthesize these perspectives by articulating a demographic approach to functional ecology.

2.11   Questions

1. Do life forms represent plant strategies?
2. Can you think of a plant species that does not fit into any of Box's life form groups in Table 2.1? The only person who has taken up this challenge so far is Peter Grubb, who suggested that the resurrection plant (e.g., *Ramonda* in the Gesneriaceae and *Myrothamnus* in Myrothamnaceae) is missing.
3. Some genera (e.g., *Festuca*) are repeated across multiple groups in Box's table of life forms. Is this a strength or a weakness of the model? Hint: consider the differences between evolutionary convergence and radiations.
4. How do Chapin's and Grime's models of plant strategies differ?
5. Do you agree with the assertion that some plant strategy models emphasize life history differences, whereas others emphasize functional differences?
6. Do avoidance, tolerance, and escape strategies exist with respect to all required resources: water, nutrients, light, and oxygen?
7. Should we aim to define plant strategies for different disturbances (such as fire and grazing) separately, or rather define plant strategies in response to disturbance more generally? Similarly, should we aim to define plant strategies in response to water, nutrients, and light separately, or rather define plant strategies in response to resource limitation more generally?
8. Do we need plant strategy models for individual ecosystems or would one model for the whole world be preferred? What trade-offs exist between these two approaches?
9. The chapter mentioned that using three correlated leaf traits (SLA, leaf area, and LDMC) to differentiate among plant strategies would be difficult. Which traits are fully independent?
10. Over time more axes have been added to plant strategy models. Is there an upper limit to the number of axes that are useful?

3

The Dimensionality of Plant Strategy Theory

I suggest that the end point will be an oligo-dimensional, somewhat untidy schema.

—Peter Grubb (1998a, p.25), "A reassessment of the strategies
of plants which cope with shortages of resources,"
Perspectives in Plant Ecology, Evolution and Systematics

3.1 Dimensionality is a blessing not a curse

Chapter 2 presented an overview of the plant strategy models that have been proposed throughout history. The objective of this chapter is to articulate the dimensionality of a general plant strategy theory used throughout the rest of the book. Each species has experienced their own evolutionary journey and has solved existential problems in ways that make them unique enough to be considered distinct species. But fundamental trade-offs have constrained what is phenotypically possible for each species. Dimensionality of the phenotype is a core concept in plant strategy theory. Data reduction methods explore what is phenotypically possible by seeking low-dimensional representations of matrices composed of N species and P traits. If each species was a fundamentally unique phenotype, then plant strategies would not exist. Plant strategy theory seeks to either categorize species into a small set of strategies or to ordinate species along continuous axes of trait variation. I emphasize the latter approach in this book. The principal question then becomes how many axes of trait variation we need to capture the important variation among species. We can continue to add dimensions by including more traits, but these will add less and less explanatory information. Can we find this asymptotic upper limit of trait dimensionality empirically? What compromises do we make by choosing a single number of trait dimensions? The intrinsic dimensionality of any multivariate set is the minimum number of parameters needed to describe it (Lee and Verleyson, 2007), and dimensional analysis of a theory identifies the base quantities, units of measure, and causal relations among the variables.

What do I mean by plant strategy "theory"? Grime often referred to his model as a theory (Campbell and Grime, 1992), but to purists, theory requires mathematical

Plant Strategies. Daniel C. Laughlin, Oxford University Press. © Daniel C. Laughlin (2023). DOI: 10.1093/oso/9780192867940.003.0003

formalism. Only a few models in the Chapter 2 were explicitly mathematical. Prediction is the goal, yet we currently lack a unified theory that accurately predicts "why a plant of this species, and not of that, is growing in a given spot" (Clapham, 1956, p. 1). Predicting the fitness of any phenotype in any environment is admittedly a lofty goal. By the end of this chapter, a quantitative framework will take shape. By the end of this book, we will have a framework for testing the existence of strategies that incorporates the minimum number of factors needed for a highly predictive, explanatory model. Failure to find any predictive relationship between traits and fitness would constitute a colossal failure of the approach, but that is hardly likely. We know that traits matter, but we still need to acquire knowledge of the contingencies that determine when and how traits determine fitness differences to develop a globally relevant strategy theory. I am hopeful that a new synthesis of empirical and theoretical approaches will emerge to galvanize progress toward this goal.

Allow me to explain a bit more about what I mean by dimensionality. Consider a point that has a dimensionality of zero (Figure 3.1). This point exhibits no variation and could represent a single strategy. Perhaps this is the Darwinian demon, also called a Hutchinsonian demon, that can achieve high fitness in all environments. Demons are useful for thought experiments precisely because they cannot exist due to ecological trade-offs (Kneitel and Chase, 2004; Laughlin, 2018). Increasing dimensionality occurs by extending outward from this singular point in any direction to create a one-dimensional vector. There are an infinite number of points along this single dimension, but we often focus on the contrast between the two ends of the continuum. For example, one dimension is often discussed as a contrast in two strategies. *r-K* selection theory is a perfect example of this—it acknowledges a continuum yet is defined by the endpoints. Again, the Myers–Briggs human personality model does the same thing: I am defined categorically as an introvert even though I am only 64% introverted. It is critical to remember that a single dimension is actually a spectrum of continuous variation, not just two extremes.

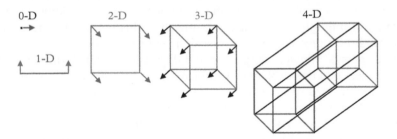

Figure 3.1 *Dimensions arise by expanding outward in an orthogonal direction. The arrows illustrate these extensions: the orange arrow expands from 0-D to 1-D, the gold arrows expand from 1-D to 2-D, the green arrows expand from 2-D to 3-D, and the black arrows expand from 3-D to 4-D. It becomes impossible to adequately illustrate the fourth dimension and beyond, but the mathematics easily extend into multiple dimensions.*

Next, extend outward from each of the two endpoints in an orthogonal (i.e., independent) direction. This process creates a plane in two dimensions. There are now four corners in the two-dimensional plane, each representing unique combinations of the two dimensions. A geometric sequence has emerged:

$$s = 2^d \qquad (3.1)$$

where s = number of categorical strategies, and d = number of continuous dimensions (Scholes et al., 1997). Extend outward in parallel from the four corners and a three-dimensional space emerges. Extend outward in parallel from the eight corners of the cube and a four-dimensional space emerges. It becomes more difficult for us to illustrate these high dimensional spaces but fortunately the mathematics scale easily into higher dimensions (Lay, 2006).

Quantifying the dimensionality of the phenotype has formed the backbone of plant strategy models and is central to determining the nature of plant strategies (Grime et al., 1997b; Westoby et al., 2002). The intrinsic dimensionality of plant traits represents the number of independent axes of functional variation among plants and is therefore a fundamental quantity in comparative plant ecology (Laughlin, 2014a; Mouillot et al., 2021). Given the vast diversity of life on Earth, ecologists are no strangers to the "curse of dimensionality," which is the general difficulty of navigating high-dimensional spaces: "What casts the pall over our victory celebration? It is the curse of dimensionality, a malediction that has plagued the scientist from the earliest days" (Bellman, 1961, p. 94). Equation 3.1 suggests that if we find evidence for four dimensions in trait space, then 16 strategies could be defined to represent the extremes. Many plant strategy models kept the dimensionality low (Tilman, 1988; Grime, 2001), much to the chagrin of ecologists who stressed a more complex view on plant strategies (Grubb, 1998a; Craine, 2009).

After decades of searching, there are many reasons why we have not reached a consensus on the dimensionality of plant strategies. A general theory has settled neither on the dimensionality of the environmental agents of selection, nor has it addressed the relationship between life history and functional traits. Moreover, the dimensionality of traits is only one component of the full dimensionality of a coherent plant strategy theory. This chapter proposes that plant strategy theory must, at a minimum, include the following quantities: a measure of fitness, demographic rates, life history traits, functional traits, temperature, disturbance regimes, limiting resources, and the densities and strategies of neighbors. Each of these components has its own dimensionality and is discussed individually. The causal relationships between these components must be assessed. The high dimensionality of plant strategy theory could be a blessing, not a curse, if it enriches our understanding of plant form and function. David Donoho (2000, p. 18) said "We have stressed the practical difficulties caused by increases in dimensionality. Now we turn to the theoretical benefits."

The test of a good theory is the accuracy of its predictions (Peters, 1991; Keddy and Laughlin, 2022). The strength of plant strategy theory will be judged by its ability to predict which phenotypes can maintain a viable population in a given environment.

For far too long, plant strategy models have relied on observations and proclamations rather than predictive frameworks. For example, it is common to hear "this sclerophyllous shrub with dense wood is stress-tolerant." Observing multiple unrelated species with the same sets of traits in similar environments may suggest that they have converged on a similar evolutionary solution (Niklas, 1997), but mere observations of evolutionary convergence do not by themselves give us the tools to predict responses to the environment. For example, if the environment changed, which it is currently doing at an unprecedented pace, can we predict how the populations will respond?

Throughout his writing, Grime often described trade-offs as "evolutionary dilemmas whereby the assumption of traits conferring fitness in one circumstance result inescapably in loss of fitness in another" (Grime et al., 1997b, p. 260), yet surprisingly he never described how to measure fitness and link it to traits. This singular phrase—"traits conferring fitness"—holds the key to synthesizing life history and functional perspectives on plant strategies. To assume that traits confer fitness because we've observed them in an environment requires a leap of faith—we have not finished the job by testing the performance of different phenotypes in multiple environments. Let's turn that leap of faith into an evidence-based predictive framework.

3.2   Plant strategy theory links traits to fitness

The theory of evolution by natural selection states that if traits are variable, heritable, and have differential effects on individual survival and reproduction, then the frequency of those traits will change in the population over time (Vincent and Brown, 2005). Individuals with advantageous traits will increase in proportion, thereby driving the population toward peaks in the fitness landscape (Wright, 1932; Niklas, 1997). But what is fitness?

Darwinian fitness is defined as "any measure of an individual's relative contribution to the gene pool of the next generation" (Niklas, 1997, p. 14) and is oriented to understanding selection of genotypes within species rather than differences among species. This is clearly a central component of fitness but lifelong reproductive output and the success of progeny is nearly impossible to measure on individual plants (Harper, 1977; Thomas, 2013). Does anyone have any idea how many seeds Methuselah, the 4,850-year-old bristlecone pine tree (*Pinus longaeva*) in the White Mountains of California, produced over its lifetime? No. Nor for that matter do we know how many of those seeds established, grew, matured, and produced their own offspring. I think the best theories are both pragmatic and operational. Theories must be grounded in concepts that can be quantified empirically otherwise they are untestable. Moreover, reproduction rate by itself is not the only important demographic rate. Consider the situation where a population of bristlecone pine trees doesn't produce a single seedling for a century, but all the trees survive. If we measure the growth rate of the population, we will conclude that the population is perfectly stable (we will soon define λ and see that it equals one in this scenario). Not a single tree reproduced, yet the population perfectly persists. This demonstrates the need to consider both rates of births and deaths if we want to understand the fitness of populations.

Population growth rate is a synthetic proxy for fitness because it integrates rates of survival and reproduction (Cole, 1954). This book emphasizes trait selection at the level of the population of a species to evaluate the diversity of plant form and function worldwide. This emphasis on population-level selection does not invoke group selection because it is analogous to modeling frequencies of alleles among genotypes (Roff, 1993). Equating fitness with population growth could lead to absurd predictions that life evolves toward ever faster growth rates over time, but this is avoided when considering the regulatory mechanisms such as density dependence or environmental deterioration (Kokko, 2021). Some evolutionary mechanisms, such as mutations and drift, can lead to deterioration of fitness. It would be incorrect to always view the traits of species as optimal matches to their environment because evolution is continually unfolding and the environment is constantly changing. We cannot assume that phenotypes have achieved peak fitness, but rather we should acknowledge that phenotypes are just fit enough to stay viable (Niklas, 1997). Fitness is also relative because a population only needs to be a little better than the competition, but no more (Bradshaw, 1987a; Garnier et al., 2016). Nevertheless, optimization by natural selection has been (and continues to be) a powerful concept for understanding plant traits in relation to their environment (Mäkelä et al., 2002; Prentice et al., 2014).

This book focuses on population growth rate as the measure of fitness for three reasons. First, measuring lifetime reproductive success of individuals is difficult or impossible for most long-lived species. Second, the long-term goal of plant strategy theory is to understand the process of environmental filtering in community assembly where it is populations that persist or go extinct in a given environment (Keddy, 1992; Grime, 2001; Kraft et al., 2015a; Keddy and Laughlin, 2022). Third, recent theory suggests that traits have stronger impacts at the population level because individual lifetime reproductive success is governed by random variation, otherwise known as luck (Snyder and Ellner, 2018). In other words, stochastic processes are stronger determinants of individual-level reproductive success, but traits are important for understanding population-level success. By focusing on populations, we can make use of stacks of theory developed by evolutionary biologists to explain drivers of population fitness.

Population growth rates of species are variable because their traits determine their fitness according to the specific abiotic and biotic context (Violle et al., 2007; Laughlin et al., 2020b). A population is well matched to an environment if it exhibits, on average, positive growth or stable population sizes over time. Likewise, a population is ill-suited to an environment if it exhibits long-term population declines. I say "on average" and "long-term" because populations naturally exhibit annual fluctuations that may not be related to longer-term trajectories.

Consider the rare Colorado butterfly plant (*Oenothera coloradensis*), a short-lived, monocarpic flower endemic to riparian habitats on the Front Range of the Rocky Mountains.[1] The number of flowering plants have been censused annually for three

---

[1] Colorado butterfly plant is only "semi-monocarpic" because a very small percentage of flowering plants survive to the next year after flowering. It defies the notion that flowering is fatal about 5% of the time (Floyd and Ranker, 1998). Colorado butterfly plant is not a great common name for this species. It is pollinated by moths, not butterflies, and some of the best remaining populations are in Wyoming, not Colorado.

decades (Heidel et al., 2019; Wepprich et al., 2022). The size of the populations fluctuates from year to year, changing from explosive population increases in some years to crashes in others. There appears to be a long-term decline on Crow Creek, but long-term increase on Diamond Creek (Figure 3.2). We need a way to quantify these differences.

The finite rate of population growth in year t (λ_t) is defined as the ratio of the number of individuals (N) in year $t + 1$ to the prior year t:

$$\lambda_t = N_{t+1}/N_t \tag{3.2}$$

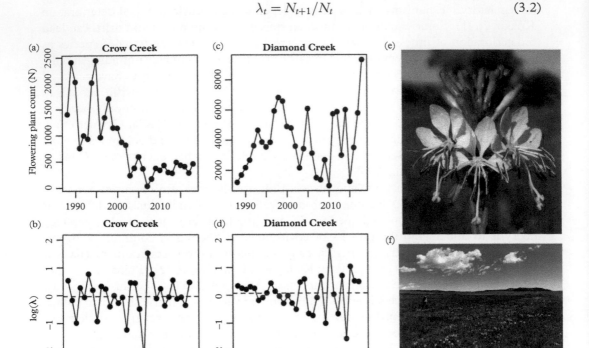

Figure 3.2 *Long-term population dynamics of the rare Colorado butterfly plant (*Oenothera coloradensis*) on two creek subpopulations (Heidel et al., 2019). (a) Censuses of the numbers of flowering plants at Crow Creek and (b) calculations of log(λ) at Crow Creek. The dashed line indicates that the average growth rate is slightly negative (log(λ) < 0) at Crow Creek. (c) Censuses of the numbers of flowering plants at Diamond Creek and (d) calculations of log(λ) at Diamond Creek. The dashed line indicates that the average growth rate is slightly positive (log(λ) > 0) at Diamond Creek. (e)* Oenothera coloradensis *flowers. (f) Wet meadow vegetation at Soapstone Prairie Natural Area in northern Colorado where* Oenothera coloradensis *grows amidst the purple* Iris missouriensis.

(a–d): Data graciously provided by Bonnie Heidel from the Wyoming Natural Diversity Database.
(e): Photo provided by Bonnie Heidel, reproduced with permission. (f): Photo by the author.

The stochastic population growth rate (λ_S) is the average growth rate of the population after taking into account environmental and demographic stochasticity (Ellner et al., 2016):

$$\log(\lambda s) = E[\log(\lambda_t)] \tag{3.3}$$

where E denotes the average expectation. Colorado butterfly plant is experiencing long-term population decline on Crow Creek ($\log(\lambda_S) = -0.037$) and long-term population increase on Diamond Creek ($\log(\lambda_S) = 0.068$).[2] What do you suppose is causing these differences in fitness?

The phenotype of Colorado butterfly plant appears to be adapted to the current environmental conditions on Diamond Creek, but something happened on Crow Creek that is eroding its suitability as habitat for this species. We don't know for sure but allow me to speculate. It turns out that willows (*Salix* spp.) have become more abundant on Crow Creek since the 1990s and willow invasion can both reduce light availability to herbaceous plants and influence the depth of the water table. The Colorado butterfly plant is somewhat shade intolerant and has acquisitive leaf and root traits that allow it to thrive in full sun if there is an adequate water supply. The species is well adapted to the open riparian habitat on Diamond Creek, but the encroaching willows on Crow Creek may have reduced the growth rates of this acquisitive species and have caused reduced opportunities for recruitment. The changing biotic and abiotic environment may have driven these changes in demographic rates, which, in combination, reduced its fitness leading to population decline.

Fitness is determined by three fundamental quantities: age or stage-specific growth, survival, and reproduction.[3] These demographic rates define the dimensionality of fitness. We explore the mathematical details of this definition of fitness in Chapter 4. All things being equal, higher rates of survival and reproduction lead to faster population growth rates—but there is a catch. Rates of survival and reproduction exhibit trade-offs (Stearns, 1992; Roff, 1993; Silvertown and Charlesworth, 2001). It is difficult to increase reproduction rates and at the same time increase survival rates to extend your lifespan—Darwinian demons be damned. The important point here is that if we can understand how traits determine growth, survival, and reproduction rates in a given environment, then we can predict its fitness in that environment. Violle and colleagues (2007) adapted Arnold's (1983) model of morphological determinants of animal fitness for plants. In their model, fitness is determined by measures of performance, which are in turn, driven by functional traits (Violle et al., 2007). In this book, I will describe two analytical frameworks to test plant strategy theory (Figure 3.3).

---

[2] You can reproduce these calculations (and many more throughout the book) using the R code and data found in my github repository at https://github.com/danielLaughlin/PlantStrategies.

[3] Demographic rates change over the lifespan of an organism, so we need to consider how these rates vary across ages and life stages (e.g., seedling, sapling, canopy tree). Note that survival and reproduction rates are sufficient for age-structured population models, but stage and size-structured population models also require information about growth rates. This is described more fully in Chapter 4 on demography.

(a) Predicting the net effects of traits on fitness

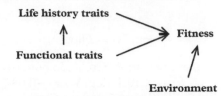

(b) Predicting fitness through demographic rates

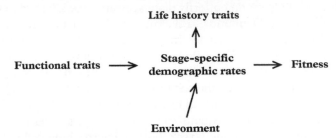

Figure 3.3 *Two analytical frameworks to test plant strategy theory. In both frameworks, demographic rates and fitness are jointly determined by the interaction between traits and the environment. (a) The most direct analytical approach is to predict fitness as a function of life history traits and functional traits, where the effects will depend on (i.e., will interact with) the environment (including both abiotic and biotic conditions). (b) The second approach assumes that traits and environments interactively affect stage-specific demographic rates, which in turn directly determine fitness. The causal relationships among functional traits, life history traits, demographic rates, and fitness is uncertain and is an active area of research. The structures shown here are only two possible causal structures among many. The uncertainty in the second approach lies in whether life history traits will remain as important predictors of fitness given that life history traits are directly calculated from the demographic rates.*

The first framework seeks to predict the net effect of traits on fitness across multiple phenotypes and environments (Figure 3.3a) (Laughlin et al., 2020b). Importantly, the effects of life history and functional traits on fitness depend on the environmental conditions that include both abiotic factors and interactions with neighbors. The details of this framework are developed in Chapters 6 through 9. This first framework doesn't ignore the complexities of demographic trade-offs but merely focuses on the object of natural selection, that is, fitness, to develop predictions for how species will respond to environmental change.

The second framework seeks to predict fitness through trait effects on demographic rates (Figure 3.3b). This framework acknowledges that traits drive trade-offs among demographic rates and seeks to build a mechanistic understanding of the indirect effects of traits on fitness. The effect of functional traits on demographic rates will, of course, depend on the environmental conditions, including both abiotic and biotic contexts. This second framework reflects the idea that life histories are an outcome of functional trait effects on demography. The details of this framework are developed in Chapters 11 and 12.

The first framework distinguishes between life history and functional traits because they are fundamentally different. Functional traits are heritable morphological, physiological, or phenological attributes that indirectly influence fitness by affecting growth, survival, and reproduction. This definition aligns well with Roff's (1993) categorization of traits. Roff distinguished four types of traits in his analysis of trait heritability: life history, morphological, physiological, and behavioral traits. If one considers phenological traits to be the plant equivalent to animal behavior traits, then the latter three categories align perfectly with the definitions of functional traits in Violle and colleagues (2007) and Garnier and colleagues (2016). Functional traits can be measured on individual plants in the field and glasshouse.

Life history traits are distinct from functional traits.[4] Life history traits are typically quantified as emergent properties of populations calculated from stage-specific demographic rates (Salguero-Gómez et al., 2016). Life history traits are estimated from population models. Lifespan is a good example. While we can measure the lifespan of a short-lived perennial, we cannot sit around and wait to measure the lifespan of a 5,000 year old tree—but we can measure a population of ancient trees and use a model to estimate the expected lifespan of a population. In short, it is very difficult to measure life history traits on individuals, so we rely on demographic models to quantify them.

Life history traits are not independent from functional traits. Plants can only achieve long lives if they are built with long-lived organs. Plants tend to generate large numbers of offspring per reproductive event only if they produce small seeds. Recent evidence suggests that functional trait effects on life history traits depend on the environmental context (Kelly et al., 2021), which indicates that more work is needed to clarify the causal relationships between environments, functional traits, life history traits, demographic rates, and fitness. Hence, this book.

Over the next few sections, I briefly summarize some proposed leading dimensions of life history traits and functional traits, and then I review how three key limiting resources, disturbance regime attributes, and temperature determine the success of plant strategies. One way to think about all this complexity is that there is a matching between the dimensions of traits and the dimensions of the environment. Our job is to discover the map that links the right trait to the right environment. The conceptual model in Figure 3.3a illustrates that if we can analytically link fitness to the right combination of traits and environments, then we can discover which traits enhance fitness in a given environment.

3.3　Life history traits potentially span three dimensions

To early ecologists and evolutionary biologists (MacArthur and Wilson, 1967), life history traits appeared to span just a single dimension. Two types of life history strategies emerged at two ends of a continuum (recall Figure 3.13). The original contrast

---

[4] Note that some authors have considered traits such as dispersal mode, seed longevity, and vegetative spread as life history traits, but these are heritable phenotypic properties, so I include them as functional traits.

between r-selected and K-selected species depended on whether or not survival in a population was density dependent as it grew in size generating a trade-off between rapid population growth and resistance to density-dependent mortality (Boyce, 1984), but this original emphasis was misapplied over time and came to be associated with early-vs-late successional plants (Pianka, 1970; Grubb, 1987). Species that exhibited short generation times, short lifespans, high fecundity, and efficient dispersal were called r-strategists given their success in open, less-crowded habitats that were periodically disturbed. Other species exhibited longer lifespans and low reproductive rates and were called K-strategists because they were successful at persisting in crowded conditions in relatively stable environments. This story was attractive to those who sought a simple explanation of plant form and function. Cadotte and colleagues (2006) provided evidence for a general trade-off between colonization ability and competitive ability, which supported some of the expectations of this interpretation of r-K selection theory.

Despite recent claims that r-K selection theory is "well accepted" (Wilson et al., 2019, p. 243), r-K selection theory is typically considered "r-K-ic" (Hutchings, 2021, p. 38) and strong doubts were cast onto r-K selection theory for several reasons (Stearns, 1992). First, the prediction that rapid population growth would exhibit a strong trade-off with resistance to density-dependent mortality did not hold up (Silvertown and Charlesworth, 2001). Second, this dimension emerged most strongly when comparisons of life history traits were made across clades (families, orders, phyla), but not among species, suggesting that these differences occurred in ancient evolutionary divergences but do not really describe evolutionary radiations within the genus level. If the one-dimensional r-K model did not hold up over time, have other life history dimensions emerged? The answer is yes, but we couldn't ask this question until recently, when stage-based models of plant populations were compiled across the globe. Stearns (1992) criticized r-K selection theory on the grounds that it ignored age-specific models of population dynamics, so it made sense to re-examine the dimensionality of plant life history traits using stage-based models.

Salguero-Gómez and colleagues (2016) compiled matrix population models to compute nine life history traits for 418 plant species across 105 families. They demonstrated that plant life history traits span at least two primary gradients: a fast–slow continuum, and a spectrum of reproductive strategies. Curiously, lifespan was not associated with the fast–slow continuum but was loaded on a third minor axis. Chapter 5 explores the dimensionality of life history traits in detail, but for now the key point is that our best empirical estimates suggest that life history traits possibly span three dimensions.

3.4   Functional traits potentially span at least five dimensions

Plants are multifaceted organisms that have evolved numerous solutions to the problem of establishing, growing, and reproducing with limited resources (Weiher et al., 1999). Our understanding of the primary axes of functional variation is rapidly expanding.

This is one of the most exciting aspects of the rapid proliferation of plant trait studies—larger syntheses are being made at larger scales. Considering that plant trait handbooks provide standardized protocols for measuring dozens of potentially important traits (Pérez-Harguindeguy et al., 2013; Klimešová et al., 2019; Freschet et al., 2021), the dimensionality of plant form and function could be a very large number! However, these traits are not all independent. Many traits are correlated, for example, leaves with high nitrogen content per unit mass exhibit fast rates of photosynthesis. Strongly correlated traits are redundant traits. Quantifying the dimensionality of plant traits is the core objective in the search for primary axes of functional variation (Laughlin, 2014a). What then is the intrinsic dimensionality of plant morphology, physiology, and phenology?

Several studies have addressed this question by analyzing global compilations of plant traits (Kattge et al., 2020). Consensus has yet to emerge from these studies, but most agree that the leading dimensions involve:

1. plant height (either at maturity or maximum height);
2. the leaf economics spectrum;
3. a collaboration gradient involving fine root diameter; and
4. a gradient in rooting depth (Díaz et al., 2016; Bergmann et al., 2020; Weigelt et al., 2021).

However, I argue that we are still missing an important gradient involving clonal growth organs, bud banks, and vegetative versus sexual reproduction (Clarke et al., 2013; Klimešová and Herben, 2015). Chapters 6 and 7 discuss these trait dimensions in more detail. Given the massive number of traits that can be measured on plants, it is hopeful that we can consider *five* leading dimensions of plant traits, each with a set of corresponding indicators. Functional traits are what make the life history traits possible. For example, it is impossible to have a long lifespan without heavy investment in perennating structures that enhance survival.

Importantly, we must also include vegetation density and the presence of other strategies in the competitive milieu (Vincent and Brown, 2005). Factors that induce changes in fitness due to increasing density of neighbors, irrespective of the strategy of the neighbors, are called density-dependent effects. Factors that induce changes in fitness due to the proportional abundance of different strategies in the competitive milieu are called frequency-dependent effects. They are distinct factors that deserve delineation but distinguishing between density-dependent and frequency-dependent selection is not straightforward. For example, increasing the density of tall woody plants would have different fitness consequences than increasing the density of annual plants on an herbaceous perennial. Density dependence was implicitly included in several plant strategy models described in Chapter 2, but frequency dependence deserves more attention. Much of our empirical understanding of the ecophysiological importance of functional traits has implicitly been developed in a density-independent framework (Chapters 6–8), but these must be integrated within a density-dependent and frequency-dependent framework (Chapter 9). We are entering slightly dangerous territory here because, by one estimation, if we account for the true number of environmental dimensions and apply Eq. 3.1, then this might require much more than 512 corresponding plant strategies (Scholes et al., 1997)!

The rest of the chapter explores the dimensionality of the primary agents of natural selection on these life history and functional trait dimensions. Grime's (2001) and Southwood's (1977) genius was their emphasis on disturbance and productivity gradients, and few ecologists would argue against their primal significance. I follow in this tradition by exploring the dimensionality of disturbance regimes and the limiting resources that determine productivity gradients. Temperature is treated separately because of its central role as a regulator of enzymatic processes (Arroyo et al., 2022). I describe the three drivers of plant evolution in the following order: resources, disturbances, and temperature. They are organized into separate sections for clarity, but this is superficial because disturbances drive resource availability, and temperature can influence disturbance regimes and resource availability—they are all interconnected (Figure 3.4).

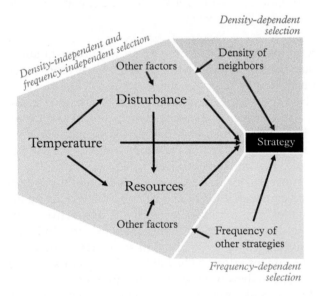

Figure 3.4 *The proposed causal drivers of plant strategies include temperature, disturbances, resources, the density of neighbors, and the frequencies of other strategies in the competitive milieu. Disturbance regimes (frequency, severity, extent, and type) and resource availability (nutrients, water, light) are affected by temperature and other factors. Resource availability is affected by the disturbance regime. Plant strategies have evolved in response to the complex interactions of density-independent and frequency-independent selection pressures (temperature, disturbance regimes, and resource availability), and density-dependent and frequency-dependent selection pressures, which can feedback to influence disturbance regimes and resource availability.*

3.5   Plants require light, water, and nutrients

Resource limitation constrains plant productivity. Plant growth is limited if plants cannot harvest light to convert CO_2 into sugar. Plants cannot convert CO_2 into sugar without an adequate water supply because diffusion of CO_2 through open leaf stomata inescapably drives the loss of water to the atmosphere. Plants cannot harvest light without enzymes and other proteins that are built with nitrogen, phosphorus, potassium, and many other mineral nutrients. This list of resources is relatively short, and all plants require them (Epstein and Bloom, 2005).

Grime argued that nutrient limitation is the key underlying factor driving the evolution of the stress-tolerant strategy. According to Grime, the similarity of traits that are common to plants living in the shade or in drylands are because of one common underlying cause: mineral nutrient limitation (Grime 2001). If you recall the discussion in Chapter 2, Craine (2009) unraveled this argument and separated the stress-tolerant strategy into four distinct strategies that have evolved in response to low nutrients, low light, low water, and low carbon (recall Figure 2.16). Carbon dioxide was at 336 ppm when I was born, and it is now well over 400 ppm. There is some evidence that canopies can reduce concentrations of CO_2 to levels that could matter to some species. Most species will grow faster if supplied with more CO_2, but CO_2 supply is very nearly constant during the daylight hours so it can't really account for differences in strategies within vegetation. Here, I consider CO_2 limitation to be linked with water limitation because water-limited plants tend to also be starved of carbon. The critical plant resources can be defined by a cube in three dimensions determined by light, water, and nutrients (Figure 3.5).

The accurate quantification of resource availability and other aspects of the abiotic environment is fundamental to this approach. Trait-based ecology has a number of excellent handbooks for standardized measurements of traits (Pérez-Harguindeguy et al., 2013; Klimešová et al., 2019; Freschet et al., 2021), but we lack similar handbooks for standardized measurements of resources and other abiotic conditions, although proposals exist for standardized greenhouse measurements (Both et al., 2015), as are shortlists of abiotic conditions for population ecology (Gibson, 2015). Some plant ecologists may not fancy themselves as soil scientists or climatologists, but we should know how to measure the conditions that matter to plants. So how should we measure light, water, and nutrients?

The light axis can be defined as photosynthetically active radiation (PAR), which is the amount of light within the 400–700 nm wavelength range that is available for photosynthesis (Larcher, 2003; Lambers and Oliveira, 2019). These values range from 0 at night to between 2,000–3,000 mmol/m^2 at midday in the summer, depending on latitude. Ideally, the light axis integrates the amount of light received over the course of the year. Plant canopies significantly reduce PAR at the ground level where recruitment occurs, so if one is interested in dynamics over time, then PAR should be measured at the ground surface.

The units for water are far trickier to define. Many studies only use precipitation because that is freely available at global scales, but precipitation totals are very different

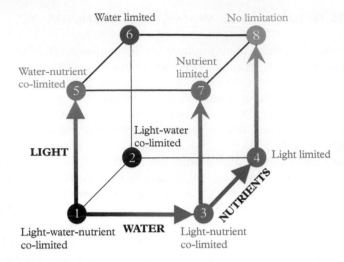

Figure 3.5 *The three-dimensional resource cube defined by variation in mineral nutrients, water, and light. Corners 1 and 2 likely do not exist in nature. This model acknowledges the existence of regions within the cube in which multiple resources are co-limiting. See text for discussion of each corner. Disturbance is not a dimension on this cube, but it can change resource availability in closed canopy ecosystems.*

from soil water availability. We could define it as gravimetric or volumetric soil water content, but these measurements mean different things to plants along gradients of soil particle size distribution (i.e., soil texture). Sandy soils have lower water-holding capacity than clay soils. If we standardize by soil texture and rooting depth, perhaps we can then define water supply in units of soil water potential, or soil water content (Vicca et al., 2012). Waterlogging leads to anoxic conditions and oxygen deprivation, which is a major source of resource limitation for plants that are ill-equipped to handle it. Too much water kills plants because they lack oxygen to fuel growth and maintenance respiration and flooded habitats have produced unique adaptations (considered in Chapter 2).

The nutrient axis is even more complex. Is it possible to collapse the great variety of essential mineral nutrients into a single axis? I can already hear my colleagues berating me for this, and it makes me squeamish, too. After all, nitrogen and phosphorus can exhibit opposite trajectories in relative abundance over timescales of soil development because as rock-derived phosphorus supplies decline over time, nitrogen supplies increase through atmospheric fixation by mutualistic bacteria (Richardson et al., 2004; Wardle et al., 2004). Molecules exhibit variable diffusion rates, some are more readily taken up by mass flow, complex organic compounds may even be a source of minerals, and most species mingle with mycorrhizal fungi to exchange carbon in return for nutrients. However, we must place limits on the proliferation of resource dimensionality and guard against reaching the extreme perspective that all is lost, and plant strategy theory is a myth to be sought only by mystics and madmen. While there are important differences between adaptations to low nitrogen versus low phosphorus, there are clear differences between infertile soil, which is often low in multiple nutrients simultaneously, and fertile soil, which often contains adequate concentrations of multiple mineral nutrients.

Defining the appropriate units to operationalize the soil nutrient axis is sure to get a crowd of soil scientists and plant ecologists into a tizzy. Given that nitrogen and phosphorus are the most limiting nutrients in terrestrial ecosystems, perhaps the sum of nitrate, ammonium, and phosphate concentrations is the most relevant pool. Or perhaps supply rates, such as the rates of mineral nutrient transformations, are preferred over pool sizes. Complications arise with this approach because some ecosystems are phosphorus limited and others are nitrogen limited (Cunha et al., 2022). Soil fertility has been modeled as a latent variable by combining the variation of multiple soil nutrient indicators together (Weiher et al., 2004; Laughlin et al., 2015b). Recently, plant ecologists have used phytometers as a way to quantify soil fertility from the perspective of multiple plant species simultaneously (Daou and Shipley, 2020; Lamontagne and Shipley, 2022). Daou and Shipley (2020) measured the relative growth rates of *Festuca rubra*, *Trifolium pratense*, *Triticum aestivum*, and *Arabidopsis thaliana* in standard climate conditions in different soil. The growth rates were combined into a single latent variable, which they called generalized fertility—the success of this modeling approach was due to the high positive correlations of growth rates of these four species across the fertility gradient. In this case, the units of soil fertility are essentially unitless, in the same way that a PCA axis is unitless.

I restrict resource limitation to three dimensions, but these dimensions are not independent. Tilman's (1988) plant strategy theory depended on an essentially one-dimensional perspective of resource limitation. He argued that the limitation of plant productivity in infertile soil permits sufficient light to reach the ground surface, but crowded vegetation in fertile soil reduces light transmission (recall Figure 3.10). I use similar logic to explore the possibilities of different combinations of resource limitation.

Let us start at the origin of the resource cube and consider the characteristics and possibilities of eight scenarios of resource limitation represented by each corner of the cube (Figure 3.5). There is of course continuous variation along all three resource dimensions, so drawing attention to the extremes is an exercise to think through their interdependences, not to deny the continuous variation in these resources. It is quite possible, indeed likely, that resource availability can lead to single resource limitation or co-limitation of plant growth (Harpole et al., 2011). Keep in mind that soil properties, hydrology, and climate are important drivers of water and nutrient availability, but the only things that can limit light are plants themselves.[5]

Corner #1 represents co-limitation of light, water, and nutrients, but this situation likely does not exist in nature. If water and nutrients limit plant growth, then the resultingly thin plant canopy will not prevent light from reaching the ground, so light cannot be limiting (Tilman, 1988). Similarly, corner #2 represents the possibility that nutrients

---

[5] This is not universally true: vascular plants have been discovered in dark caves in China (Monro et al., 2018). But for a general theory, this assumption is sufficient.

are abundant, but water and light are limiting. It is unlikely for a plant canopy to close with insufficient water, so this scenario is also not plausible (Craine, 2009).

Can light and nutrients be co-limited, as in corner #3? Given an adequate water supply, plants can develop into a canopy that is dense enough to limit light even if nutrients are limited because, unlike light, nutrients can be stored in plant tissues (Craine, 2009). Of course, exceptions exist: 1.5% PAR (a reasonable amount of light) can reach the forest floor in the extremely nutrient-poor caatinga forests (Coomes and Grubb, 1996). If neither water nor nutrients limit plant growth, then the canopy will close, and light will become limited (corner #4). Of course, all it takes is one tree to fall to create a gap large enough for light to become abundant at the ground surface, so these environments dynamically move between corners #3 and #7 and between corners #4 and #8 due to spatiotemporal variation in tree falls that drive forest gap dynamics. Given the chronic depletion of light beneath the forest canopy, this is where the highest diversity of shade-tolerant species evolved.

In dry lands, water and nutrients are often co-limited (Grime, 1994; Hooper and Johnson, 1999). Nutrient uptake is limited in dry soil because rates of both mass flow and diffusion decline (Craine and Dybzinski, 2013). When water and nutrients co-limit productivity light will not be limiting (corner #5). Plant productivity would still be limited if water was the only limiting resource (even if nutrients were available), so light would still be abundant in corner #6.

The last two situations are only possible after disturbances, which are fully described in Section 3.6. Grime treated disturbance and productivity as independent dimensions, but resource availability is clearly driven by disturbances. If water is not limiting, then the canopy will likely close, and light will be limited whether nutrients are limiting or not. The only way to remove light limitation in the understory in these scenarios is to open the canopy through a disturbance (corners #7 and #8). Disturbances also likely affect the availability of nutrients through decomposition and increased rates of mineralization, but disturbances will have less of an impact on water availability since that is driven more by hydrology and weather.

The resource cube includes regions where multiple resources co-limit plant growth (Harpole et al., 2011). Microeconomics theory has been applied to ecophysiology to explore how plants may optimize their growth in the presence of multiple limiting resources (Wright et al., 2003). For example, if production is limited by two resources, the least-cost combination of inputs that optimizes production for a given total cost is the location on the equiproduct curve that has the same slope as and is tangential to the equal-cost line (see Wright et al., 2003 for in-depth discussion of this theory). This "least cost theory" was developed to show that plants in dry environments are expected to, and in fact do, operate at higher leaf nitrogen concentrations but lower stomatal conductance (Wright et al., 2001). Subsequently, least cost theory was used to account for the competing costs of maintaining both water flow and photosynthetic capacity, and the model accurately predicted variation in leaf traits across climatic gradients (Prentice et al., 2014). Higher carboxylation capacity together with tighter stomatal regulation of transpiration in species with high leaf nitrogen per area allows them to achieve higher water-use efficiency (Querejeta et al., 2022).

The density and identity of other strategies within the vegetation also drives variation in resource availability (Figure 3.6), but I wait to describe these effects in detail until Chapters 4, 8, and 9. Resource competition is "the process by which two or more individuals acquire resources from a potentially common, limiting supply" (Craine and Dybzinski, 2013, p. 834). Competition for light is considered asymmetric (or one-sided) because light is supplied from above and therefore "larger individuals obtain a disproportionate share of the resources (for their relative size) and suppress the growth of smaller individuals" (Weiner, 1990, p. 360). In contrast, competition for nutrients and water is often considered size symmetric. In other words, having deeper roots alone does not permit an individual to obtain a disproportionate share of belowground resources. Rather, plants compete for nutrients by maximizing their root length (or mycorrhizal hyphae length) to preempt nutrients from being obtained by neighbors. Competition

Density independent   Density dependent   Frequency and density dependent

Figure 3.6 *Illustration of density-independent, density-dependent, and both frequency- and density-dependent selection, in relation to the availability of light, nutrients, and water. Light (indicated by the golden rays) is supplied by the sun from above the canopy and can be usurped by taller individuals; hence, competition for light is size asymmetric. Nutrients (indicated by the molecules of ammonium, NH_4^+) are supplied in patches belowground and tend to be greatest near the soil surface. Water (indicated by the blue gradient) is most abundant near the soil surface, but deeper water layers exist in some habitats that can be tapped into by deep rooted plants. Competition for nutrients and water is generally thought to be size symmetric. Density dependence occurs when the increasing density of neighbors (either conspecific or heterospecific neighbors) induces negative effects on demographic rates. Frequency-dependent effects occur when the magnitude of the competitive effect depends on the traits of the neighbors.*

for water is less understood: it might also be driven by supply preemption, but often the best competitors tolerate the extreme negative tensions in their xylem that occurs when soil water potential is low. Plant strategies may be able to partition the water supply if one strategy specializes at finding water from deep soil layers and others capture recently fallen precipitation in the shallow soil layers.

3.6   Disturbance regimes account for type, frequency, severity, and extent

This section discusses the dimensionality of disturbance regimes. Disturbances remove plant biomass (Grime, 1979) and there is a wide array of adaptations to disturbance. Life history theory captured the imagination of ecologists at about the same time when disturbances came to be seen as central drivers of ecological dynamics (Pianka, 1970; Connell, 1978; Huston, 1979; Pickett and White, 1985). Disturbance and its associated release from negative density dependence by freeing up space for colonization was central to r-K selection theory, where r-strategists rose to prominence in recently disturbed open ground and K-strategists came to dominate in stable, undisturbed conditions. The habitat duration axis in Grime's CSR model is loosely defined as frequency of disturbance, where the ruderal strategy is adapted to rapidly colonizing open ground. But much has been learned about disturbance regimes in the last four decades that requires us to reconsider how disturbances have shaped the evolution of plant strategies.

We can no longer talk about disturbance as a single dimension—it would be foolish to do so because there are many types of disturbances (Table 3.1). Walker and Willig (1999) made the delightful insight that the classical elements of earth, water, air, and fire[6] capture the variation in abiotic disturbance types. Earlier work by Jackson (1968), which also emphasized these classical elements, was influential in Australia and a precursor to the vital attributes model (Noble and Slatyer, 1980). Earth (or tectonic) disturbances include earthquakes, landslides, erosion, and volcanic eruptions. Atmospheric disturbances include hurricanes,[7] tornados, dune drift, treefalls, and uprooting. Water disturbances include drought,[8] floods, and glaciers. Finally, fire is one of the most important disturbances to global vegetation. Biotic disturbances, on the other hand, can be separated into non-human and human disturbances. Non-human disturbances include herbivory, invasion, and animal activities. Human disturbances include agriculture, forestry, mineral extraction, military activity, transportation infrastructure, and urban development. Some of these disturbances are sudden events ("pulse" disturbances) and others are more subtle yet pervasive and extensive ("press" disturbances) (Collins et al., 2011).

---

[6] "*Earth my body, water my blood, air my breath, and fire my spirit*" is a favorite chant in pagan circles.

[7] Hurricanes are called "cyclones" if you live near the South Pacific or Indian Oceans and "typhoons" if you live near the western Pacific Ocean.

[8] Extreme climatic events are often viewed as a disturbance because they can be traced to a specific seasonal event, are distinguished from chronic water limitation, and cause widespread removal of biomass. Recent hotter droughts under climate change seem to be outside the historic natural range of climatic variation.

Table 3.1 *The major types of disturbance on Earth.*

Element	Primary disturbance
Earth (tectonic)	Earthquake
	Landslide
	Erosion
	Volcano (a'a, pahoehoe, lahar, tephra)
	Rock outcrops
Air	Hurricane
	Tornado
	Dune drift
	Treefall
	Uprooting
Water	Drought
	Flood
	Glacier
Fire	Fire
Biotic—Non-human	Herbivory
	Invasion
	Animal activity
Biotic—Human	Agriculture
	Forestry
	Mineral extraction
	Military activity
	Transportation infrastructure
	Urban development

Reproduced from Walker and Willig (1999) with permission.

The effects of volcanoes are qualitatively different than the effects of treefall and herbivory. Volcanic eruptions can incinerate all biological material leaving nothing but rock and ash in their wake. However, species like *Metrosideros collina* that occur near active volcanoes have acquired tolerance of sulfur dioxide emissions by closing stomata (Winner and Mooney, 1980). Tree blow downs can be destructive, but they leave behind legacies of biological material that drive forest recovery. The former are often classified as drivers of primary succession and the latter as drivers of secondary succession (Walker and Willig, 1999; Walker and del Moral, 2003; Prach and Walker, 2020). Beyond this distinction there are several attributes of disturbance regimes that require elaboration and quantification.

Disturbance types vary in their frequency (rate of occurrence over time), their magnitude (extent of effects in space), their intensity (energy release), and their severity (effects on the biota). Are these attributes independent or are they correlated and redundant? Walker and Willig (1999) plotted the dimensionality of disturbance types in a three-dimensional space. The ordinal scores used to rank the spatial extent, frequency, and severity of disturbances were illuminating. The first thing to note is that disturbance intensity and severity were so strongly and positively correlated that intensity was ignored because it was redundant with severity. In other words, intense disturbances (e.g., tall flame lengths, extreme wind speeds) produce severe effects (e.g., high mortality of plants). If we know the intensity, then we can usually infer the severity.[9]

The next thing to consider is how severity and frequency are associated. Frequent disturbances tend to be less severe than infrequent disturbances. Mammal herbivory is extremely frequent, but rarely deadly. On the other hand, disturbances that occur infrequently can be extremely severe. Volcanic eruptions are infrequent, but often deadly. Severity and frequency are therefore inversely correlated attributes of disturbances that span a gradient from frequent, low-severity disturbances to infrequent severe disturbances (Figure 3.7a) (White and Jentsch, 2001).

Disturbances also vary considerably in spatial extent. Some disturbances like volcanic eruptions, regional drought, or forestry operations tend to occur over large spatial extents, whereas others, like animal activities, are extremely localized. How is extent related to the frequency–severity dimension? If we subject Walker and Willig's (1999) ordinal scores of these disturbance attributes to dimensionality reduction using principal components analysis, we see that extent is orthogonal to the frequency–severity dimension (Figure 3.7b). Disturbance regimes are at least two dimensional. In other words, disturbances that are frequent and low severity exhibit a range of spatial extents, and those that are infrequent and high severity also exhibit a range of spatial extents.

Are some disturbance regimes more likely to promote natural selection on traits than others? Harper (1977, p. 627) argued that frequent, low-severity disturbances are more likely to lead to natural selection on traits than infrequent, high-severity disturbances. The chronic yet low levels of predation on plants by herbivores is likely one of the most important disturbance agents of natural selection and has led to a battery of evolved defenses (Agrawal and Fishbein, 2006).[10] Frequent fire regimes are also associated with unique adaptations such as smoke-induced germination requirements (Dixon et al., 1995) or enhanced thickening of the corky outer bark in woody plants (Uhl and Kauffman, 1990; Laughlin et al., 2011a). Infrequent fire regimes and stand-replacing disturbances are less likely to select for unique adaptations like these, but rather the

---

[9] There are exceptions. Smoldering fires are extremely low intensity yet can kill stands of trees.

[10] It makes one wonder what plant strategies can survive the recent devastating outbreaks of spongy moth (*Lymantria dispar dispar*) herbivory outside of its native range.

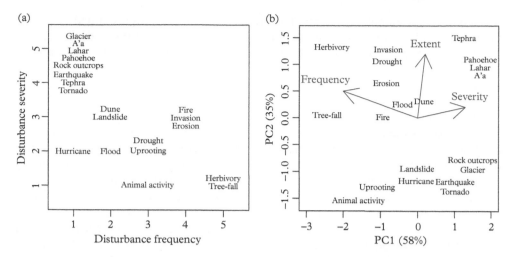

Figure 3.7 *(a) Disturbance severity and frequency are negatively correlated among disturbance types. Herbivory and tree-fall are frequent, low severity disturbances whereas glaciers and volcanic eruptions are infrequent—high severity disturbances. (b) The two dimensions of disturbance regimes illustrated by a principal components analysis. Disturbance types that are clumped together have the same scores and are separated only for readability. Ordinal scores were estimated using expert opinion and provided to the author by Lars Walker, with thanks (Walker and Willig, 1999; Walker and del Moral, 2003).*

innate ability to resprout and regenerate vegetatively is an important feature of plants found in these disturbance regimes (Bellingham and Sparrow, 2000). Spatial extent of disturbances is unlikely to select for physiological strategies directly because plants only respond to their immediate environment, not the environment hundreds of meters away (Grime, 2001). But a mechanism for long-distance dispersal is required for plants to be able to colonize large, denuded landscapes (Nathan, 2006; Gillespie et al., 2012), so dispersal traits will enhance the likelihood of invasion into large disturbances.

The correlations among frequency, severity, and extent have been evaluated across disturbance types, but do these correlations hold when we examine variation within each type individually? Fire ecologists describe fire regimes that vary in frequency and severity (Agee, 1996; Laughlin and Fulé 2008). High-severity fire regimes include chaparral, fynbos shrublands, *Eucalyptus* woodlands, and subalpine spruce-fir forests that burn relatively infrequently, but when they burn, they burn hot, whole crowns are consumed, and fire rages across the landscape driven by enormous flame lengths. Low-severity fire regimes include open woodlands and grasslands, where fires may be flashy in their consumption of fine fuels, but the fires are generally not deadly. So, within individual disturbance types we can also see a gradient from high severity/low frequency fire regimes to low severity/high frequency fire regimes. It is reasonable to assume that this gradient exists within other disturbance types as well.

Grime treated disturbance and site productivity as orthogonal factors but didn't emphasize how resource availability declines with succession. In general, disturbances strongly affect resource availability. Disturbance events have been defined as a "relatively abrupt change in resource availability or ecological structure or function, often associated with the conversion of live to dead biomass" (Burton et al., 2020, p. 885). Disturbances release a pulse of nutrients from decaying plant tissue and increase light availability at the ground level (Jentsch and White, 2019). Disturbance types will differ in their relative impact on resource release. For example, a tree blow down event will open the canopy and increase light and nutrients in a single patch (Lusk and Laughlin, 2016). Canopy disturbances have strong positive effects on light and nutrient availability, but the effects on water are less clear. While removing vegetation does increase water content in the soil because less water is transpired, water availability is also strongly tied to precipitation regimes and local hydrology, which could effectively swamp out any draw down by vegetation. Fires have complex effects on soil nutrient availability. A low-severity forest fire may provide a short-lived pulse of available nitrogen, but a high-severity fire may combust so much organic matter in the soil that available nitrogen is reduced (Certini, 2005). Nitrogen availability increases while phosphorus availability decreases over long-term soil pedogenesis following primary disturbances (Wardle et al., 2004; Peltzer et al., 2010). Given these complex effects of disturbances on resources, we must anticipate that disturbance regimes will have both direct effects on plant strategies that drive selection on lifespan and reproductive rate, as well as indirect effects mediated through resource availability that drive selection on physiological adaptations (Figure 3.4).

3.7 Temperature regulates all biological processes

Temperature is a complicated environmental factor that covaries with many abiotic gradients, including latitude, elevation, and precipitation, among others. From enzymatic rates to cosmic searches for extraterrestrial life, temperature has long been known to be the key factor regulating biological processes. Some planets are too close to their stars and would be too hot for life (as we know it), whereas others are too cold; Goldilocks planets are "just right." Temperature is clearly one of the most powerful controls on the distribution of vegetation globally (Collinson, 1988; Akin, 1991; Breckle, 2002). Woodward and Kelly (1997, p. 48) called temperature the "central controller." All biological processes are rate limited, and temperature is the primary driver of these rates. The uptake of water and nutrients is temperature sensitive because of changes in the permeability of membranes as well as sensitivity to enzymatic reactions to temperature. Leaf growth in cold temperatures are limited by wall extensibility and in hot temperatures are limited by water balance (Woodward and Kelly, 1997).

Process models of enzymatic rates demonstrate that rates peak at optimum values (Arcus and Mulholland, 2020; Arroyo et al., 2022). These optimum temperatures are specific to each biochemical enzyme. When temperatures exceed the optimum, the enzymatic rates then fall because higher temperatures begin to denature the enzymes. Rates of net photosynthesis and respiration exhibit such non-linear relationships with temperature. The C_3 photosynthetic pathway achieves maximum rates at 18°C and the C_4 photosynthetic pathway achieves optimum rates at 28°C (Figure 3.8) (Duffy et al., 2021). Interestingly, the enzymes involved in total ecosystem respiration have optimum

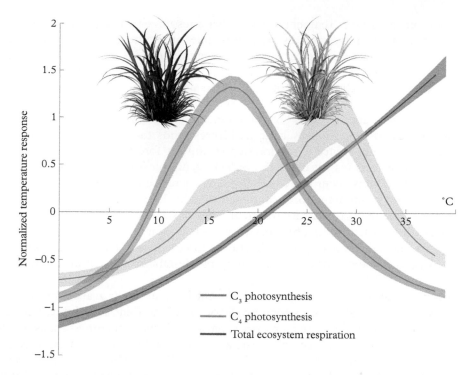

Figure 3.8 *The normalized (scaled to unit variance) global temperature response of C₃ photosynthesis (green) exhibits an optimum at 18° C, and C₄ photosynthesis (yellow) exhibits an optimum at 28° C. Total ecosystem respiration (brown) exhibits an optimum at 62° C (Liang et al., 2018), well beyond the range limits of this figure. Shaded areas represent the 90% confidence interval of projections.*
Data from the FLUXNET database reproduced from Duffy et al. (2021).

temperatures at 62°C (Liang et al., 2018), so as rates of photosynthesis decline with increasing temperatures, respiration rates continue to rise.

Mean annual temperature is the most common measurement used in large-scale models of species and phenotype distributions, but plant fitness is likely more related to minimum and maximum temperatures than the mean. But these measures are so strongly correlated that the mean is often as good a predictor as the extremes. Moreover, these data are typically modelled at the 1-km scale, which does not capture important microclimatic variation (Zellweger et al., 2020). Soil temperature could also be a useful way to include temperature as a predictor of plant fitness, and global datasets of soil temperature are now being developed (Lembrechts et al., 2020).

Plant strategy models have treated temperature in a variety of different ways: Grime subsumed temperature into the productivity gradient—cold-limited sites were dominated by stress-tolerant species (Grime, 2001); Tilman considered that temperature was a driver of belowground resource availabilities—cold-limited sites were dominated by unproductive open canopy vegetation where light was abundant and soil resources were not (Tilman, 1988). Temperature is neither a resource that can be consumed nor a disturbance, but resources and disturbance regimes are both affected by temperature.

Disturbance regimes vary across global temperature gradients (Burton et al., 2020). Cryoturbation is the mixing of soils by freeze-thaw of frozen soil water—this physical disturbance is most likely to occur in cold deserts and tundra biomes where temperatures are low (Figure 3.9). Bioturbation—the mixing of soils by small mammals, worms, ants, termites, etc.—appears to be limited to tundra and open grassland biomes. Extensive disturbance to vegetation by large herbivores tends to occur in cold biomes as well, including but not limited to tundra, steppe, and dry forests. Extensive disturbance by herbivorous insects (especially bark beetles) occurs mostly in colder montane and subalpine forests. Fires occur more broadly—anywhere the fuel can sufficiently dry— and are common in montane forests, grasslands, and desert scrub. Extreme drought occurs most often in grassland, dry forest, and desert scrub biomes. Storms that cause widespread disturbance most often occur in regions where temperatures are warm enough to support the development of hurricane force storms. Landslides are generally limited to warmer biomes where large rain events destabilize the soil in mountainous regions.

Temperature also drives variation in water, nutrient, and light availability. While it is unlikely that temperature is a strong predictor of soil nutrient availability, temperature is related to several ecosystem processes that are driven by microbes. Rates of litter decomposition and root turnover in the soil are processes that are also strongly related to temperature (Gill and Jackson, 2000; Fierer et al., 2005). The relative nitrification potential of soil exhibit peak rates of nitrification by ammonia oxidizing bacteria (AOB) at 30°C and by ammonia oxidizing archaea (AOA) at 40°C (Norton and Ouyang, 2019). Taken together, these may explain the low nutrient availability in some cold ecosystems, and the weak yet increasing relationship between mean annual temperature and foliar $\delta^{15}N$, an indicator of soil nitrogen supply rate (Craine et al., 2009).

The relationship between temperature and soil water potential is even more complicated and involves the relationship between temperature and precipitation. Hot and dry environments (e.g., hot deserts) exhibit high rates of evapotranspiration and have chronically dry soils. But hot and wet environments (tropical rain forests) constantly resupply the soil through precipitation, so despite the high evapotranspiration rates, the soil water potential typically remains high. Tropical dry forests experience soil water deficits during the dry season.

Light limitation also exhibits a complicated relationship with temperature due to the interaction with precipitation. Recall that light only becomes limiting in productive environments that support a dense canopy. The relationship between temperature and net primary production (NPP) is generally positive (Figure 3.10) (Del Grosso et al., 2008), although it is dependent on sufficient precipitation. Leaf area index (LAI) is a measure of the amount of leaf area per unit land area and is positively correlated with NPP (Waring and Schlesinger, 1985). Therefore, light will be most limited in habitats with high LAI in warmer and wetter environments.

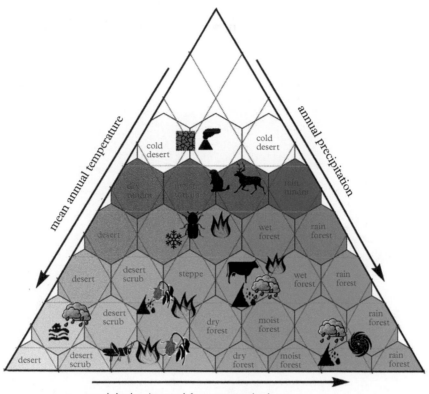

precipitation / potential evapotranspiration

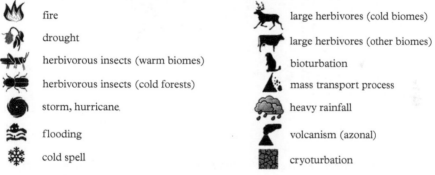

Figure 3.9 *Common disturbance types vary among climatically determined biomes as shown using the Holdridge climate zones, which are defined by mean annual temperature, annual precipitation, and the precipitation-to-potential evapotranspiration ratio.*

Reproduced from Burton et al. (2020) with permission.

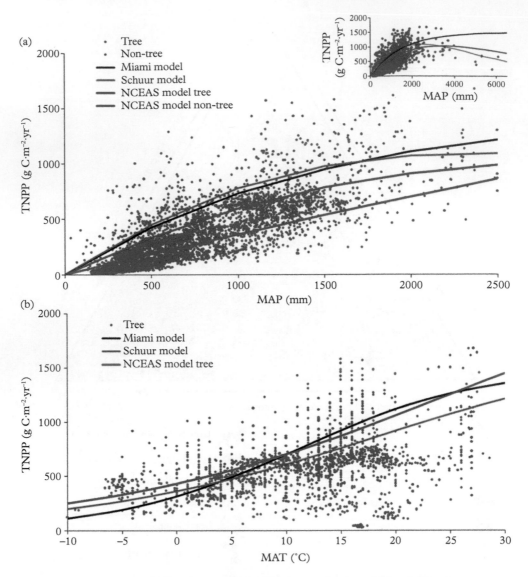

Figure 3.10 *(a) Total net primary production (TNPP) for tree-dominated (blue dots) and non-tree-dominated (red dots) systems compared with mean annual precipitation (MAP). Lines represent predictions from different models: the Miami, Schuur, and NCEAS model described in Del Grosso et al. (2008). The inset expands the x-axis to include a much higher limit of MAP to show how production declines at the highest precipitation amounts. (b) TNPP for tree-dominated systems as a function of mean annual temperature (MAT) in the Miami, Schuur, and NCEAS models.*
Reproduced from Del Grosso et al. (2008) with permission.

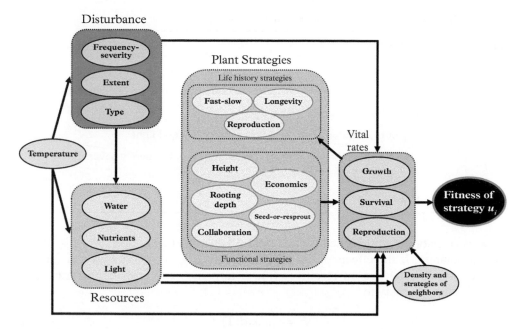

Figure 3.11 *The proposed dimensionality of each major construct and possible causal relationships between drivers of fitness in plant strategy theory. This expands Figure 3.3b into its component parts. The goal of the theory is to predict the fitness, or log(λ), of any plant strategy (u_i) using its traits given the availability of resources, the disturbance regime, the temperature, and the density and frequency of other strategies in the community. Feedbacks over time between these factors are likely but are not illustrated for simplicity. The efficacy of the strategy for improving fitness entirely depends on the abiotic and biotic conditions of the site; therefore, any statistical model for predicting fitness must include interaction effects between traits and environmental conditions.*

3.8   Toward a formal plant strategy theory

Those who prefer simple approaches like Grime's CSR model will not like the expansion of productivity and disturbance into multiple dimensions. I stress here that I do not make these expansions lightly. We must carefully consider the benefits of increasing the complexity of any theory. On the other hand, those of you who prefer high-dimensional models may think that I have not gone far enough! I hope my explanations are reasonable for why I guard against the indefinite expansion into intractable numbers of dimensions.

Figure 3.11 summarizes the proposed dimensionality and causal structure of plant strategy theory. The goal of any useful plant strategy model is to use traits to predict fitness in a given environment. In anticipation of a mathematical formalism, we can present the general structure of the arguments that should be included in an empirical fitness function F:

$$F = f(u_i, \boldsymbol{u}, \boldsymbol{x}, \boldsymbol{r}, \boldsymbol{d}, T), i = 1, \ldots, S \qquad (3.4)$$

Table 3.2 *Definitions of the arguments in the fitness function (Eq. 3.4).*

Variable	Definition
u_i	The focal strategy, a scalar or vector of trait values
u	Other strategies, vector, or matrix of other trait values in the community
x	Population sizes of each strategy
r	Vector of resources, including water, nutrients, and light
d	Vector of disturbance attributes, including type, frequency-severity, and extent
T	Temperature
S	Number of strategies

where each parameter is defined in Table 3.2 following Vincent and Brown's (2005) general notation. If the effects of traits (**u**) and densities (**x**) are manifested through their local-scale influence on resources (*r*), disturbances (*d*), and temperature (*T*), then this could simplify to

$$F = f(u_i; r, d, T) \qquad (3.5)$$

These equations are discussed in Chapters 8 and 9. The factors in these equations comprise the minimum number of dimensions that we need to contemplate on our quest to understand plant strategies.

This chapter attempts to synthesize previous work and to work toward an operational plant strategy theory that it is empirically testable. It serves much like a trail guide for the rest of the book, but trail guides cannot replace the experience of exploring the trail itself. A rich body of life history theory and comparative functional ecology awaits. It is time to build our understanding of plant strategies from the ground up. To do so, there is no better place to start than plant demography and life history theory.

3.9 Chapter summary

- The intrinsic dimensionality of any multivariate dataset is the minimum number of parameters needed to describe it, and dimensional analysis of a theory identifies the base quantities, their units of measure, and causal relations among them.
- Plant strategy theory needs to define the necessary and sufficient causal factors that affect fitness across environmental gradients. The goal of the theory is to predict the likelihood of successful establishment, growth, and survival of any phenotype in a particular environment.
- The strength of plant strategy theory will be judged by its ability to predict whether a phenotype can maintain a viable population in a given environment.

- The singular phrase—"traits conferring fitness"—holds the key to synthesizing life history and functional perspectives on plant strategies.

- Population growth rate is determined by age/stage-specific rates of growth, survival, and reproduction.

- Two predictive frameworks are described in this book. The first framework seeks to predict the net effect of traits on fitness across multiple phenotypes and environments. The second framework seeks to predict fitness through trait effects on demographic rates.

- Life history traits are functions of demographic rates, and they span at least three dimensions: a fast–slow continuum, reproductive strategies, and lifespan.

- Functional traits are morphological, physiological, or phenological traits that potentially impact fitness. Our understanding of the dimensionality of functional traits is rapidly growing, but at least five dimensions are important: an economics spectrum, height, rooting depth, belowground collaboration with fungi, and a seed-or-resprout spectrum.

- Plant strategies have evolved in response to the complex interactions of density independent and frequency independent selection pressures (i.e., temperature, disturbance regimes, and resource availability), and density-dependent (i.e., vegetation density) and frequency-dependent selection pressures (i.e., the presence of other strategies).

- Three categories of abiotic factors are distinguished: temperature, disturbance regimes, and resources. Disturbances influence resource availability and temperature affects both resource availability and disturbance regimes.

- Resources can be conceptualized as a three-dimensional cube representing variation in light, water, and mineral nutrients, where different corners of the cube represent either abundance, limitation, or co-limitation of these resources.

- Disturbance types can be arrayed along two dimensions: a gradient of frequency and severity, and a gradient of disturbance extent. These two dimensions can vary within a disturbance type as well.

- Temperature is the ultimate regulating factor because it governs non-linear reaction rates for all biological processes.

- The central challenge of plant strategy theory is determining how these fundamental abiotic dimensions determine which sets of traits optimize population growth rate.

3.10 Questions

1. Are life history traits a direct result of functional traits, or are there examples of life history traits that are not determined by the phenotype?

2. Are life history traits just the demographic consequences of functional traits in a given environment, or are they attributes of species that also affect fitness independently?

3. What compromises are made when trying to determine the dimensionality of plant traits? Can this dimensionality only be determined empirically, or can theory play a stronger role?

4. What other units could define the dimensions on the resource cube? How should the nutrient axis be operationalized?

5. Do strategies evolve in response to average resource levels or in response to spatial and temporal variation in resource levels driven by disturbances?

6. How do disturbances (other than fire) exhibit variation in frequency, severity, and extent?

7. Is space a limiting resource? If so, is time a limiting resource?

8. How have human activities altered habitats with respect to the habitat's location in the resource cube?

9. Can cooperative resource sharing (e.g., mycorrhizal networks that link resource use among individuals across multiple species) be considered a strategy that is under selection?

10. Does the framework described in this chapter allow for a strategy to exhibit routine flexibility as described by Peter Grubb's "switching" and "gearing down" strategies?

11. How could parasitic and carnivorous plants fit into this framework of plant strategies?

Part II

Demography and Life History

Part II establishes demography as a central component of plant strategy theory, as the existence of a successful strategy depends on positive demographic outcomes that lead to population persistence. These chapters explain how to compare life history strategies of species based on their different schedules of births and deaths.

Chapter 4 describes how populations increase or decline in size and how to empirically quantify population growth rates using age- and stage-based matrix models and integral projection models.

Chapter 5 explores the theory that underpins density-independent population growth and the demographic trade-offs that constrain life history strategies. Plant life history strategies are proposed to span at least three independent spectrums of life history traits.

Part II

Demography and Life History

4

Plant Demography

Disproportionately less attention has been devoted by the geobotanists to the life cycle and population biology of species than to their assembly into communities, notwithstanding that a knowledge of the demographic properties of species is essential for a full understanding of community composition and dynamics.

—James White (1985b, p. 2), "The population structure of vegetation,"
The Population Structure of Vegetation

While trait variation may influence the fate of populations, luck often governs the lives of individuals.

—Robin Snyder and Stephen Ellner (2018, p. E90),
"Pluck or luck: does trait variation or chance drive variation in lifetime reproductive success?" in *The American Naturalist*

4.1 Plant strategies prove themselves through positive demographic outcomes

Functional ecologists may wonder why we need two whole chapters on population models and life history theory in a book on plant strategies. My reasoning is straightforward. All successful strategies must exhibit positive demographic outcomes that permit the persistence of a population. Plant strategies that survive and reproduce will persist, but strategies that die and fail to reproduce will go extinct. I echo Silvertown's proclamation that any viable theory of plant strategies "must ultimately be founded on demography" (Silvertown et al., 1992, p. 135).

The first step in testing plant strategy theory is determining what phenotypes exist by analyzing the variation and covariation of functional and life history traits. However, the next step is arguably even more important—to establish that population growth rates of different strategies change across environments, which is often assumed, yet rarely tested. Observing a trait in an environment and assuming that the trait is optimized for

Plant Strategies. Daniel C. Laughlin, Oxford University Press. © Daniel C. Laughlin (2023). DOI: 10.1093/oso/9780192867940.003.0004

that environment is not strong evidence for the importance of that trait. A strong test would take the next step to determine whether traits affect demographic performance in that environment. For this reason, I emphasize demography.[1]

Life histories describe schedules of births and deaths that affect age at first reproduction, lifespan, and other key events. In many ecological studies of plant communities, life histories of plants are sometimes casually described as simple discrete categories: annuals reproduce and die in one year, biennials reproduce and die in their second year, and perennials reproduce continually for many years before they eventually die. But variation in plant life histories is so much more than these simple categories (Salguero-Gómez et al., 2016; Paniw et al., 2021). Natural selection has generated variation in these schedules among plant species, yet trade-offs among life history traits constrain which combinations of traits are possible. Some plants live for only a few weeks, others live for millennia. The former invest more into growth and reproduction, the latter invest more in survival.

The primary objective of this chapter is to describe how demographers measure population growth rates and use models to learn about the life history traits of plants. We first explore schedules of births and deaths from an age-based perspective, then we generalize to stage-based population models, which are often more useful for plants. We examine population models for real species to calculate life history traits to prepare ourselves for Chapter 5, where we explore the constraints and trade-offs among life history traits. This introduction to plant population demography covers the essentials for plant strategy theory, but I refer the eager demographer to other more in-depth treatments (Ebert, 1999; Caswell, 2001; Gibson, 2015; Ellner et al., 2016; Salguero-Gómez and Gamelon 2021).

4.2   Population growth rate is a matter of birth and death

All populations have the potential to exhibit exponential population growth. If we assume a closed population that exhibits no migration, then we only need to add the number of births in time t ($Births_t$) and subtract the number of deaths in time t ($Deaths_t$) to project population size in time $t+1$ (N_{t+1}) if we know the population size in time t (N_t):

$$N_{t+1} = N_t + Births_t - Deaths_t \tag{4.1}$$

The per capita (i.e., per individual) birth and death rate are defined as the number of births per individual ($b = Births_t/N_t$) and the number of deaths per individual

---

[1] Some of my colleagues may disagree with me. Those who think measuring fitness is too difficult or even esoteric will prefer the simpler observational approach that has dominated the literature over the last several decades. But I argue that doing the difficult work of linking traits to demographic rates and fitness would constitute the most rigorous test of plant strategy theory.

$(d = Deaths_t/N_t)$. Rearrangement shows that $Births_t = bN_t$ and $Deaths_t = dN_t$. Substituting these into Eq. 4.1 we get

$$N_{t+1} = N_t + bN_t - dN_t, \tag{4.2}$$

and factoring simplifies it to

$$N_{t+1} = (1 + b - d)N_t \tag{4.3}$$

The multiplier $(1 + b - d)$ is one way to express λ_t, which we defined in Eq. 3.2 as the finite (also known as discrete) rate of population growth. Therefore,

$$N_{t+1} = \lambda_t N_t. \tag{4.4}$$

We can also express these rates in continuous (or instantaneous) time, rather than over discrete time steps. Recalling the definitions of per capita birth b and death rates d from Eq. 4.2, the rate of change in population size over any instant can be expressed as

$$\frac{dN}{dt} = bN - dN, \tag{4.5}$$

and factoring simplifies this to

$$\frac{dN}{dt} = (b - d)\, N. \tag{4.6}$$

The multiplier $(b - d)$ can be rewritten as r, which has come to be known as the intrinsic rate of natural increase, also called the Malthusian parameter of population increase (Fisher, 1930). In contrast to λ, r is a per capita rate of population change because if we substitute r into Eq. 4.6

$$\frac{dN}{dt} = rN, \tag{4.7}$$

and solve for r, we see that

$$r = \frac{1}{N}\frac{dN}{dt}. \tag{4.8}$$

In summary, λ is a measure of population growth rate over a discrete time interval, whereas r is a measure of per capita population growth rate expressed in units of "individuals per individual per unit time." The distinction is important, but the parameters are directly related through inverse expressions:

$$\lambda = e^r, \tag{4.9}$$

$$r = log\,(\lambda) \tag{4.10}$$

where *log* is the natural logarithm (Hutchings, 2021).

Silvertown and Charlesworth (2001) emphasized that if population models ever get too confusing, just remember this simple fact: population sizes increase with births and lack of deaths and decrease with deaths and lack of births. All else being equal, if the birth rate is greater than the death rate, then a population will increase exponentially according to Eqs. 4.4 and 4.7. If $\lambda < 1$ or $r < 0$, then population size declines geometrically or exponentially, respectively, and if $\lambda > 1$ or $r > 0$, then population size increases geometrically or exponentially.

These models are based on a full census of the population (N), with no information about the age or size structure of the population. One of the key insights obtained by the first population demographers was that changes in population size are driven by the age-specific schedules of births and deaths because heterogeneity among individuals in a population contribute to variation in population growth rates. Through a bit of calculus, it was discovered that r could be written as an expression of these age-specific schedules in the Euler–Lotka equation (Stearns, 1992; Roff, 1993):

$$1 = \int_{x=\alpha}^{x=\omega} e^{-rx} l_x m_x \, dx \qquad (4.11)$$

where $x =$ age, $\alpha =$ age at maturity, $\omega =$ age at last reproduction, $l_x =$ cumulative survivorship to age x, and $m_x =$ per capita reproduction at age x. Note that age is a property of an individual. The other four parameters are properties of a population because they are aggregated properties of the individuals.

The discovery of this equality was perhaps one of the greatest mathematical discoveries in the history of ecology. Any equation that begins with "one equals" is bound to pronounce something spectacular. In this case, the intrinsic rate of increase (r) of a population in a stable age distribution (i.e., the proportion of individuals among different age classes are constant) can be solved if we know the age-specific survivorship (l_x) and age-specific reproduction (m_x) of a population. In other words, the age-specific survivorship and reproduction rates determine the asymptotic, or long-term average, r. If we can understand how functional traits determine age-specific survivorship and reproduction rates across environmental gradients, then we can predict the performance of any population.

4.3   Life tables can be used to compute life history traits

For the elegance of the Euler–Lotka equation to sink in, let us see how it works with a hypothetical short-lived perennial plant, *Hierba rapida*. Table 4.1 lists the l_x and m_x vectors in the life table for *Hierba rapida*. This is called a cohort life table because I followed the fates of 1,000 individuals from birth to death. The l_x vector always starts at 1.0 for age class zero and declines monotonically. The m_x vector represents per capita reproduction, or the average number of offspring per individual.

Table 4.1 *Cohort life table for* Hierba rapida, *a hypothetical short-lived perennial plant, starting with 1000 individuals.*

Age (x)	N	l_x	p_x	m_x	$l_x m_x$
0	1000	1.000	0	0	0
1	300	0.300	0.3	0	0
2	100	0.100	0.33	6	0.6
3	50	0.050	0.5	7	0.35
4	30	0.030	0.6	6	0.18
5	10	0.010	0.33	3	0.03
6	5	0.005	-	0	0

N = number alive at age x; l_x = age-specific survivorship, computed as $l_x = N_x/N_0$; p_x = transition survival probabilities, computed as $p_x = l_{x+1}/l_x$; m_x = age-specific fecundity, average per capita number of offspring at age x; $l_x m_x$ = product of l_x and m_x, the number of recruits per survivor.

The Euler–Lotka equation has no closed form solution (i.e., an exact solution cannot be found using a finite number of standard mathematical operations), but fortunately, we can approximate the solution using numerical methods.[2] I encourage you to reproduce these calculations for yourself using the book's R scripts at https://github.com/danielLaughlin/PlantStrategies. To make things easier, let us rewrite Eq. 4.11 as a discrete sum rather than as a continuous integral, and set the equation equal to zero so that we can find the roots of the function:

$$0 = \left(\sum_{x=\alpha}^{x=\omega} e^{-rx} l_x m_x \right) - 1 \tag{4.12}$$

Our estimate of r for this species based on these age-specific schedules of survival and reproduction is $r = 0.056$. This relatively small positive value of r indicates that the population is growing slowly. Now let us forget for a moment about keeping track of the numbers of individuals in each age class and think just about total population size, N. We can use r to project population size into the future. The discrete time expression of a population that is growing exponentially can be found by integrating the differential Eq. 4.7:

$$N_t = N_0 e^{rt} \tag{4.13}$$

where N_0 is population size at the start of the projection and N_t is the predicted size at time t. Using our estimate of $r = 0.056$, if we start with 12 individuals, then in 10 years the

---

[2] By numerical methods I mean by brute force. If the value is too high, you reduce it. If the value is too low, you increase it until you find a stable solution.

population will increase to 21 individuals. In this example $\lambda = 1.058$. We can interpret this to mean that the population is growing at a rate of 5.8% per year. If λ was 0.98, then the population would decline by 2% per year.

There are two life history traits that we can extract from the life table. Age at maturity (α) is the age at which *Hierba rapida* becomes reproductively active and age at last reproduction (ω) is the oldest age at which it produces offspring. *Hierba rapida* starts to reproduce at age 2 and stops reproducing at age 5 (Table 4.1). Note that these life history traits form the beginning and endpoints of the integral (or sum) in the Euler–Lotka equation.

We can also calculate the net reproductive rate (R_0) directly from the life table. R_0 is the average number of offspring per individual that lives to the average life expectancy:

$$R_0 = \sum^{\omega} l_x m_x \tag{4.14}$$

Therefore, the net reproductive rate of *Hierba rapida* in Table 4.1 is $R_0 = 1.16$. The most important thing to see here is that the product $l_x m_x$ is ultimately more important than l_x or m_x alone for determining r or R_0. This also means that we could equivalently use ages 0 and ∞ as endpoints in the integral, because $l_x m_x$ would equal zero for $0 < x < \alpha$ and $\omega < x < \infty$.

We can also compute generation time (T), the average time between the birth of an individual and the birth of its offspring. There are several ways to compute this, which are nearly equivalent.[3] We use the following expression:

$$T = \frac{\log R_0}{\log \lambda} \tag{4.15}$$

In our example then $T = 2.67$. Note that you can also estimate r without numerical methods using the following expression:

$$r = \frac{\log_e R_0}{T} \tag{4.16}$$

Try it for yourself. The estimate is remarkably similar to the brute force solution.

Life tables are foundational to the study of population dynamics, but it is not entirely obvious how life tables can be used to project population sizes while keeping track of the total counts in each age class. Matrix population models are convenient for projecting the numbers of individuals in each age group into the future, and life tables can be directly translated into matrices (Stearns, 1992; Caswell, 2001). These matrices that are based on age are called Leslie matrices, after Robert H. Leslie, the Oxford scholar who discovered this convenient translation (Leslie, 1945). To do so, we can first illustrate the model graphically to make it easy to see how the conversion is made. Table 4.1 includes

---

[3] An alternative expression to compute generation time is $T = \frac{\sum x l_x m_x}{\sum l_x m_x}$.

Table 4.2 *Projection matrix A for* Hierba rapida *calculated from its life table. The first row represents per capita fecundities.*

		"From age x"					
		1	2	3	4	5	6
"To age x"	1	0	2.0	3.5	3.6	1.0	0
	2	0.30	0	0	0	0	0
	3	0	0.33	0	0	0	0
	4	0	0	0.50	0	0	0
	5	0	0	0	0.60	0	0
	6	0	0	0	0	0.33	0

NB: In this case, this table is also called a Leslie matrix because it is based on ages.

the transition survival probabilities p_x, which are computed as $p_x = l_{x+1}/l_x$. The fecundity transitions are computed as $F_x = p_x m_x$ because this was a post-breeding census; see Ellner and colleagues (2016) for a discussion of the important distinction between pre-breeding and post-breeding censuses.

These transitions can be arranged into a 6 × 6 matrix **A** in Table 4.2 where the columns indicate "from age x" and the rows represent "to age x." One-year old plants survive to age 2 with probability 0.30. In other words, 0.3 of the individuals that are one year of age survive to age 2. Two-year old plants survive to age 3 with probability 0.33, and so on up through each annual transition. Note that the diagonal elements are zero in the age-based Leslie matrix because it is impossible for an individual to stay the same age if it survives. The per capita fecundities are listed in the first row because any individual between the ages 2–5 can create offspring. Figure 4.1a illustrates the Leslie matrix in graphical form and all arrows in the diagram can be matched to their associated element in the matrix.

It is helpful to check whether we did our conversion correctly. To do so, we can compute λ directly from this matrix using the tools of linear algebra (Lay, 2006). It is beyond the scope of this book to provide the tools to conduct eigenanalysis, but the R scripts provide standard functions that will help you do so.[4] The dominant eigenvalue of $\mathbf{A} = \lambda$. For a $p \times p$ matrix, p eigenvalues exist. The dominant eigenvalue is a real number and is greatest in absolute value than all the other eigenvalues. In the case of a matrix population model, it is directly interpretable as the asymptotic population growth rate (λ).

---

[4] For those of you who must know now: λ is an eigenvalue of the matrix **A** if $\mathbf{Av} = \lambda\mathbf{v}$, and the vector **v** is an eigenvector of **A** belonging to the eigenvalue λ. Eigenvalues λ of **A** are roots of the characteristic polynomial equation: $\det(\mathbf{A} - \lambda I) = 0$, where "det" is the determinant and I is the identity matrix. See Caswell (2001) for details.

(a)

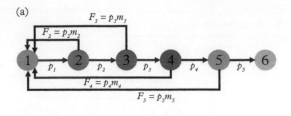

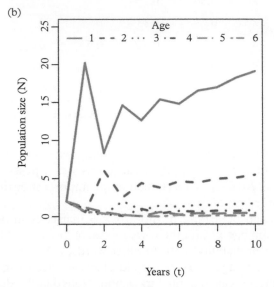

Figure 4.1 *(a) Graphical representation of the Leslie matrix derived from the life table of the hypothetical plant* Hierba rapida. *(b) Projection of population sizes for each age in the population of* Hierba rapida *starting with two individuals in each age class. Note the transient dynamics in the first few years before the size classes settle into the stable age distribution.*

The dominant eigenvalue of the matrix in Table 4.2 is 1.057, which matches the estimation of λ that we made from the life table.

This Leslie matrix can project the numbers in each age class through time using matrix multiplication. To estimate the numbers of individuals in each age class in time $t + 1$, pre-multiply the vector of numbers of individuals in each age class in time t (n_t) by the Leslie matrix **A**:

$$n_{t+1} = \mathbf{A}n_t \tag{4.17}$$

For example, let us start with two individuals in each age class (totaling 12 individuals) and project the population forward (Figure 4.1b). We first see transient dynamics (i.e., reordering of the relative abundances of each age class) in the first few years, then the population settles into its stable age distribution by year 10. Recall that an age distribution is stable when the proportions of each age class are constant over time and each age class increases at rate λ. Note that the total count of 27 total individuals in 10 years differs slightly from our original estimate of 21 individuals because the original estimate only applied the census-based exponential model (Eq. 4.13) and did not account for age-specific rates of survival and reproduction.

4.4   From ages to stages

We started this chapter by evaluating the life history of a short-lived perennial because short lifespans make the analysis and inspection of life tables tractable. But trees can live for millennia. A life table for *Pinus longaeva,* the longest known living non-clonal organism, would exceed 5,000 rows! Moreover, ages can be difficult to ascertain for long-lived plants.[5] Fortunately, reproduction and survival rates are more strongly related to size than age in plants (Harper, 1977; Bierzychudek, 1982; Caswell, 2001), where small individuals generally have lower probabilities of surviving and reproducing than large individuals. This has led many plant demographers to the conclusion that models based on size are superior, but note is it also possible to model ages and stages simultaneously (Caswell and Salguero-Gómez, 2013).

Stages can represent different sizes (e.g., small, medium, large) or stages in their life cycles (e.g., seedling, flowering plant, dormant). In perennial species, plants start as seeds, seeds germinate into immature plants that can persist in vegetative form for years, and flowering stages can mature late in life and continue to produce offspring for centuries. It is not only more convenient to classify plants into stages, but also more appropriate. The rows and columns in these projection matrices represent stages and are called Lefkovitch matrices, after Leonard P. Lefkovitch, who first pioneered their use (Lefkovitch, 1965). These matrices have proved to be absolutely central to our understanding of plant life histories. Russian plant ecologist Tikhon Alexandrovich Rabotnov was one of the first to analyze plant demography on the basis of stages (Rabotnov, 1985), and now an open-access global compilation of these matrix population models called COMPADRE is available for widespread use (Salguero-Gómez et al., 2015). The basics of Lefkovitch matrix models are demonstrated using one of these matrix models estimated for an iconic prairie wildflower.

Echinacea angustifolia (narrow-leaved coneflower) is a long-lived perennial forb that grows in the dry prairies in the Great Plains of North America (Figure 4.2). The genus name derives from the Greek "echinos," meaning hedgehog, because of its spiny ripe head of seed. It has many of the same tongue-tingling properties as the cultivated *Echinacea purpurea,* which is grown commercially to make the beloved bitter tea that is used as a cold and flu remedy. *Echinacea angustifolia* was one of the most important medical herbs for Indigenous tribes on the North American plains as a remedy for coughs,

---

[5] Donald Currey, a graduate student at the time, was using tree cores to study historical climate dynamics and could not extract his increment borers from one ancient pine called Prometheus. With permission from the Forest Service, he cut the tree to accurately cross date the tree rings, only to discover that he cut down the oldest known living non-clonal organism on the planet (Currey, 1965).

(a)   (b)

Figure 4.2 *(a) Echinacea angustifolia (narrow-leaved coneflower) in a xeric prairie in Bear Butte State Park, South Dakota, USA. (b) Life cycle diagram illustrating all the possible transitions between five stages. Each transition is color-coded by demographic rate according to whether it contributes to survival, growth, or reproduction, which are later used to create a demographic triangle.*

(a): Photo by Paul Rothrock, reproduced here with permission under a CC BY-SA license (https:// swbiodiversity.org/seinet/).

colds, sore throats, toothache, and even snakebite (Kindscher, 1992), and is still harvested in the wild by herbalists today. The medicinal properties are found primarily in the taproots of this species, causing wild populations to be susceptible to overharvesting by digging too many large mature plants.

Hurlburt (1999) studied the demography of this wild herb and created population models using five discrete stages: seedlings, small rosettes, medium-sized flowering plants, large-sized flowering plants, and a dormant stage (Table 4.3). The projection matrix **A** for this species is a 5 × 5 square matrix. An important property of stage-based Lefkovitch matrices that distinguishes themselves from age-based Leslie matrices are that the elements in the diagonal of the matrix are non-zero. This permits an individual to stay in the same stage year after year. For example, a large plant will remain a large plant with a probability of 0.90.

Some plants don't wake up in the spring and delay emergence from the same rootstock. This dormancy can last up to three years by some estimates (Steshenko, 1976). Dormancy is not a well-understood demographic phenomenon but is likely a very important part of a successful strategy for surviving fire, drought, or other harsh conditions (Shefferson et al., 2014). Figure 4.2 is a life cycle diagram that illustrates the transitions between each of these stages, and the arrows represent transitions between stages.

Table 4.3 *Projection matrix A of transition probabilities and fecundities among five stages in the life cycle of* Echinacea angustifolia.

		"*From this stage*"				
		Seedling	Small	Medium	Large	Dormant
"*To this stage*"	Seedling	0	0	0.10	0.27	0
	Small	0.52	0.69	0.02	0	0.36
	Medium	0	0.17	0.71	0.02	0.38
	Large	0	0.01	0.20	0.90	0.26
	Dormant	0.02	0.02	0.04	0.05	0

NB: In this case, this table is also called a Lefkovitch matrix because it is based on stages.
Data from Hurlburt (1999).

Stage-based models are much more flexible than age-based models because they can allow for multiple possibilities. Some individuals can skip a stage if they grow really fast. Some individuals can shrink and get smaller the next year—this happens more often than you might think in plants (Salguero-Gómez and Casper, 2010). Seedlings either transition to small rosettes with probability 0.52 or go dormant with probability 0.02 (Table 4.3). Small rosettes either stay as rosettes with probability 0.69, or they grow into medium-sized flowering plants at a rate of 0.17. Some small-rosettes jump a stage and become large flowering plants at an unlikely rate of 0.01, and some go dormant at a rate of 0.02. Medium-sized flowering plants produce seedlings, regress into small non-flowering plants, stay the same size, grow into large flowering plants, or go dormant. Large flowering plants produce new seedlings, regress to medium sized plants, stay the same stage, or go dormant. Finally, dormant plants either emerge as small rosettes, medium-sized flowering plants, or large flowering plants, but never stay dormant for more than one year. That is an interesting result in and off itself, because it is theoretically possible for plants to remain dormant for multiple consecutive years.

It is hard to describe just how much information is contained in this matrix. For a thorough treatment of matrix models, see Caswell (2001) and Ebert (1999). Let us start with population projection. Hurlburt (1999) estimated that there were 220 seedlings, 1245 small rosettes, 330 medium flowering plants, 155 large flowering plants, and 195 dormant individuals in one of her study populations. We can apply Eq. 4.17 the same way for stage-based Lefkovitch models. Figure 4.3 shows that there is some transient reordering of the relative abundances of each stage in the beginning of the projection, but after about 10 years the stages reach their stable stage distribution and then continue to increase at a steady rate.

The right eigenvector of **A** can be used to compute the stable stage distribution, w (Caswell, 2001). In R, the eigen() function by default computes the right eigenvector of any square matrix. If you extract the first eigenvector of **A**, the stable stage distribution

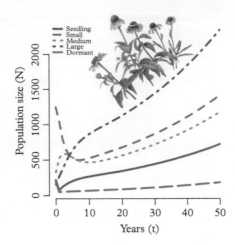

Figure 4.3 *Projection of population sizes for each stage in the population of* Echinacea angustifolia *starting with 220 seedlings, 1245 small rosettes, 330 medium flowering plants, 155 large flowering plants, and 195 dormant individuals. Note the transient dynamics in the first few years after which the projections settle into the stable stage distribution.*

is computed by relativizing the vector by taking the absolute value of each element and dividing each element by the sum of the vector. This new vector will then sum to unity and each element represents the proportion that each stage contributes to the population size when stable. The stable stage distribution of A is $w = (0.12, 0.24, 0.20, 0.40, 0.03)$.

Once the stable stage distribution is reached, we can see the stages projected to be most abundant. Ranked in descending order, large plants will comprise 40% of the population, small plants will comprise 24%, medium plants will comprise 20%, seedlings will comprise 12%, and dormant plants will comprise 3% of the population. These number are reflected in the rank order in the projected population sizes in Figure 4.3. This model is "deterministic" because it converges on a single rate of increase, as opposed to a "stochastic" model, which allows other factors to change the rate of increase annually. It is unlikely that a deterministic value of λ can predict long-term population dynamics, but it is a useful measure of the average potential of the population at the time the data was collected (Crone et al., 2011). The dominant eigenvalue of A is 1.024, which means that the population is projected to increase 2.4% annually once it has reached the stable age distribution.

Another important vector called the reproductive value can be extracted from the projection matrix. Reproductive value has been interpreted many ways (at least three, according to Hal Caswell) but let us consider a general definition that the reproductive value of a stage represents the long-term contribution of reproduction to the population. The left eigenvector of A (Caswell, 2001) is used to compute reproductive value, v. To do this in R, one must conduct eigenanalysis on the transpose of A, denoted A^T. If you extract the first eigenvector of A^T, the reproductive value is computed by relativizing the vector by taking the absolute value of each element and dividing each element by the sum of the vector, then divide each element by the first element so that the first element equals one. You can reproduce these results using the book's R scripts as well as the functions in the R package rage (Jones et al., 2022). The reproductive value of **A** is $v = (1.0, 1.9, 3.1, 3.5, 2.7)$, and note that all v_i values are relative to v_1. Unsurprisingly, the largest size class contributes the most to reproduction, but you might be surprised to see that seedlings and dormant plants also contribute to reproduction. How can this be the case when they have zero fecundity? The answer is because some of those seedlings and dormant plants grow up into reproductive individuals, thus contributing to long-term population growth.

Just like the life table and the Leslie matrix, we can use the Lefkovitch matrix to compute the net reproductive rate ($R_0 = 1.9$), generation time ($T = 26.3$), age at maturity ($\alpha = 5$ years), and longevity ($\omega = 99$ years) for *Echinacea angustifolia*. No one has ever observed an individual *Echinacea angustifolia* plant live for 99 years, but we can use the logic of probability to estimate how long an individual could survive based on the survival probabilities encoded in the matrix (Caswell, 2001). It is amazing that we can make such estimations by recording the survival rates of a few plants over a few years. It is equally amazing that such a small unassuming herb will outlive most of us.

So far in this chapter we have used a life table to solve for r (and by extension λ) using the Euler–Lotka equation, to estimate the net reproductive rate, to estimate generation time, and to translate the life table into an equivalent Leslie matrix. We used the Leslie matrix to project population sizes while tracking each age class. Then we explored a stage-based matrix population model for a long-lived perennial plant and projected its population into the future. Now let us scale up and see how individual population models can tell us something about the variation in life histories across multiple species simultaneously.

4.5   Silvertown and Franco's demographic triangle

Plant demography started out as a discipline that shared a primary interest in quantifying population dynamics of single species. When you read the classic demography papers (and you really should) you will find that they are nearly all focused on single species. Many of these papers demonstrated that population growth rates of a species varied across different sites and local populations. For example, one classic study documented that *Dipsacus sylvestris* (teasel) population growth rates (λ) varied from 0.275 to 2.605 (Werner and Caswell, 1977)! These studies demonstrated that fitness varied across environmental contexts, but a single-species focus does not allow us to discover which traits are driving those fitness differences.[6]

---

[6] Single-species demographic studies are critical in conservation of rare and managed species, but the day-to-day data collection of single species populations can get tiring. I routinely heard stories from fellow graduate students that they started hating their study species because they were tired of counting them! I became a community ecologist because I wanted to learn about whole communities of species, and so my early research

The plant population ecology literature took a turn in the early 1990s when a sufficient number of population studies became available for synthesis. We can thank John Harper for stimulating international interest in plant population biology in the 1970s. One thing that clearly motivated Harper was his firm belief that plants are superior to animals for studying population dynamics because "plants stand still and wait to be counted" (Harper, 1977, p. 515).

It was Jonathan Silvertown and Miguel Franco, however, who pioneered a new field of comparative plant demography. They were interested in making comparisons across multiple species to learn something deeper about the ecology and evolution of life histories. Silvertown wrote, "It is easy to get lost in the details of demographic studies and to lose site of the bigger picture" (Silvertown and Charlesworth, 2001, p. 172). Elasticity analysis provided just the tool to compare plant population dynamics to find grander patterns.

The projection matrix **A** can be pushed and prodded to learn something about the sensitivity of λ to stage-specific survival and reproduction. This prodding is called perturbation analysis. We want to learn which specific transitions in the matrix **A** exert the largest changes in population growth rate. We will denote each cell in **A** as a_{ij} to keep track of each row i and column j. Conceptually, perturbation analysis is equivalent to varying one element in **A** while holding all other elements constant to determine the effects of this element on λ. Mathematically, we compute the partial derivative of λ with respect to a_{ij}, $\partial \lambda / \partial a_{ij}$. Each a_{ij} has a corresponding sensitivity value, s_{ij}:

$$\frac{\partial \lambda}{\partial a_{ij}} = s_{ij} = \frac{v \cdot w^T}{\langle v, w \rangle}, \tag{4.18}$$

where v and w are the left and right eigenvectors of **A** and $<v, w>$ is their scalar product.

These sensitivities are difficult to compare between those that affect survival probabilities (which are bound between 0 and 1) and those that affect fecundity (which can be numbers > 1). Therefore, de Kroon and colleagues (1986) proposed the elasticity metric (e_{ij}) that sum to one:

$$e_{ij} = \frac{a_{ij}}{\lambda} \times s_{ij}. \tag{4.19}$$

Elasticities are interpreted as the relative "contributions" of each of the a_{ij} to λ (Caswell, 2001). They are valuable tools in the demographic toolbox for understanding which demographic rates contribute the most to population growth. This is especially important for the conservation of rare species threatened with extinction. Figure 4.4 illustrates the elasticities of the matrix model for *Echinacea angustifolia*. It is evident that the elasticities along the diagonal are the largest, indicating that persistence (stasis) within stage classes is the most important contributor to population growth (Hurlburt, 1999).

was focused on the community and ecosystem levels. It wasn't until later when I recognized the centrality of demography to a functional perspective on community ecology.

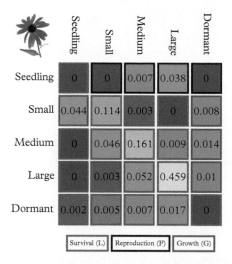

Figure 4.4 *Matrix of elasticities for* Echinacea angustifolia. *The background color of the cells denote the importance of the transition to population growth rate, increasing in importance from blue to gold colors. The boundary of each cell is also color-coded to distinguish among the three demographic rates (survival, reproduction, and growth) used to locate the species in the demographic triangle.*

The largest elasticity by far is associated with persistence within the largest size class ($e_{large, large} = 0.459$), suggesting that if survival rates within this class fall for some reason, then population growth will decline quickly. Recall that this species is prized for its medicinal roots: when herbalists dig up too many large plants, then local populations may be threatened with extinction. To maintain viable populations, limits must be placed on the number of large individuals that can be harvested.

Elasticities can also be used to compare species life histories by comparing the relative importance of each demographic rate (growth, survival, and reproduction) with respect to their influence on population growth rate (Silvertown and Franco, 1993; Silvertown et al., 1993). In this framework, elasticities for transitions that represent different demographic processes are combined and summarized into three types (Figure 4.4). First, recruitment of seeds into the seed bank and recruitment of seedlings or juveniles from current seed production collectively represent fecundity (F). Second, retrogression (i.e., shrinkage) due to plants decreasing in size, reverting from a flowering to a vegetative state or becoming dormant, and stasis (i.e., survival within the same stage from year to year) collectively represent survival (L). Third, clonal growth and progression to later stage classes collectively represent growth (G). Some authors prefer to include clonal growth within fecundity.

Silvertown made these calculations for dozens of species and plotted the results in a ternary space that he called the "demographic triangle" (Figure 4.5a) (Silvertown and Franco, 1993; Silvertown et al., 1993; Franco and Silvertown, 1996, 2004). The space is triangular because the elasticities sum to one (recall the discussion in Section 2.5: if you

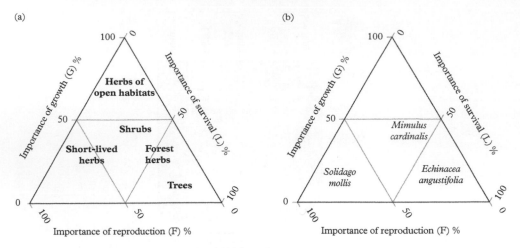

Figure 4.5 *(a) Silvertown and Franco's demographic triangle and the approximate location of different plant life histories within the ternary space. (b) Locations of three herbaceous perennials within the demographic triangle. Elasticities were computed from matrix population models in the COMPADRE database.*

(a): Reproduced from Silvertown and Franco (1993) with permission.

know two values then the third is a given). Short-lived semelparous herbs cluster toward the bottom left where reproduction is all important. Semelparous perennial herbs of open habitats are located toward the top of the triangle where growth and reproduction are most important. Iteroparous herbaceous perennials in forests tend toward the right due to lower fecundity. Long-lived trees tend to cluster in the lower right corner, where survival is most important.

To make this type of analysis more concrete, let us quantify the relative importance of growth, survival, and fecundity using the elasticities for *Echinacea angustifolia*. In Figure 4.4, transitions related to growth (G) include many of the elements in the lower diagonal of the matrix, because these record transitions from smaller to larger size classes. Transitions related to survival (L) include elements on the diagonal because these record stasis, and a few elements on the upper diagonal because these record shrinkage from larger to smaller size classes. Survival also includes the transitions into a dormant state recorded in the last row of the matrix. Fecundity (F) includes all elements in the first row except the first.

Summing the elasticities within each of the three categories gives an indication of the importance of that demographic rate to population growth. For *Echinacea angustifolia*, survival sums to 0.78, growth sums to 0.18, and fecundity sums to 0.04. These numbers suggest that survival is the most important and that recruitment is the least important component of the life cycle for maintaining a stable population of *Echinacea angustifolia*. This is logical given that it is a long-lived perennial forb with a high importance of survival and stasis, but it seems to disagree with Silvertown's model, which places it near the life history of a tree (Figure 4.5b). *Echinacea angustifolia* bucks the original generalities, suggesting that other long-lived perennial forbs from dry or fire-prone habitats may also behave more like woody plants with respect to their life history.

We can conduct the same analysis on other perennial species found in the COMPADRE database. *Solidago mollis* (soft goldenrod) is a short-lived perennial herb in the Asteraceae (aster family). Goldenrods are known for their showy golden capitulescences (clusters of inflorescences) that brighten trails in late summer and fall. The matrix for this model was drawn from a study in the mixed grass prairie of Kansas, USA (Dalgleish et al., 2010). Summing the elasticities places this species in the extreme lower left corner of the demographic triangle, indicating that it behaves like a short-lived herb (Figure 4.5b). *Mimulus cardinalis* (scarlet monkeyflower) is a short-lived perennial herb in the Phrymaceae (lopseed family) that is native to western North America. It can be found growing naturally in seeps and streambanks and is cultivated widely for its showy red flowers. The matrix for this model was drawn from a study in the Sierra Nevada mountains of California (Angert, 2006). Summing the elasticities places this species in the middle right region of the demographic triangle (Figure 4.5b). There are important limitations to this elasticity ternary plot approach (Shea et al., 1994; Takada et al., 2018), but these ideas are important because they jump-started the modern journey toward linking comparative functional ecology with demography.

4.6 From stages to ages

We started this chapter by considering age-based models built from life tables and have now extended our arsenal to include stage-based models that are more flexible and appropriate for most plants. However, it is not easy to see how the stage-based matrix can be used to learn about age-specific survival and reproduction vectors. It turns out that we can also estimate l_x and m_x by making a life table from the projection matrix using methods developed by—you guessed it—Hal Caswell (2001). Fortunately for us, these methods have been translated into R functions in the Rage package (Jones et al., 2022).

The authors of the COMPADRE database split the full **A** matrix into three sub-matrices: **U**, **F**, and **C**. The **U** matrix includes the transition probabilities between classes that represent survival, growth (or "development") and shrinkage (or "de-development"). The **F** matrix includes the fecundities, the average number of recruits per individual in each size class. The **C** matrix summarizes clonal growth (sometimes called asexual reproduction) and is not included if the original study did not measure it. The projection matrix **A** is the matrix sum of the component matrices:

$$\mathbf{A} = \mathbf{U} + \mathbf{F} + \mathbf{C}. \tag{4.20}$$

By splitting **A** into its components, we can conveniently estimate life tables for each species.

Let us add a few more matrix models to our growing list of species in this chapter. *Asclepias meadii* (Mead's milkweed) is a rare species of milkweed native to the tallgrass

Table 4.4 *Comparison of life history traits for five plant species ranked in order of increasing generation time.*

Species	Generation time (T)	Age to maturity (α)	Net reproductive rate (R_0)	Degree of iteroparity (D)
Solidago mollis	1	1	1.8	−0.3
Mimulus cardinalis	5	2	23.5	1.5
Asclepias meadii	14	6	0.03	6.0
Echinacea angustifolia	26	5	1.9	3.7
Silene acaulis	176	24	5.3	3.8

prairies in the midwestern USA. This perennial herb in the Asclepiadaceae (milkweed family) reproduces rarely from seed and is listed as threatened mostly due to habitat loss. The matrix for this model was drawn from restoration projects in the tallgrass prairie of Illinois and Indiana, USA (Bell et al., 2003). *Silene acaulis* (cushion pink) is a long-lived cushion plant in the Caryophyllaceae (pink family) that grows in alpine and arctic tundra. Cushion plants retain their warmth in the extreme high elevation environment by growing low to the ground. The matrix for this model was drawn from a study in the Wrangell mountains in Alaska, USA (Morris and Doak, 1998). Using the equations from earlier, we can reproduce the calculations of five life history traits listed in Table 4.4 for the five herbaceous species described in this chapter.

Recall from Section 3.3 that Salguero-Gómez and colleagues (2016) computed life history metrics for 418 plant species worldwide to examine life history strategies and found two main axes. The first axis represented variation along the fast–slow continuum, which is a contrast of slow species (long generation times and later reproductive maturity) and fast species. Note that the first two columns in Table 4.4 are positively correlated because these traits represent variation along the fast–slow continuum. *Solidago mollis* and *Mimulus cardinalis* have fast life histories and *Echinacea angustifolia* and *Silene acaulis* have slower life histories. The degree of iteroparity (D) is a continuous life history trait that ordinates species along a spectrum from monocarpic to polycarpic reproduction schedules. The second independent axis of variation represented variation in reproductive strategies as a contrast between semelparous (monocarpic) and iteroparous species. Note that D is not correlated with the first two life history traits that are associated with the fast–slow continuum.

4.7   Seed banks can be modeled as discrete stages

The life cycle of plants can include a dormant stage when seeds downregulate respiration rates and halt embryonic development. It can be difficult to believe, but dormant seeds are living plants, and despite their near invisibility to the casual observer, seeds should count in the census of a population! One census of the Mojave Desert annual *Linanthus parryae* indicated that the ratio of seeds-to-plants in the population was 30:1 (Epling et al., 1960). Researching the flux of seeds into and out of the soil is hard, but ignoring these demographic rates will only continue to cloud our understanding of plant life histories.

Seed dormancy varies strongly among species, and it has proven extremely difficult to predict which species develop seeds with a long dormancy phase (Doak et al., 2002). According to John Harper, "some seeds are born dormant, some acquire dormancy, and some have dormancy thrust upon them" (Harper, 1977, p. 65). In other words, some seeds are innately dormant for a variety of reasons, including seed coat impermeability and chemical inhibition, whereas dormancy in other seeds is enforced by the lack of suitable light, water, or temperature requirements for germination (Rees, 1997a; Baskin and Baskin, 1998). Dormancy is therefore a mechanism of escape from unfavorable environmental conditions through time. It is a truly remarkable evolutionary development and some seeds have proven to be exceptionally long lived.

Dormant seeds contribute to a reserve of propagules called a soil seed bank. Note that seed banks should only be added to a model if seeds can live longer than the period of sampling (e.g., one year) or more; otherwise, including them is a fatal modeling error. Adding a seed bank stage to a population model adds another row and column to the transition matrix. However, obtaining accurate estimates of those parameters is not so easy. Recent advances in coupling seed bank and germination experiments provide a useful framework (Paniw et al., 2017). Consider a simple population model for an annual plant using a matrix of two stages: seeds and plants (Table 4.5). The probability of a seed staying in the seed bank is the product of the probability of seed survival (s_s) and the probability of not germinating ($1 - g_o$). The probability of a seed emerging from the seed bank is the product of the probability of seed survival (s_s) and the probability of germinating (g_o). The number of seeds entering the seed bank is the product of three quantities: the per capita number of seeds produced (f), the survival rate of the reproductive plants from germination to reproduction (s_{ad}), and the probability of new seeds not germinating ($1 - g_n$). The transition rate of plants in year t to plants in year $t + 1$ is the product of three quantities: the per capita number of seeds produced (f), the survival rate of the reproductive plants from germination to reproduction (s_{ad}), and the germination rate of seeds in the seed bank (g_n).

Seed viability tends to decay over time, but it does not follow a predictable negative exponential rate of decay, making generalization across species tricky (Rees and Long, 1993). In general, underestimating seed survivorship will reduce population stability and overestimating seed survivorship will increase population stability. Based on reviews of the literature and simulation analysis of the 2×2 matrix model with a seed bank in Table 4.5, an important generality has emerged: "Species with shorter and more variable aboveground lifespans are most likely to rely on seed banks for population persistence. In addition, the importance of seed banks to population persistence is likely to diminish with increasing population size" (Doak et al., 2002, p. 332).

Ecologists have long recognized the importance of seed banks to plant population dynamics, but the inclusion of seed banks into population models has been slow to materialize (Adams et al. 2005; Nguyen et al., 2019), likely because estimating the

Table 4.5 *Two-by-two matrix population model with a seed bank stage.*

	Seed	Plant
Seed	$(1 - g_o)\, s_s$	$s_{ad}\, f\, (1 - g_n)$
Plant	$g_o\, s_s$	$s_{ad}\, f\, g_n$

f = fecundity; s_{ad} = survival from germination to reproduction; g_n = germination rate of newly produced seeds; g_o = germination rate of seeds in the seed bank; s_s = seed survival.
Reproduced from Doak et al., (2002) with permission.

rate of seed inputs into the seed bank and outputs from the seed bank is a formidable task. Paniw and colleagues (2017) used experimentally buried seed bags to estimate some of these difficult rates (Figure 4.6). They used the experiments and observations to estimate four key transitions: the rate of new seed input into the seed bank (*goSB*), the rate of new seeds that germinate immediately the following year (*goCont*), the probability of staying in the soil seed bank (*staySB*), and the rate of emergence from the seed bank (*outSB*). Hopefully these new approaches increase the inclusion of dormant stages of the life cycle in plants to improve our understanding of plant population dynamics.

4.8 Population models can incorporate density dependence

So far, we have examined the growth rate of populations in isolation, but individuals do not grow in a vacuum. Individuals struggle to survive and reproduce within a milieu of interactions with individuals of the same species (conspecifics) and with individuals of other species (heterospecifics). Population size is regulated by density-dependent mortality, which increases the per capita death rate, and density-dependent fecundity, which decreases the per capita birth rate (Silvertown and Charlesworth, 2001). Competition experiments have produced mixed results about competition, but they agree on one thing: individual plant growth responds strongly to density, and "members of a species tend to be their own worst enemies" (Niklas, 1997, p. 13).

In single-species populations, the relationship between mean biomass per plant and density is often modeled as a hyperbolic function:

$$w = w_m(1 + aN)^{-1} \tag{4.21}$$

where w = mean biomass per plant, w_m = the maximum potential biomass per plant, a = the area necessary to achieve w_m, and N = density (Watkinson, 1980). Thus,

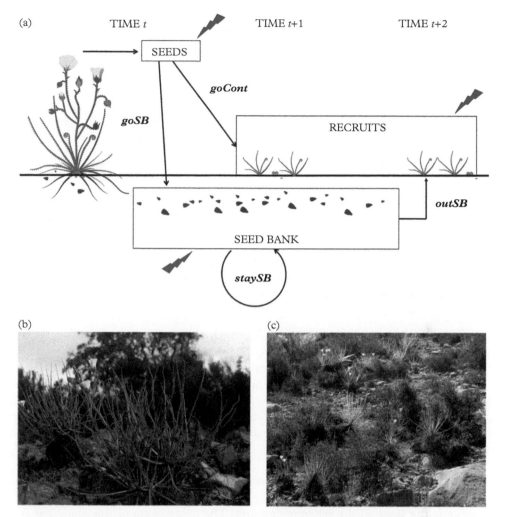

Figure 4.6 *(a) New seeds either contribute to new recruits the following year (*goCont*), or they enter the soil seed bank (*goSB*). Once in the soil, they can either stay in the seed bank (*staySB*), eventually emerge as new seedlings (*outSB*), or die in the soil. Red lightning bolts represent mortality of seeds or seedlings before establishment. (b) This approach was used to model the population dynamics of the spectacular dewy pine (*Drosophila lusitanicum*), a carnivorous short-lived subshrub. (c) Most carnivorous plants live in wet habitats, but this species grows in the herriza—a dry, fire-prone Mediterranean heathland.*
(a): Reproduced from Paniw et al. (2017) with permission. (b, c): Photos by Fernando Ojeda reproduced here with permission.

mean biomass per plant declines non-linearly with increasing density (Figure 4.7a) and the relationship of w to density is called the competition-density effect (Firbank and Watkinson, 1990).

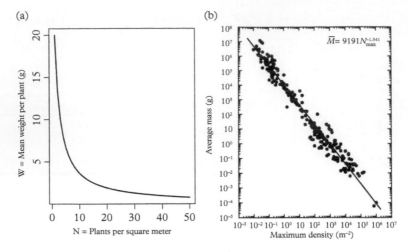

Figure 4.7 *(a) The general relationship between plant density and mean mass per plant, following Eq. 4.21. (b) Empirical relationship between maximum density and average mass among plants.* Reproduced from Enquist et al. (1998) with permission.

As density rises and as crowding intensifies with plant growth, plants die and density declines over time—a phenomenon called self-thinning, which describes how density and biomass change over time in the same place. This sets an upper bound on the competition-density effect, which describes the negative relationship between mean plant biomass and density at one moment in time across a range of places (Silvertown and Charlesworth, 2001). The trajectory of populations undergoing self-thinning approach a line of constant slope. The relationship between mean plant biomass and density is described by the equation

$$w = cN^{-b} \tag{4.22}$$

where $-b$ is the slope of the self-thinning line. The exact value of this slope has been vigorously debated. For a very long time, the exponent was thought to be $-3/2$, based on the logical ratio of plant volume (a cubic linear dimension, l^3) to the area occupied by the plant (a square linear dimension, l^2), attributed to Yoda (1963).[7]

However, many now argue that the exponent is $-4/3$, which is derived from constraints on metabolic rates with body size in both plants and animals (Enquist et al., 1998). The empirical data are generally supportive of this revised exponent. Figure 4.7b illustrates the strong negative linear relationship between log maximum density and log average biomass among plants ranging in size from the tiniest duckweeds (*Lemna* spp.) to the most massive coastal redwoods (*Sequoia sempervirens*). The scaling exponent of this line is -1.341, which is statistically indistinguishable from $-4/3$. Since space is finite,

---

[7] An important phenomenon, this is.

a large number of small individuals in a given space will eventually be reduced to a small number of large individuals.

Given that plant size has strong effects on individual survival and reproduction, density must exert strong control over population dynamics. The negative effects of density dependence have been shown time and time again. For example, negative density dependence results in reduced annual rates of germination, seedling survival, and seedling growth of the canopy tree *Miliusa horsfieldii* in Huai Kha Khaeng Wildlife Sanctuary in western Thailand (Figure 4.8). The effects of density dependence were incorporated into an individual-based model to simulate population dynamics and viability to demonstrate the importance of frugivore dispersal to tropical forest composition and dynamics (Caughlin et al., 2015).

Figure 4.8 *Photos of* Miliusa horsfieldii *(Annonaceae) in Huai Kha Khaeng Wildlife Sanctuary, Thailand. (a) Tagged seedling; (b) Reproductive tree with fruits; (c) Canopy tree. Consistent negative effects of conspecific seedling density on (d) germination, (e) seedling survival, and (f) seedling growth rates of* Miliusa horsfieldii. *Dashed lines represent the central moment of the Bayesian posterior distribution and light grey lines represent uncertainty in the estimate.*

Reproduced from Caughlin et al. (2015) with permission.

Density dependence has also been incorporated into matrix population models by expressing the elements in the **A** matrix as functions of density. For example, Silva Matos and colleagues (1999) modeled the population dynamics of the subcanopy palm *Euterpe edulis* in a semideciduous forest at the Municipal Reserve of Santa Genebra in the state of São Paulo, Brazil. They found that the transition rate from seedlings into larger size classes (g) were density dependent, so they incorporated density dependence into their matrix model by using the following function for this transition rate:

$$g = g_m(1 + aN_1)^{-1}, \tag{4.23}$$

where g_m is the maximum mean transition rate at low density, N_1 is the initial density of seedlings, and a is a parameter that measures the strength of the reduction in g with increasing density. Notice the strong similarity in form between this function and Eq. 4.21. Density dependence can be incorporated into individual functions for each transition in a matrix or density effects can be integrated into size-dependent vital rate (i.e., demographic rate) regression models within integral projection models.

4.9   Integral projection models consider size as a continuum

The delineation of the breakpoints between plant size classes is sometimes arbitrary. This led population ecologists down a path toward the development of the integral projection model (IPM) (Ellner and Rees, 2006; Rees et al., 2014; Ellner et al., 2016), which treats size as a continuous variable. IPMs are not to be confused with the other IPMs: integrated pest management, integrated population models, or the fusion of the latter with an IPM to create an integrated integral projection model (IPM^2) (Plard et al., 2019). Be aware of this acronym and the context of its use.

The key feature of the IPM is that it treats size as a continuous variable, which allows demographers to use regression models and the stacks of probability theory associated with them. They have been shown to perform similarly to matrix models, although users should be aware of their inherent differences (Doak et al., 2021; Ellner et al., 2022). IPMs can combine multiple stages (seed banks) together with continuous stages (size). Quite frankly, they can do just about anything provided you synthesize all the various functions correctly. The resulting projection matrix of an IPM can be analyzed using all the matrix techniques discussed earlier. I refer interested readers to a book and some recent articles that describe this method in great detail (Merow et al., 2014a; Rees et al., 2014; Ellner et al., 2016) and to the R packages IPMpack (Metcalf et al., 2013) and ipmr (Levin et al., 2021) to facilitate model implementation.

I introduce this framework for constructing population models because they are inherently more flexible and because any covariate can be incorporated into the vital rate regression models. So rather than treating size as the only predictor of survival and reproduction, we can easily incorporate other predictors such as traits, density, traits of neighbors, or environmental conditions (recall Eq. 3.4). Understanding vital rate regression models for IPMs lays the groundwork for the later discussion of trait-based demographic rates.

The core feature of an IPM is the kernel, which is a function that describes how the size distribution of individuals in one year changes in the following year by keeping track of individual growth, survival, and reproduction. The time step can be any interval, but we will stick with one year for consistency. The advantage of the IPM is that we can leverage the power (and tractability) of regression models to explore how size impacts growth, survival, and reproduction. We will split the kernel, $K(z', z)$, which maps the size distribution at time t (denoted as z) to the size distribution in time $t + 1$ (denoted as z') into a growth/survival kernel P, and a fecundity kernel F. Recall that the A matrix can be split into its components: the survival/growth matrix U and a fecundity matrix F (Eq. 4.20). Similarly, the IPM kernel $K(z', z)$ is composed of two sub-kernels:

$$K(z', z) = P(z', z) + F(z', z). \qquad (4.24)$$

The survival/growth kernel $P(z', z)$ describes the probability that an individual survives, and if it does survive, whether it grows or shrinks. The fecundity kernel $F(z', z)$ describes the number of recruits created by reproductive individuals and their size class distribution.

To motivate the construction of an IPM, we build the kernels using a sample dataset of a long-lived perennial grass *Festuca arizonica*. This C_3 bunchgrass is native to the southwestern USA and is found in montane grasslands and forests. Margaret Moore has been mapping 1-m^2 chart quadrats since 2002, and I demonstrate how to build an IPM using this long-term demographic data that is freely available (Moore et al., 2022). The spatial data I use here has already been processed by algorithms to produce the demographic response data that is needed for the model (Stears et al., 2022a). To build an IPM, you need a dataset of individual plants during a transition from year t to $t + 1$. In other words, each row is an individual plant, and information on size in year t, whether it survived to year $t + 1$, and if it survived, then its new size in year $t + 1$. The dataset can also include whether it flowered, how many seeds it produced, and whether it is a new recruit to the population.

The core features of the IPM are the vital rate regression models, where growth, survival, and fecundity are all functions of plant size (Merow et al., 2014a). The survival function, $s(z)$, depends on the size of an individual at time t and predicts the probability of survival to time $t + 1$. We estimate $s(z)$ by modeling survival (1 = survived, 0 = died) in year $t + 1$ as a function of size in year t using a logit link function and binomial error distribution (Figure 4.9a).

The growth function, $g(z' \mid z)$, describes the probability density of size z' that an individual of size z grows between year t and year $t + 1$ if it survived to year $t + 1$. This function is estimated by modeling size in year $t + 1$ as a function of size in year t using a Gaussian error distribution (Figure 4.9b). The survival/growth kernel is the product of the survival and growth functions:

$$P(z', z) = s(z) g(z' \mid z). \qquad (4.25)$$

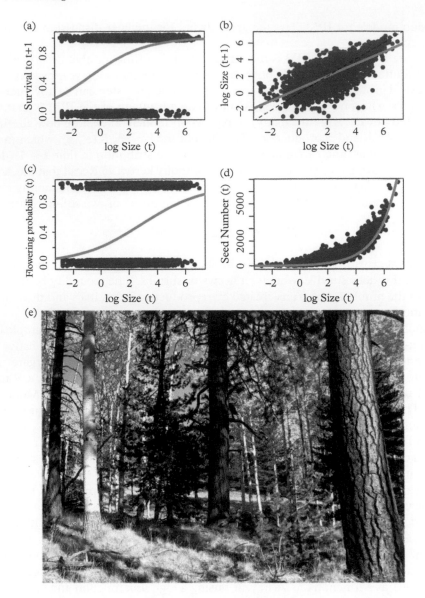

Figure 4.9 *Vital rate regression models for Arizona fescue (*Festuca arizonica*) as a function of log size (measured as basal area) in year* t: *(a) probability of survival, (b) growth rate model, (c) probability of flowering, and (d) seed production. (e) Open montane forest is prime habitat for Arizona fescue, which can be seen setting seed in the lower left of the photo, San Francisco Peaks, Flagstaff, Arizona, USA.*
Photo by the author.

For a typical perennial plant, the fecundity kernel $F(z', z)$ is the product of four ingredients:

1. The size-dependent flowering probability ($p_{flower}(z)$) (Figure 4.9c).
2. Size-dependent seed production conditional on flowering ($f_{seeds}(z)$) (Figure 4.9d).
3. The establishment probability of each seed produced (p_{estab}), which is often estimated as the ratio of number of recruits in year $t + 1$ to the number of seeds produced in year t.
4. The size distribution of recruits at time $t + 1$ ($f_{recruit\ size}(z')$):

$$F(z', z) = p_{flower}(z) f_{seeds}(z) p_{estab} f_{recruitsize}(z').$$ (4.26)

Using these functions, we can compute a projection matrix (Figure 4.10), which is oriented in a different direction compared to matrix population models given the tradition within the IPM literature of viewing the demographic rates as regression models. In other words, the *y*-axis increases from bottom to top. Compared to a matrix model, the rows are inverted such that fecundity is seen in the bottom of the matrix (as opposed to the top row), and the growth/survival kernel reaches highest densities from bottom left to top right (as opposed to top left to bottom right). Note that this matrix has all the same properties as the matrix models discussed earlier. For example, we can use eigenanalysis to compute λ. In this case $\lambda = 0.986$.

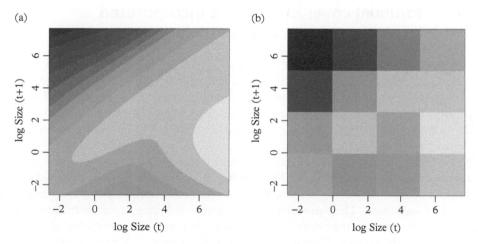

Figure 4.10 *Integral projection models and matrix population models are two ends of a continuum between infinitesimally small size classes to large size classes. All projection matrix models for* Festuca arizonica *were created using the IPM framework but were discretized using (a) 500 size classes, and (b) 4 size classes. Dark colors are low transition probabilities and light colors are high probabilities. The probabilities of surviving and growing are generally along the diagonal from lower left to upper right. The highest probabilities of reproducing are seen near the lower right where larger individuals in time* t *create offspring of small sizes in time* t + 1.
Inspired by Merow et al. (2014a).

Meshpoints are the number of rows and columns in the matrix. If a smaller number of meshpoints are used then we can see that IPMs and matrix projection models are two ends of a spectrum (Merow et al., 2014a): Figure 4.10a illustrates the full kernel using 500 meshpoints with smooth changes in probability densities over the size distribution, whereas Figure 4.10b illustrates kernels using four meshpoints using the same data—these are more akin to a standard matrix population model.

Evidence shows that projections from matrix population models for long-lived trees are sensitive to the number of size classes and that too few categories underestimate longevity. Zuidema and colleagues (2010) showed that the projected population growth rates stabilized using IPMs with 100–1000 size classes and recommended using size classes between 0.1–1.0 cm width to obtain reliable output. Another major challenge for long-lived plants is modeling recruitment. Tree census monitoring rarely count the smallest seedlings, and many tree IPMs model recruitment as the transition into a minimum diameter size class (often 2.54 cm diameter at breast height). These are rather old seedlings by this time, so there is considerable uncertainty about specific yearly recruitment rates in these cases. IPMs have been used to study the life history and population dynamics of long-lived trees in tropical forests (Metcalf et al., 2009; Zuidema et al., 2009; Kuswandi and Murdjoko, 2015; Needham et al., 2018; Zucaratto et al., 2021), temperate trees (Needham et al., 2016), shrubs from South Africa (Merow et al., 2014b), and tree cholla in the Southwest USA desert (Elderd and Miller, 2016).

4.10   Additional covariates can be incorporated into vital rate regression models

Size is a continuous variable in the vital rate regression models, but other covariates can influence demographic rates. There are a few covariates that are of central importance to plant strategy theory: traits, environments, neighborhood density, and traits of neighbors. While incorporating these covariates is technically possible for matrix population models, IPMs make the process remarkably easy. I describe the utility of this approach by way of two examples.

Westerband and Horvitz (2017) were interested in how photosynthetic rate affected growth rates across a gradient in light availability in two perennial herbs, *Heliconia tortuosa* and *Calathea crotalifera*, in tropical forest understories at the Las Cruces Biological Station in Costa Rica. They modeled size in year $t + 1$ as a function of interactions between size in year t, canopy openness (a measure of light availability), and A_{max} (maximum photosynthetic rate). They found that, as light increased, λ increased for *Calathea* but decreased for *Heliconia* (Figure 4.11). Further, as A_{max} increased, λ did not change for *Calathea* but increased for *Heliconia* in low light. They concluded that photosynthetic rate could affect λ through its effects on growth, but that the magnitude of this effect varies across species. This study is an excellent example of how traits and resource levels can be used to understand drivers of demographic rates other than size.

Struckman and colleagues (2019) built an integral projection model for the common milkweed, *Asclepias syriaca*, a perennial herb that is common in the Eastern USA. The

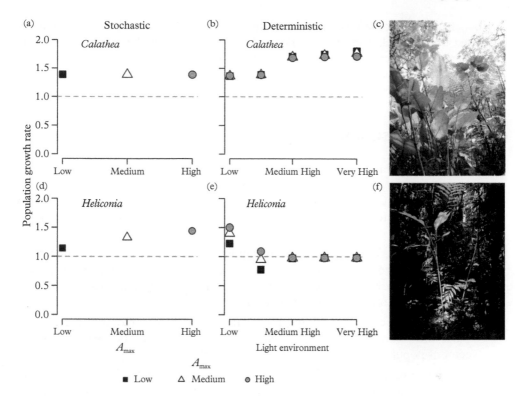

Figure 4.11 *Estimated population growth rates (λ) for (a–c)* Calathea crotalifera *and (d–f)* Heliconia tortuosa.

Reproduced from Westerband and Horvitz (2017) with permission. Photos of Costa Rican study species provided by Andrea Westerbrand, with permission.

authors were primarily interested in how chemical defenses influence the population growth rate of this key food plant for monarch butterflies. They measured traits related to leaf nutritional quality, structural composition, morphology, and a secondary metabolite called cardenolides. Secondary metabolites are by-products of metabolic pathways in plant cells and a considerable body of work has investigated how these are used as a defense against insect herbivory (Agrawal et al., 2012). They found that plants with higher levels of cardenolides in the leaves had lower structural strength and reduced sexual and clonal reproduction. This led to a trade-off between defense and reproduction that affected λ.

The ability of incorporating additional covariates into the vital rate regression models permits powerful tests of the effects of traits on demographic rates. These two examples show their usefulness in the context of understanding intraspecific trait variation on population dynamics. Later in the book, I extend the approach to multiple species to develop mechanistic models that explain how trait effects on fitness are mediated through demographic rates.

This chapter demonstrated how population growth rates and life history traits are estimated using population models and explored variations of the models to account for density dependence and dormant life stages. Chapter 5 discusses life history theory, which explores why many of the life history traits exhibit trade-offs and how these trade-offs constrain the possible combinations of life history traits in plants.

4.11   Chapter summary

- All successful plant strategies must exhibit positive demographic outcomes. It is not sufficient to merely observe a species in a habitat and assume its traits confer fitness in those environmental conditions. Strong tests of plant strategy theory must demonstrate that phenotypes can predict demographic outcomes in a given environment.

- The Euler–Lotka equation shows us that fitness is a function of age-specific rates of survival and reproduction. Age-specific demographic rates in life tables can be used to compute life history traits including age to maturity, lifespan, net reproductive rate, generation time, and degree of iteroparity.

- Life tables can be translated into Leslie matrices that summarize transition probabilities and fecundities of each age class. These matrices can be used to project a vector of population sizes of each age class into the future.

- Stage-based models based on size classes are often more useful for modular organisms like plants given that survival and reproduction are more strongly related to size than age. Plants not only can survive and reproduce, but also, they can shrink, the plant can remain dormant belowground, and their seeds can remain dormant in the soil seed bank. Lefkovitch matrices are stage-based population matrix models.

- We can determine the relative importance of each vital rate on its impact on population growth rate by computing vital rate elasticities, which quantify the impact of a vital rate on population growth rate. Silvertown's demographic triangle can be used to compare species' life histories by computing elasticities for growth, stasis, and fecundity.

- Given the desire to interpret stage-based models in a life history theory that is grounded in ages, methods were developed to convert stage-based models into age-based life tables to compute life history traits.

- Seed banks can be modeled as discrete stages in population models by knowing the probability of entering, remaining in, and emerging from the soil seed bank.

- Most demographic theory that focuses on intrinsic population growth rate operates under an assumption of density independence because the Euler–Lotka equation does not account for population regulation by itself or other species, but plant growth is strongly regulated by density-dependent processes. Models can accommodate this important source of variation by allowing transition probabilities and demographic rates to be functions of local neighborhood density.

- Integral projection models have recently been developed that treat size as a continuous variable rather than needing to bin plants into arbitrary size classes. This flexible approach allows us to estimate population growth rates by building vital rate regression models that include any number of covariates as predictors. This approach is necessary to explore how traits determine individual demographic rates across species.

4.12   Questions

1. Place each of the transition probabilities and fecundities in Table 4.3 onto each of the arrows in the life cycle diagram on Figure 4.2. You should also be able to do the inverse: create a matrix from any life cycle diagram.
2. Confirm that the stable stage distribution of **A** in Table 4.3 is $w = (0.12, 0.24, 0.20, 0.40, 0.03)$.
3. Confirm that the reproductive value of **A** in Table 4.3 is $v = (1.0, 1.9, 3.1, 3.5, 2.7)$.
4. Select two matrix population models from COMPADRE and compare the life history traits of the two species, including net reproduction rate, generation time, degree of iteroparity, and age to maturity.
5. Build an integral projection model for *Muhlenbergia montana* using the dataset provided and compare the model to the one presented for *Festuca arizonica*. You can find the data and R code at https://github.com/danielLaughlin/PlantStrategies.

5

Life History Theory Applied to Plants

The evolution of life history traits and their plasticity determines the population dynamics of interacting species. Its explanatory power, barely tapped, could reach as far as communities. There is much to be done.

—Stephen Stearns (1992, p. 9), *The Evolution of Life Histories*

There is much to be done. There is also a real hope that we may be getting somewhere.

—Mark Westoby and colleagues (2002, p. 148), "Plant ecological strategies: some leading dimensions of variation between species," in *Annual Review of Ecology and Systematics*

5.1 Population growth rate is an integrative proxy of average fitness in a population

Population models deepen our understanding of the evolution of life histories (Hutchings, 2021). Life histories are the age- and stage-specific schedules of reproduction and mortality. Life history theory delineates what is demographically possible and a core component of plant strategy theory is understanding the constraints on life histories and drivers of population trajectories. It has been argued that evolutionary biologists should become demographers (Metcalf and Pavard, 2007)—I would add that functional ecologists should take the same advice (Salguero-Gómez et al., 2018; Laughlin et al., 2020b).

Life history theory is based on the understanding that natural selection has been the principal driving force in the diversification of life on Earth. This theory is grounded in two important ideas (Stearns, 1992; Roff, 1993). First, natural selection tends to maximize the fitness of populations by selecting the fittest genotypes. Second, trade-offs limit the possible combinations of traits expressed by the genotype. Lamont Cole (1954) is often credited to be the first to focus attention on selection pressures on population growth rate, rather than reproductive rate alone, but we can trace it further back to Fisher (1930), Wright (1932), Andrewartha and Birch (1954), and of course conceptually back to Charles Darwin and Alfred Russel Wallace (Darwin and Wallace, 1858; Darwin, 1859). Birch (1960, p. 10) wrote, "Natural selection will tend to maximize

Plant Strategies. Daniel C. Laughlin, Oxford University Press. © Daniel C. Laughlin (2023). DOI: 10.1093/oso/9780192867940.003.0005

r for the environment in which the species lives." This is not to say that natural selection is driving species toward perfection. It is not the "survival of the fittest," but rather the "survival of the tolerably more fit" (Niklas, 1997, p. 19). If traits exhibit variation, if traits are heritable, and if traits have differential effects on survival and reproduction, then the genotypes with superior traits will live more often and produce more offspring, which inevitably leads to increased rates of population growth compared to genotypes with underperforming traits. Population growth rate is a "global" mean proxy of fitness because it involves all the life history components simultaneously, in contrast to "local" measures of fitness, such as survival or reproduction alone, which can influence each other due to trade-offs driven by resource allocation.[1]

The Euler–Lotka equation (Eq. 4.11) elegantly states that *r* is a function of l_x and m_x. Recall that *r* is the intrinsic rate of natural increase of a population at its stable age distribution. This equation ignores density dependence. Density dependence is a major regulator of population size, and the logistic equation of population growth was developed to account for this regulation by introducing a parameter that represents the carrying capacity of the population (K), the maximum population size that can be supported by the resources in a given environment.[2] K is absent from the Euler–Lotka equation. Parameters that represent frequency dependence are also absent. Modeling the determinants of population growth rate in the absence of neighbors is one empirical way to quantify the fundamental niche (Laughlin et al., 2020b), which represents the range of environments suitable for a species in the absence of species interactions. The fundamental niche is typically much larger than the realized niche, which is the range of environments where the species actually occurs, and it is reduced in size by competitive interactions with itself or others (recall Figure 1.6). Despite its central placement in ecology textbooks, ecology has not satisfactorily quantified the fundamental niche of plants (Austin, 1990). The fundamental niche may be even larger than we realize. Simply take a stroll through an outdoor botanical garden and you will discover trees from all over the world often growing in a surprisingly different climate than where they originated (Grubb, 1985; Chen and Sun, 2018). Consider this rather extreme example: the natural range of the Queen of the Andes (*Puya raimondii*) is restricted to locations above 3000 meters in Peru and Bolivia, yet it can grow and flower in a coastal botanical garden in San Francisco! One study showed that realized niches underestimate fundamental climatic limits by demonstrating species are establishing in sites with an average of 3°C (range: 0–11°C) higher minimum temperatures than in their native ranges (Bocsi et al., 2016). The same is true for edaphic gradients. According to Warming (1909), "plants

---

[1] Fitness components in their most fundamental form are the demographic rates of survival and reproduction. Demographic rates are synonymous with vital rates. Some consider growth to be a third fitness component, but others subsume it within survival. Others have a more liberal view of fitness components: for example, Roff (1993, p. 347) considers offspring size as "undoubtedly an important fitness component."

[2] In my opinion K, or carrying capacity, is a troublesome parameter in population models. While *r* is based on real phenomena like births and deaths, K is a phenomenological simplification to account for density dependence and the ever-fluctuating, difficult-to-quantify resources, too. Other parameterizations of density dependence exist, such as the *r*-α model, which solves the serious problem of positive infinite growth rates when $r > 0$ and $N > K$ as starting conditions in the *r*-*K* logistic growth model (Mallet, 2012).

are evidently in general, tolerably impartial as regards soil, if we except certain chemical and physical extremes (abundance of common salt, of lime, or of water), so long as they have not competitors." This is one of the greatest problems for plant strategy theory to solve: disentangling the effects of the abiotic environment versus the effects of species interactions on fitness.

5.2   A brief digression on what else natural selection could be optimizing

Population growth rate is not the only possible feature of a plant that could be optimized by natural selection. Shugart (1997, p. 24) noted that

> in many cases, it is not clear what natural selection is optimizing. There are several obvious choices ranging from optimizing (or maximizing) survival, fitness, efficacious use of energy, water or nutrients, occupancy of space, etc. Further, it is possible that two sympatric species functioning in the same environment may, owing to their past evolutionary and biogeographical history, be optimized through natural selection toward different goals.

Being optimized toward different goals is not necessarily a contradiction. Being *r*-selected means that the intrinsic population growth rate is optimized in the absence of heterospecific and conspecific neighbors. Being *K*-selected means that the phenotype is optimized to maintain large population sizes to achieve competitive dominance (Roughgarden, 1976). Traits that optimize *K* would be the central focus. Growth rate, body size, and resource-use efficiency would be logical traits to study when the concern is maintaining long-term competitive advantage in a crowd (Roff, 1986; Kozłowski and Wiegert, 1987).

Many physiological ecologists have focused on traits that optimize net rate of plant growth. For example, Givnish (1982) developed a game theoretical model of leaf height strategies among forest herbs under the assumption that natural selection favored plants whose form optimizes net carbon gain since that phenotype would be best equipped to compete for more light, water, and nutrients. Many other studies assumed that natural selection favors traits that optimize the net rate of growth (Orians and Solbrig, 1977; Mooney and Gulmon, 1979). Bloom and colleagues (1985) discussed the analogies between economic decisions in business to resource acquisition strategies in plants. Their interest was in explaining fitness—the contribution of genes to future generations—but their primary focus was on ways that plants optimize primary productivity, and they argued that the quality and quantity of reproduction was strongly correlated with productivity. This landmark paper was a revolutionary approach to the study of plant strategies because it clarified how resources can co-limit plant growth and how different strategies of resource uptake, storage, and utilization have evolved to balance the limitation of multiple resources. Anten (2016) listed several other possible traits that could be optimized as fitness proxies, including leaf nitrogen content, leaf area index (LAI, the amount of leaf area per unit land area), stomatal conductance or leaf photosynthetic capacity.

The bottom line is that if optimizing a trait does not improve fitness, then the population that possesses that trait could decline or even go extinct. Fitness is the common currency that sets the "evolutionary play in the ecological theatre" (Hutchinson, 1965). Understanding the demographic consequences of trait variation is the most promising path toward understanding life history evolution and plant strategies. The fundamental niche can be quantified by determining the drivers of fitness in the absence of competition, whereas studies aimed at quantifying the realized niche will need to consider density dependence and frequency dependence. The invasion exponent λ_i is a quantity that represents the potential growth rate of a population when rare in the presence of competitors (Metz et al., 1992; Caswell and Takada, 2004; McGill and Brown, 2007; Roff, 2008). It is the growth rate of a population when the size of the focal population is low and when the population sizes of competitors are at equilibrium (Chesson, 2000). Chapters 8 and 9 describe empirical and theoretical methods to find the traits that optimize fitness in the presence of neighbors, but the rest of this chapter emphasizes theories developed under the assumptions of density independence.

5.3   Optimizing the product of survival and reproduction

The Euler–Lotka equation is clear that the product of l_x and m_x is more important than each function individually. That is to say, the product function $l_x m_x$ is under selection, not l_x or m_x alone. If natural selection optimizes the product, what predictions can we make about selection on l_x and m_x? Understanding these fundamental selection pressures on l_x and m_x is key to understanding the observed variation in life history strategies. Let us take another look at the life history schedule of our hypothetical perennial plant *Hierba rapida* in Table 4.1. Each individual in this population produced on average 1.2 offspring throughout their life, allowing the population to increase slightly ($r = 0.056$, $\lambda = 1.058$).[3]

What happens to the net reproductive rate (R_0) and per capita population growth rate (r) if we fix l_x and lifespan, and shift the m_x curve to the left or right? Figure 5.1 plots the $l_x m_x$ curve together with l_x and m_x individually so you can observe the effects of shifting m_x on the $l_x m_x$ function. If we shift the m_x curve to the left to simulate earlier reproduction (Figure 5.1b), then R_0 more than doubles to 2.9, leading to big increases in population growth rate ($r = 0.78$, $\lambda = 2.18$). However, if we shift the m_x curve to the right to simulate delayed reproduction (Figure 5.1c), then R_0 is cut in half to 0.59, and the growth rate plummets ($r = -0.15$, $\lambda = 0.86$). More offspring are produced if reproduction occurs early because there are more individuals to reproduce early in life. All things being equal, natural selection should favor earlier reproduction.

---

[3] These numbers are only true if the population is at a stationary equilibrium.

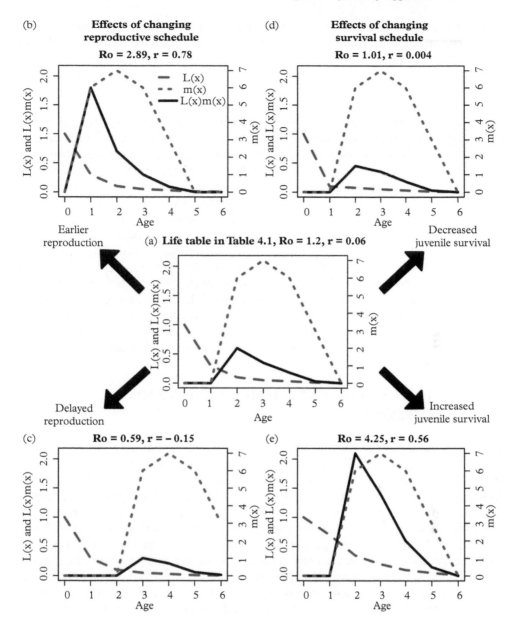

Figure 5.1 *Effects of changing reproductive and survival schedules on the net reproductive rate (R₀) and per capita population growth rate (r). Note that (a) is in the center. (a) All changes are shown with respect to the life table of our hypothetical perennial plant* Hierba rapida *(Table 4.1), illustrated in the center panel. (b) Effects of earlier reproduction while holding survival rate constant. (c) Effects of delayed reproduction while holding survival rate constant. (d) Effects of decreased juvenile survival while holding reproduction rate constant. (e) Effects of increased juvenile survival while holding reproduction rate constant.*

What if we fix the rate of reproduction and vary survivorship? A steeper survivorship curve simulates decreased juvenile survival (Figure 5.1d), which leads to a reduced net reproductive rate ($R_0 = 1.01$) and reduced population growth rate ($r = 0.004$, $\lambda = 1.004$). However, if we increase juvenile survival rate (Figure 5.1e), this increases net reproductive rate ($R_0 = 4.25$) and population growth rate ($r = 0.56$, $\lambda = 1.75$). All things being equal, natural selection should favor increased juvenile survival.

Based on these simulations, fitness is more sensitive to changes in mortality and fecundity in early age classes than older classes, so all else being equal natural selection should favor earlier reproduction and increased juvenile survival (Stearns, 1992; Silvertown and Charlesworth, 2001). Small changes in survival and reproduction have large effects early in life but have small effects later in life because net reproductive rate is the sum of the $l_x m_x$ function. This should favor the evolution of the earliest possible age of reproductive maturity and the concentration of reproduction at that age. Early maturation means that individuals have a higher probability of surviving to reproductive age. Early maturation is associated with higher r and R_0, which shortens generation times. As Stephen Stearns (1992, p. 124) wrote, "Early maturation is easily understood. It is delayed maturity that calls for explanation." Why do we not we live in a world that is overrun by species that reproduce as early as possible and have low juvenile mortality rates? How can it be that long-lived trees dominate our landscapes? Why do some herbaceous species, for example, the long-lived *Silene acaulis* (Chapter 4), delay reproduction for a few decades? The answers to these questions can be found in the study of life history trade-offs.

5.4   Trade-offs make some trait combinations impossible

Trade-offs between life history traits are generated because resources such as water, nutrients, and light are limiting, which has led to genetic and physiological constraints on which life history trait combinations are possible. Plants must divide up their limited pool of resources and allocate them to different needs: growth requires the construction of new tissues, survival requires expending energy through maintenance respiration, and reproduction requires the construction of nutrient-rich tissue. Stearns (1992, p. 161) described the "general life history problem" as a quandary for how to optimize the allocation of limited resources to growth, survival, and reproduction from birth all the way to death. Even Theophrastus recognized that "when the trees leaf very luxuriantly, they are more likely to be sterile, and whenever they bear copiously, they are more likely to leaf poorly, as though Nature could not satisfy both elements of the tree's growth but must spend her resource first on one and then the other" (quoted in Arber, 1950, p. 20).

The allocation of resources toward reproduction inevitably comes at a high cost. El-Kassaby and Barclay (1992) demonstrated this trade-off clearly by evaluating the cost of reproduction in Douglas fir (*Pseudotsuga menziesii*). They quantified the annual growth of the trees by measuring the width of the growth rings and related these to annual cone production. They found a consistently negative relationship between these two quantities, indicating that individuals that allocated more resources to cone production grew less than species that produced fewer cones. If reproduction decreases growth rates, then early reproduction could also lead to higher mortality rates. In another classic study, Law and colleagues (1977) demonstrated that the inappropriately named perennial grass *Poa annua* exhibited higher mortality in populations with large reproductive output. Trade-offs between reproduction and both growth and survival lead to trade-offs between reproduction now versus reproduction later because reproduction depends on previous growth and survival (Silvertown and Charlesworth, 2001). Improvement of fitness components early in life can sometimes lead to the deterioration of fitness components later in life. Rüger and colleagues (2018) described a trade-off among tropical trees where some species prioritize adult growth and survival at the expense of recruitment and juvenile growth and survival, but benefit from extended fecundity through long lifespans.

Taken together, delaying reproductive maturity might lead to increased overall fecundity later in life. If a plant waits to reproduce it will be larger at first reproduction and can produce more seeds (Metcalf et al., 2008). The benefits of early maturation are shorter generation times and higher survival to reproductive maturity because the juvenile period is shorter. In contrast, the benefits of delayed maturation are higher initial fecundity because longer growth leads to larger body sizes, lower juvenile mortality rates, and higher fecundity through extended lifespans (Stearns, 1992). Among wet tropical rain forest tree species in La Selva, Costa Rica, the longest living trees exhibited the lowest recruitment rates (Lieberman et al., 1985).

A variety of shapes in the l_x and m_x function have been described, but do plants conform to those that have been proposed for animals? Probably not. Pearl and Miner (1935) described three basic types of mortality rates that lead to their three well-known survivorship curves (Figure 5.2a). Some species exhibit low mortality rate when young but then a fast rate of morality when old (Type I), some experience a more or less constant rate of mortality throughout the life cycle (Type II), and others suffer high mortality early in life but low mortality later in life (Type III). In reality, these curves represent more of a continuum and can be quantified by Keyfitz's entropy (Salguero-Gómez et al., 2016). True Type I survivorship curves are rare in plants for the simple fact that high rates of seed mortality are ubiquitous. The closest one might get to a Type I curve would be a long-lived monocarpic perennial (Noodén, 1988b), where individuals of a population live to a certain age and die quickly after reproducing, such as synched populations of semelparous century plants (*Agave* spp.) and Hawaiian silverswords in the genera *Argyroxiphium, Dubautia,* and *Wilkesia,* but this still doesn't consider the high mortality rates of seeds in these species.

Tracking survival of long-lived plants is challenging. Chart quadrats are maps of vegetation plots where the locations of individual plants are mapped on coordinate paper. Chart quadrats were originally proposed by Weaver and Clements (1929) as a way to track vegetation through time in great detail (Figure 5.2b), but their potential use for demography wasn't unlocked until Peter Adler developed software to track individual plant growth and survival (Lauenroth and Adler, 2008). White (1985a) reviewed the installation of these quadrats throughout the western USA, leading to a revival of their

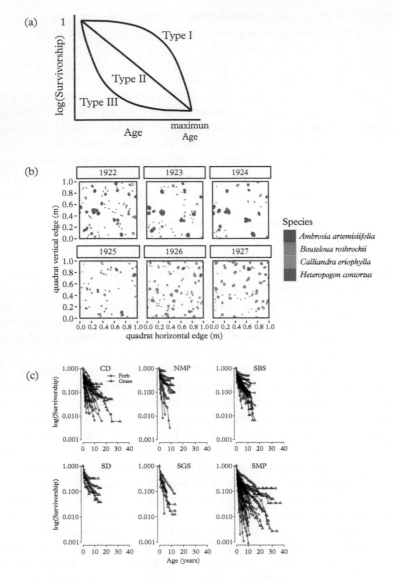

Figure 5.2 *(a) The three hypothetical survivorship curves. (b) Example of a digitized chart quadrat map collected over six consecutive years from a quadrat at the Santa Rita Experimental Range near Tucson, Arizona. Four species are included here: two grasses (*Bouteloua rothrockii *and* Heteropogon contortus*) were mapped as polygons, and two forbs (*Ambrosia artemisiifolia *and* Calliandra eriophylla*) were mapped as points. (c) Survivorship curves for 109 herbaceous perennials from the Jornada Experimental Range, New Mexico (CD), Fort Keough, Montana (NMP), US Sheep Experiment Station, Idaho (SBS), Santa Rita Experimental Range, Arizona (SD), Shortgrass steppe long-term ecological research station, Colorado (SGS), and mixed-grass prairie in Hays, Kansas (SMP).*

(a, b): Figure reproduced from Stears et al. (2022a) with permission using the open access dataset of Anderson et al. (2012). (c): Reproduced from Chu et al. (2014) with permission.

use. Using data from chart quadrats from six different grassland habitats in the western USA, Chu and colleagues (2014) plotted log-survival versus age for 109 herbaceous perennial species. Forbs and grasses followed Type III survivorship curves, but forbs approached Type II curves in many instances (Figure 5.2c). Type I curves were never observed in this study. Type II curves appear linear because risk of mortality is constant throughout life, whereas Type III curves are more concave because risk of mortality declines with age.

The reproduction function m_x is thought to exhibit three basic shapes: uniform, asymptotic, and triangular (i.e., hump-shaped) (Roff, 1993). Fecundity in plants tends to increase with age and size both within and between species (Primack, 1979; Watkinson and White, 1986; Wenk and Falster, 2015; Wenk et al., 2018), but can level off with age. Many matrix population models of trees often use a constant fecundity term across all size-classes (Salguero-Gómez et al., 2015), which would suggest that a uniform shape for m_x is common in woody plants, but this is often a methodological choice for convenience because it is so difficult to measure fecundity across size classes for trees. In some cases, growth may slow with age, and in these circumstances, fecundity could plateau at a certain age. Some species of plants have been shown to exhibit a triangular reproduction function (Harper and White, 1974), which would indicate physiological decline and senescence of fecundity with age. Some recent studies have indicated that plants can increase fecundity with age (Jones et al., 2014; Jones and Vaupel, 2017), whereas others have shown that senescence is common in trees (Qiu et al., 2021). Section 5.8 explores senescence as it relates to lifespan in more detail.

5.5 Reproduction can either be fatal or it can recur throughout the lifespan

The contrast between the annual and perennial life history in plants is well known. All gardeners know the difference between a plant that dies and must be replanted every year versus those that resprout year after year. A related, but not identical, concept is variation in reproductive strategies from semelparity to iteroparity, the second dimension of life history traits (Salguero-Gómez et al., 2016). The reproductive strategy of semelparity (also called monocarpy in plants) is distinguished by the fact that reproduction is fatal. The plant reproduces once then dies.[4] In contrast, iteroparity (also called polycarpy in plants) is the ability to reproduce more than once throughout the lifespan. All annual plants are semelparous, but not all semelparous plants are annuals (Figure 5.3). Many semelparous plants live for dozens of years. For example, the blue agave (*Agave tequilana*), belovedly cultivated for the production of tequila, lives for several decades. Then,

---

[4] It can be difficult to remember what semelparity means. Iteroparity is easy because reproduction is "iterated" throughout life. But semelparity has a wonderful etymology. Semele was a mortal lover of the Greek god Zeus and was carrying his child Dionysus. When Zeus appeared to her in his chariot engulfed in thunder and lightning Semele was consumed in flames, but Zeus was able to save the child. Reproduction was fatal for Semele, as it is for all semelparous (monocarpic) plants (Willson, 1983).

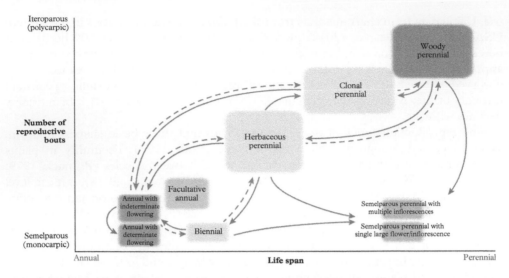

Figure 5.3 *Life history diversity in plants arrayed along axes of lifespan and number of reproductive bouts. The green boxes for herbaceous perennials, clonal perennials, and woody perennials should probably be longer in vertical extent given their great flexibility in number of reproductive bouts. Solid arrows indicate the most prevalent direction of evolutionary transitions, and dashed arrows indicate when transitions have been shown to go in the opposite direction.*

Reproduced from Friedman (2020) with permission.

when it sends up its mighty 5-meter-tall stalk of yellow flowers to reproduce, it rapidly senesces and dies.

Cole (1954) notably pioneered the optimization of *r* as a measure of fitness, but his seminal paper is also famous for a paradox. He proposed that an annual plant can attain the equivalent population growth rate of a perennial species simply by increasing its number of offspring by one. This surprising result raised important evolutionary questions about why perennials could persist at all. It is an important question because the woody habit and perennial life history are ancestral in angiosperms, and annuals evolved numerous times from perennial ancestors (Barrett et al., 1997; Friedman, 2020).[5] Perennials became precocious (i.e., early reproductive maturation) through natural selection. Annuals are not just plants that are unable to perennate—they also lost their perennating structures because it was selectively advantageous to do so.

Let us consider Cole's argument. Recall that the size of the population *N* in time $t + 1$ can be estimated as the product of the finite rate of increase (λ) and the current population size (N_t):

$$N_{t+1} = \lambda N_t. \tag{5.1}$$

---

[5] But perenniality has also been shown to evolve from annual ancestors, the most famous example being the Hawaiian silverswords in the genera *Argyroxiphium, Dubautia,* and *Wilkesia,* a perennial radiation from annual ancestors (Carlquist et al., 2003).

Let R_a represent the number of annual recruits, and R_p represent the number of perennial recruits. The model for annual population growth (assuming no juvenile mortality) is expressed as

$$N_{t+1} = R_a\, N_t. \tag{5.2}$$

The population will increase if $R_a > 1$. We can interpret R_a in the same way we interpret λ. The iteroparous perennial population model (assuming no adult mortality), differs by the addition of N_t to the product:

$$N_{t+1} = R_p\, N_t + N_t, \tag{5.3}$$

which simplifies to

$$N_{t+1} = (R_p + 1)\, N_t \tag{5.4}$$

If λ for the annual and perennial are to be equivalent, then $R_a = R_p + 1$. Therefore, the annual type only needs to increase its fecundity by 1 to exhibit equal fitness. This paradox stymied evolutionary biologists for a few decades, but the flaw in the analysis was that mortality rates were ignored, and according to Stearns "leaving out mortality rates when doing life history theory is a bad idea" (1992, p. 187).

Charnov and Schaffer (1973) solved the paradox by including mortality, which led to the important insight that returns on investment by retaining mature adults in a perennial population are greater than producing a single extra seed in annuals (Burghardt and Metcalf, 2017). To see this clearly, let S_m = mature survival, and S_j = juvenile survival. Annual plant growth rate is then equal to juvenile survival times the number of recruits:

$$\lambda_a = S_j R_a, \tag{5.5}$$

which leads to the annual population model:

$$N_{t+1} = (S_j R_a)\, N_t. \tag{5.6}$$

Perennial plant survival is equal to juvenile survival times the number of recruits plus mature plant survival:

$$\lambda_p = S_j\, R_p + S_m, \tag{5.7}$$

which leads to the perennial population model:

$$N_{t+1} = (S_j\, R_p + S_m)\, N_t. \tag{5.8}$$

For the two to have equivalent fitness,

$$S_j\, R_a = S_j\, R_p + S_m. \tag{5.9}$$

Solving for R_a we see that

$$R_a = \frac{S_j R_p}{S_j} + \frac{S_m}{S_j},$$

(5.10)

which simplifies to

$$R_a = R_p + \frac{S_m}{S_j}.$$

(5.11)

This result suggests that the advantage of one life history strategy over the other depends on the ratio of mature to juvenile survival (S_m/S_j), also called the adult-to-offspring survival ratio. Iteroparous perennials are favored when juvenile survival is low or unpredictable, whereas semelparous annuals are favored when juvenile survival is high and when adult survival is low (Stearns, 1992; Friedman, 2020). Put another way, increasing adult mortality or decreasing juvenile mortality enhances the value of juveniles and favors a shift toward annual semelparity. The second axis of variation in plant life histories is a continuous gradient from semelparity to iteroparity (Salguero-Gómez et al., 2016), which we now know has been driven by the relative importance of juvenile and mature survival rates. The ratio of mature to juvenile survival also plays an important role in Pausas and Keeley's (2014) model of evolution of resprouting and seeding strategies (recall Figure 2.5).

But survival and reproduction are not selected in isolation because selection acts on the summed product of the $l_x m_x$ function. Charlesworth (1980) extended Charnov and Schaffer's model and found that semelparous and iteroparous life histories have equal fitness when

$$\frac{R_i}{R_s} = 1 - \frac{S_m}{\lambda},$$

(5.12)

where R_i and R_s refer to iteroparous and semelparous reproduction rates, respectively. This function states that the ratio of iteroparous to semelparous reproduction decreases with decreasing λ and increasing mature plant survival. Roff (1993) described the following example to make this result clear: when mature survival rates are low ($S_m = 0.1$) and finite rate of increase is high ($\lambda = 2$), then semelparous organisms need to only produce 1.05 times the number of offspring than iteroparous organisms to maintain equal fitness. But if mature survival rates are high ($S_m = 0.9$) and the population is stationary ($\lambda = 1$), then semelparous organisms are predicted to need 10 times the number of offspring than an iteroparous organism to have equal fitness.

Empirical results support these theoretical predictions. Reproductive effort is a measure of investment in reproductive organs. In plants, reproductive effort can be quantified as the ratio of propagule mass to total biomass at maturity (Harper and Ogden, 1970). Reproductive effort is generally higher in annuals than perennials (Wilson and Thompson, 1989): in annual grasses it varied between 44% and 66%, whereas in perennial grasses it varied between 0.2% to 31%. Importantly, this relationship held when controlling for differences in plant size between annuals and perennials. Roff (1993) also summarized estimates of reproductive effort and showed that it is much higher in annual and semelparous species than in perennials (see Table 5.4 in Roff, 1993). A comparison of wild and domestic species found that plants with longer lifespans allocate more to storage and perennating tissues than to seed mass (Vico et al., 2016).

What about biennials? "Biennial" is a common word that we throw around casually for plants that take approximately two years to reproduce then die, but the lifespan of these so-called biennials is highly plastic and varies widely. Few species exhibit high fidelity to a lifespan of exactly two years. It can be frustrating for students who frequently set down to categorize species in their study site into tidy groups such as annual, biennial, and perennial, only to find that some species are listed in regional floras as being "annual/biennial/perennial"! Such plasticity is likely driven by environmental conditions, where lifespan is shorter in unproductive habitats and longer in productive habitats. Interestingly, Robin Hart (1977) extended the models of Charnov and Schaffer (1973) to evaluate how biennial reproduction compares to annuals and perennials. Hart demonstrated theoretically that biennials need to produce four times as many offspring than perennials and two times as many offspring than annuals to have equal fitness with each of them. Empirical evidence supports these theoretical claims because, according to Hart's measurements, biennials produce on average four to five times as many seeds as both annuals and perennials. This is likely made possible because biennials store more energy and are larger in size than annuals, and because biennials do not have to invest in as much perennating structures and so can invest more in reproduction (Roff, 1993).

Annuals have committed to risky behavior, so any means of increasing the likelihood of seed production would likely be selectively advantageous. Self-fertilization is much more common in annual than perennial species and dioecy is more common in perennial than annual species, suggesting that separating the sexes and permitting self-fertilization allows annuals to enhance seed set despite the risk of increased inbreeding depression. The sequence of this evolutionary event is not well understood because self-fertilization may actually drive the evolution of an annual life history, not the other way around (Friedman, 2020).

Reproductive strategies are associated with different habitats (Roff, 1993). Iteroparous perennials tend to be favored where juvenile mortality is high, including mesic productive habitats where light becomes limiting under the shade of canopy trees. Semelparous annuals are favored in open vegetation that is either kept open through disturbance or in arid habitats where light is naturally less limiting if they time their germination properly to escape unfavorable seasons of moisture limitation.

Little data on reproductive schedules exist for perennial plants, but schedules appear to be quite variable across species (Wenk and Falster, 2015). Reproductive allocation (RA), sometimes called reproductive effort, is a measure of how much energy is apportioned to reproduction versus growth and maintenance, ranging from 0 to 1. RA is difficult to estimate precisely but can be done by partitioning annual biomass production into reproductive versus non-reproductive tissue, where RA can be expressed as the ratio of dry mass of reproductive tissue to the total dry mass (Abrahamson and Gadgil, 1973).

RA = 0 for juvenile plants, but increases as soon as age or size at maturation is reached and a variety of different RA schedules are theoretically possible (Figure 5.4). Wenk and colleagues (2018) quantified RA for 14 dominant woody plants in Kuring'gai Chase National Park in Australia and showed that many iteroparous woody perennial plants allocate all surplus energy to reproduction when reproductive allocation includes seeds and all accessory reproductive structures and when reproductive allocation is calculated after accounting for energy used to replace lost biomass. This study also demonstrated that both leaf mass per area (LMA) and maximum height were positively correlated with age at which RA = 0.5, indicating that functional traits can predict reproductive allocation schedules.

5.6   Modularity displaces longevity from the fast–slow continuum

All plants age. Aging is simply the inevitable passage of time (Shefferson et al., 2017; Nathan, 2021). But, unlike most animals, some plants appear to escape the vagaries of old age and continue to reproduce for millennia. Some plants appear to have achieved potential immortality. Plant lifespan spans orders of magnitude: some plants live for a few weeks, others live for thousands of years (Thomas, 2013). How do plants live for so long? Estimates of age in clonal plants exceed that of non-clonal plants by an order of magnitude (Table 5.1), suggesting that plant clonality, dormancy, and modularity (i.e., the ability to indeterminately repeat the basic building blocks of plant organs, discussed in Chapter 6) play a role in extending lifespan. These lifespan estimates are lifespans of genets (genetic individuals), and interestingly ramets (clonal individuals within a genet) of clonal plants exhibit greater senescence than sexually reproducing species (Salguero-Gómez, 2018). Underground trees (geoxyles) that propagate vegetatively belowground can live for exceptionally long times, like the species *Jacaranda decurrens* from the Brazilian Cerrado that has been estimated to be as old as 3,801 years (Alves et al., 2013)! Theories of plant senescence are catching up to classical theories of senescence that were derived largely for unitary organisms like animals (Salguero-Gómez et al., 2013; Shefferson et al., 2017).

Senescence in plants can refer to both organ-level decay and whole-plant degeneration, so we need to be clear in defining senescence. Organ-level senescence in plants is well studied—the transformation of leaf color in autumn from green to golds, purples, and crimsons is a vivid example. The lifespan of individual organs is much shorter than the genetic individual: roots and leaves last for anywhere between weeks to several years (Mencuccini and Munné-Bosch, 2017). Plants have the unusual capacity to recycle nutrients from these decaying organs and reroute them to where they are needed most for future growth and reproduction. Nutrient economics are relevant to plant strategy theory, but I limit my use of the word senescence in this book to whole plant senescence, unless clearly stated otherwise. Senescence is the physiological decay leading to increasing mortality and decreasing fertility with age (Noodén, 1988a; Shefferson et al., 2017);

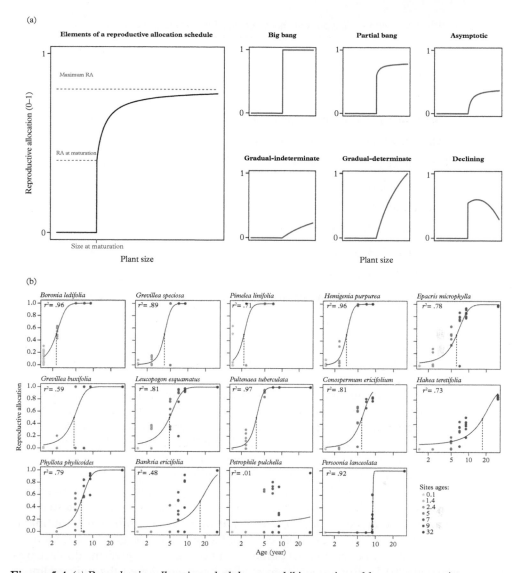

Figure 5.4 *(a) Reproductive allocation schedules can exhibit a variety of forms across species. (b) Variation in empirical reproductive allocation schedules in 14 woody species from the Kuring'gai Chase National Park, Sydney, Australia. Species are arranged in order of increasing age to reproductive maturity. Vertical dashed lines indicate the age at which RA = 0.5.*

(a): Reproduced from Wenk and Falster (2015) with permission. (b): Reproduced from Wenk et al. (2018) with permission.

Table 5.1 *Maximum lifespans of the longest-lived non-clonal and clonal plants.*

Species	Age (yrs)
Non-clonal plants	
Bristlecone pine (*Pinus longaeva*)	5,000+
Patagonian cypress (*Fitzroya cupressoides*)	3,600+
Giant sequoia (*Sequoiadendron giganteum*)	3,200
Huon pine (*Dacrydium franklinii*)	2,200+
Common juniper (*Juniperus communis*)	2,000
Clonal plants	
King's lomatia (*Lomatia tasmanica*)	43,000+
Huckleberry (*Gaylussacia brachycerium*)	13,000+
Creosote (*Larrea tridentata*)	11,000+
Quaking aspen (*Populus tremuloides*)	10,000+

Adapted from Thomas (2013) with permission.

thus, senescence is linked directly to schedules of age-specific birth and death rates in a population.

All theories of senescence start with the fact that selection gradients on age-specific survival and fertility decline with age (Charlesworth, 2000; Caswell and Shyu, 2017). In other words, given the triangular shape of the $l_x m_x$ curve (Roff, 1993), decreasing mortality and increasing fertility late in life has little effect on fitness (Figure 5.1). The disposable soma theory proposes that many organisms experience physiological decline late in life because it does not pay to invest in the maintenance of the soma (i.e. the mortal body) when resources should be spent on maintaining and replicating the immortal germ line (Kirkwood, 1977). Organisms must invest something into the soma so that it doesn't fall apart and die before it reproduces, but there is little to be gained in maintaining a soma that is bound to die from external factors. There is greater selection for reproduction and growth over survival of the soma because old organisms don't reproduce (Kirkwood, 2017). But some recent studies have questioned the universality of senescence and have stimulated renewed interest in the assumptions of these theories for all taxa (Baudisch et al., 2013; Jones et al., 2014; Roper et al., 2021). The problem with this theory for iteroparous plants is that plants do not differentiate soma and germ cells early in life, and the modular growth of plants may allow plants to escape senescence by replacing old tissue with new modules—if plants maintain the soma then they can continue to reproduce late into their lives (Roach, 1993; Pedersen, 1999; Gremer et al., 2017).

This discussion must distinguish between semelparous and iteroparous species. Semelparous (monocarpic) plants senesce because reproduction is fatal (Burghardt and

Metcalf, 2017). The size at flowering among semelparous species is therefore an optimization problem: the larger the plant gets, the more seeds it can produce, but extending life to grow larger comes with greater risk of death from environmental factors that would prevent the plant from reproducing at all (Metcalf et al., 2003; Rees et al., 2006). The physiological processes that lead to death have been especially well studied in monocarpic crop plants (Noodén, 1988a), and now we know that oxidative stress increases once resources from leaves are diverted into fruit production (Mencuccini and Munné-Bosch, 2017). In fact, we can delay mortality in semelparous species by simply removing their inflorescences! The lesson here is that models of plant survival spanning multiple taxa must treat semelparous and iteroparous species differently because cause of death is exogenous for iteroparous species but can be either endogenous or exogenous for semelparous species.

Among unitary organisms such as animals, there is a clear positive relationship between age at maturity (α) and maximum longevity (ω), which are the traits that define the limits of the integral in the Euler–Lotka equation (Eq. 4.11). Technically, ω represents age at last reproduction. In animals, this can be considerably younger than lifespan. Many plants that reproduce early tend to die young and plants that delay maturity tend to live longer lives, but the relationship between age at maturity and longevity is not fully resolved among plants. The relationship was positive according to Silvertown and colleages (2001), but Loehle (1988a) showed that the relationship was rather strong among angiosperms but much weaker among gymnosperms (Figure 5.5). Nevertheless, it is still generally thought that these two traits evolved together because early reproduction negatively affects subsequent survival. Reproduction is also related to the probability of reproductive success. If there is high variation in reproductive success, then it would pay off to reduce reproductive effort to live longer and reproduce more times. This "bet-hedging" would allow the plant to increase the number of offspring by "sampling" a higher number of conditions throughout life (Stearns, 1992). This would certainly describe the behavior of Great Basin bristlecone pine (*Pinus longaeva*) that grows in the White Mountains of California (Table 5.1, Figure 5.5b). Successful reproduction in this highly variable subalpine environment is rare, so bristlecone pine spreads out the likelihood of reproduction by attempting to reproduce over a long lifespan (Silvertown and Charlesworth, 2001).

Some scholars doubt that senescence occurs at all in some iteroparous species. In a recent review, the majority of studies that tracked ages and demographic rates in herbaceous perennial plants found no evidence for age effects on mortality or fecundity (Dahlgren and Roach, 2017). Moreover, the largest analysis of demographic models found no evidence for senescence in 93% of studies (Baudisch et al., 2013; Gremer et al., 2017; Roper et al., 2021). The latter analysis used stage-to-age methods described in Chapter 4, so they are less robust than tracking age directly, but the overwhelming evidence is strongly suggestive.

Consider the geophyte *Borderea pyrenaica* that grows in the Central Pyrenees. This species can live for > 300 years, making it one of the longest living non-clonal herbs known to science (García and Antor, 1995). It resprouts from a bulb year after year, and bud scars on the below-ground bulb can be counted to estimate individual plant age

Figure 5.5 *(a) Relationship between average age at reproductive maturity and average longevity in angiosperms and gymnosperms. Examples of long-lived conifers in North America include the following: (b) Great Basin bristlecone pine (*Pinus longaeva) *in the White Mountains, Inyo National Forest, California, USA. (c) Giant sequoia (*Sequoiadendron giganteum) *in Long Meadow Grove, Sequoia National Forest, California, USA. (d) Coastal redwood (*Sequoia sempervirens) *in the California Federation of Women's Clubs Grove, Humboldt Redwoods State Park, California, USA. An albino redwood can be seen near the middle of the image.*

(a): Reproduced from data in Loehle (1988a) with permission. (b, c): Photos by the author. (d): Photo provided by Michael Remke with permission.

(Figure 5.6). Remarkably, there appear to be no age-related declines in physiological performance (Morales et al., 2013) and mortality rates may decrease with age in this species (García et al., 2011).

Evidence for senescence in long-lived woody plants is more controversial. Trees differ from herbaceous plants because large old trees accumulate dead tissue. Put more poetically, "The tree depends for its very form on the continual accumulation of dead

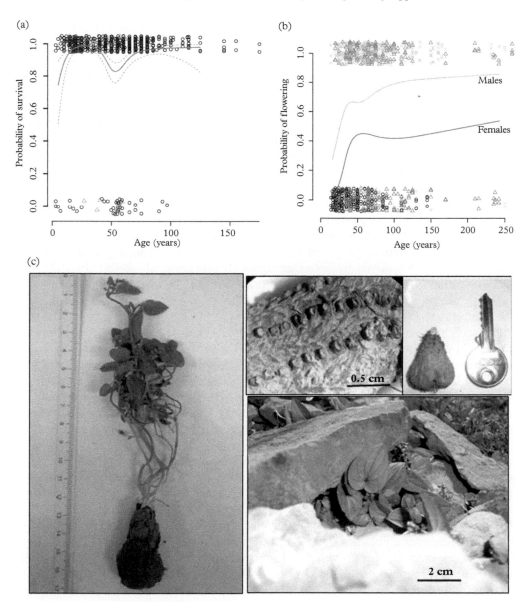

Figure 5.6 *Relationship between (a) age and probability of survival and (b) age and probability of flowering in* Borderea pyrenaica *in the Central Pyrenees. (c) Photographs of excised individual of* Borderea pyrenaica *with scale bars.*

(a, b): Figures reproduced from García et al. (2011) with permission. (c): Reproduced from Morales et al. (2013) with permission.

tissue—it is a thin layer of living sepulcher enclosing a growing corpse" (Harper, 1977, p. 218). The accumulation of dead tissues may lead to deterioration in demographic rates (Harper and White, 1974). This is exactly what was found in one recent global study (Qiu et al., 2021), which presented evidence for fecundity declines in 80% of the studied species from temperate and tropical forests. These results are important for understanding which tree size-classes contribute the most to recruitment in forests. Long-lived iteroparous plants, such as trees, tend to experience a constant background rate of mortality, and this is generally believed to be caused by environmental factors. There is a clear need for additional forensic data about what kills plants (Shefferson et al., 2017; McDowell et al., 2022). Unfortunately, trees are dying at faster rates all over the world because of climate change (van Mantgem et al., 2009; Allen et al., 2010; Anderegg et al., 2013). The rapid increase in mortality rates in trees across the world may be giving demographers the opportunity they need to determine what kills trees (Hammond et al., 2019). In 2002, we witnessed one of the most drought-tolerant pines in North America, the piñon pine (*Pinus edulis*) exhibit widespread die-off in the Southwest USA driven by hotter droughts (Breshears et al., 2005). In 2021, we witnessed the most drought-tolerant tree, one-seed juniper (*Juniperus monosperma*), exhibit widespread die-off, sending shockwaves through the ecological community in the southwestern USA.

How do iteroparous plants escape senescence? Scholars of senescence speculate that this is related to the modular nature of plants and their ability to drink from the proverbial fountain of youth through the production of new meristems. Individual plants can be viewed as populations of individual buds (Harper, 1977; White, 1979). This modular structure allows plants to compartmentalize sections of their bodies that cease to function (Shigo, 1986). There are three possible mechanisms for how plants may escape senescence. First, plants have multiple meristems that can be formed throughout life and many plants exhibit indeterminate modular growth. These features can allow plants to escape senescence because senescence would require the simultaneous death of all meristems, which is unlikely (Dahlgren and Roach, 2017). Second, many plants can remain dormant belowground in unfavorable years, which could permit the plant to reset and rejuvenate demographic rates (Shefferson et al., 2014; Gremer et al., 2017). In other words, plants can age and not grow.[6] Plants live altogether differently, and it can be difficult for animals to think about plants that can simply wait a year to grow. Third, continuous somatic growth through clonal propagation of new meristems could lead to negative senescence, defined as the increase of fertility and decline in mortality rates with age (Vaupel et al., 2004). Circumstantial evidence for this can be seen in the fact that clonal plants exhibit some of the longest lifespans known on the planet (de Witte and Stöcklin, 2010) (Table 5.1). Senescence may also be staved off in the many species of plants that undergo somatic mutations within individual plants (Whitham and Slobodchikoff, 1981).

---

[6] Silvertown and Charlesworth (2001) called juvenile plants that age but don't grow "oskars", after Oskar in *The Tin Drum* by Günter Grass (1959). Oskar decided he'd rather stay young than grow up only to become a grocer.

So far, we have only considered the lifespan of adult plants. However, the lifespan of seeds is just as important as the lifespan of adults. Doak and colleagues (2002) described two extremes of life history strategies in plants. At one end, are the long-lived plants with short-lived seeds, reproductive adults that have low mortality rates, catastrophe-prone seedlings, and slow reproductive maturation. He likened these plants to vertebrate animal life histories. The other end is typified by short-lived adults with highly buffered, long-lived seeds. Theory suggests that there are trade-offs between the costs of seed dormancy and the benefits of population buffering against environmental variation (Ellner, 1985a, b; Rees, 1997a). The cost of seed dormancy is manifested by the negative effects of delayed reproduction on current population size. Greater seed dormancy should be selected in variable environments, such as deserts with high interannual variation in rainfall, and also in habitats consisting of dense vegetation, such as grasslands, where periodic disturbance opens up the canopy (Ellner, 1986). Rees (1993) found a negative yet weak empirical relationship between adult longevity and seed longevity, suggesting that short-lived species exhibit the greatest potential for seed dormancy. This life history trade-off between sexual reproduction (seeding) and adult survival (resprouting) appears to be a central dimension of plant strategies (Bellingham and Sparrow, 2000; Klimešová and Klimeš, 2007; Pausas and Keeley, 2014). On the one hand, long-lived plants have higher bud bank densities, which allows them to resprout and persist even amid strong environmental variation. On the other hand, plants with long-lived seeds rely on emergence from the seed bank in favorable years for population persistence. We explore this trade-off in more detail in Chapter 6.

5.7 Three life history strategies explain global variation in plant demographic rates

Life history traits do not evolve independently. We have seen that trade-offs constrain which life history trait combinations are possible. Salguero-Gómez and colleagues (2016) computed nine life history traits for 418 plant species across 105 families compiled in the COMPADRE database and subjected the traits to principal components analysis. Rather than finding a single dimension, they found strong evidence that the intrinsic dimensionality of life history variation was (at least) two dimensional (Figure 5.7), and unlike *r-K* selection theory these results were robust after accounting for phylogenetic structure (Ackerly, 2009). The first dimension represented variation along a "fast–slow continuum." Species on the slow end exhibited long generation times, older ages to sexual maturity, low mean reproduction, and slow growth. Species on the fast end exhibited short generation times, young ages to sexual maturity, high mean reproduction, and fast growth.[7] Similarly, a fast–slow continuum has been proposed to explain life history variation among trees in a tropical forest where gap formation drives population dynamics. The fast end of the continuum is characterized by

---

[7] Low-density population growth rates, that is, potential rates of increase when rare, would be an important trait to include in a future analysis of projection matrices. These rates are discussed more fully in Chapters 8 and 9.

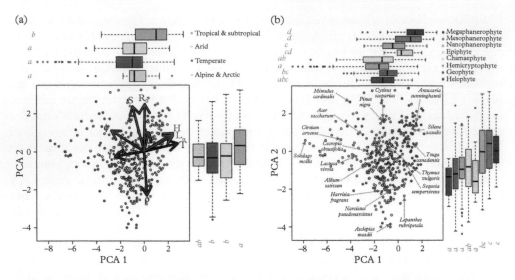

Figure 5.7 *A global analysis suggests that at least two dimensions capture the variation in life history traits: the fast–slow continuum (PCA1) and reproductive strategies (PCA2). The following life history traits were included: those relating to turnover time (*T* = generation time), longevity (*H* = survivorship curve type, *L_α* = age at sexual maturity), growth (*γ* = progressive growth, *ρ* = retrogressive growth), and reproduction (*φ* = mean sexual reproduction, *S* = degree of iteroparity, *R_0* = net reproductive rate, *L_ω* = mature life expectancy).*
Reproduced from Salguero-Gómez et al. (2016) with permission.

shade-intolerant, fast-growing trees with high mortality rates, versus the slow end, which is characterized by shade-tolerant, slow-growing trees that exhibit low mortality rates (Loehle, 1988a; Condit et al., 1996; Enquist et al., 1999; Rüger et al., 2018). While the results in Figure 5.7 represent our best current understanding of life history trade-offs at a global scale, there are key data gaps and biases in the database (Römer et al., 2021), and it must be acknowledged that trade-offs could differ within life forms or across disturbance regimes. For example, the growth-survival trade-off appears to be less frequently observed in disturbance-prone forests (Russo et al., 2021).

The second dimension describes variation in "reproductive strategies" that are unrelated to mean reproduction rate. Species at one end of the spectrum reproduce more frequently throughout their life cycle (iteroparous) and experience less shrinkage (i.e., reduction in size from one year to the next) because, in plants, being big leads to greater reproduction. Species on the other end reproduce less frequently and exhibit higher rates of shrinkage. The authors included net reproductive rate (R_0) on this axis, but R_0 is very likely a function of the environment the plant was measured in (much like λ), rather than an intrinsic property of the species. In other words, if we were to measure R_0 in an unsuitable environment for a species, it would be a far lower number. Nevertheless, this axis was interpreted as a gradient in degree of iteroparity and retrogression. If one restricts the analyses to trees, the secondary trade-off is a stature–recruitment trade-off, where some species prioritize adult growth and survival at the expense of recruitment and juvenile growth and survival (Rüger et al., 2018; Kambach et al., 2022). This demonstrates how dominant life history trade-offs may differ across life forms.

Most life history theory indicates that lifespan should be correlated with age at reproductive maturity (Silvertown and Charlesworth, 2001), but plant lifespan was not correlated with the fast–slow continuum in the COMPADRE analysis. Salguero-Gómez and colleagues (2016) used the Kaiser rule to determine the dimensionality of the dataset and the only eigenvalues > 1.0 were for the first two dimensions, so they limited their discussion to two axes. However, hidden within the third dimension was something remarkable that is (in my opinion) an important feature of plant life histories that distinguish themselves from animal life histories. Lifespan loaded onto a third dimension, which suggests that plant lifespan is independent of age at maturity. This indicates that early reproduction is likely a weaker constraint on longer lifespan in plants, possibly due to the modular construction of plants. Indeed, the correlation is weakest among gymnosperms that achieve exceptionally long lifespans (Figure 5.5). Plant lifespan may be independent of age to sexual maturity if plants can defy the degenerative process of senescence (Vaupel et al., 2004; Jones and Vaupel, 2017). Another life history trade-off between demographic buffering versus lability might also exist (Hilde et al., 2020; Le Coeur et al., 2022), suggesting that other dimensions of life history variation remain to be discovered.

Some plant ecologists are satisfied with a life history perspective on plant strategies, but life history traits alone are not sufficient for understanding plant adaptations to environmental conditions. A demographer might say that everything that is worth knowing is wrapped up in the patterns of births, deaths, migration, and dispersal, but a functional ecologist would disagree. Consider the century-long debate about whether life forms, which share similar life histories, are useful as plant strategies. The Raunkiaer life form system was criticized precisely because the system could not distinguish between trees that grew in dry versus wet climates, or grasses that grew in fertile versus infertile soil. Frequencies of life forms vary somewhat across climate gradients (Raunkiaer, 1934), but each life form is found across broad climate gradients (Salguero-Gómez et al., 2016). For this very reason, Box (1981) was compelled to devise 77 different life forms to account for preference to different environments within similar life forms. The same argument can be made for life history strategies. We can easily find two trees that each attain a lifespan of 300 years but exhibit very different morphological and physiological traits. These two trees occupy different habitats because their functional traits optimize fitness in different environmental conditions. Their lifespans tell us less about habitat preference, but we could predict their optimal habitats if we knew that one tree had small, tough leaves and the other had large, thin leaves. Understanding variation in life histories is not the end game of plant strategy theory. The end game is to use the information encoded in both life history and functional traits to predict fitness across environmental gradients. Next, we turn to the traits that have been the focus of plant strategy models that emphasize ecological function (Garnier et al., 2016). This chapter described what is demographically possible for plants, and Chapter 6 covers the phenotypic traits that turn these possibilities into realities.

5.8   Chapter summary

- Life history theory assumes that natural selection optimizes fitness and that trade-offs among fitness components constrain the possible combinations of life history traits.

- The product function $l_x m_x$ is under selection, not l_x or m_x alone. Fitness is more sensitive to changes in mortality and fecundity in early age classes than older classes, so natural selection should favor earlier reproduction and increased juvenile survival.

- Trade-offs between life history traits are generated because resources such as water, nutrients, and light are limiting. Plants must divide up their limited pool of resources and allocate them to different needs: growth requires the construction of new tissues, survival requires expending energy through maintenance respiration, and reproduction requires the construction of nutrient-rich tissue.

- Type II and III survivorship curves are the most common curves in plants, indicating that rates of survival and reproduction do not inevitably decline with age in plants.

- The reproduction function m_x is thought to exhibit three basic shapes: uniform, asymptotic, and triangular (i.e., hump-shaped).

- The advantage of one life history strategy over the other depends on the ratio of mature to juvenile survival. Iteroparous perennials are favored when juvenile survival is low or unpredictable, whereas semelparous annuals are favored when juvenile survival is high and when adult survival is low.

- Little data on reproductive schedules exist for perennial plants, but schedules appear to be quite variable across species. LMA and maximum height were positively correlated with age at which reproductive allocation equals 0.5, indicating that functional traits can predict reproductive allocation schedules.

- There is some doubt that senescence occurs at all in iteroparous species. In a recent review, the majority of studies that tracked ages and demographic rates in herbaceous perennial plants found no evidence for age effects on mortality or fecundity, and researchers have speculated that this is due to the modular nature of plants.

- Some ecologists appear to be satisfied with a purely life history perspective on plant strategies, but life history traits alone are not sufficient for understanding plant adaptations to environmental conditions.

5.9   Questions

1. All things being equal, why would evolution tend to select for earlier age at reproduction?
2. What conditions select for long lifespan?
3. Do plants escape senescence?
4. Can long-lived clonal plants achieve immortality?

Part III

Comparative Functional Ecology

Part III establishes the field of comparative functional ecology as the central framework underlying differences in plant strategies by exploring the vast variation and covariation in plant functional traits worldwide.

Chapter 6 examines the whole plant phenotype as an integration of traits across roots, clonal growth organs, bud banks, stems, leaves, seeds, and flowers. The multidimensional phenotype is proposed to span at least five independent trait spectrums.

Chapter 7 establishes that gradients of resource limitation including water, nutrients, and light, gradients in disturbance regimes including type, severity and frequency, and gradients in temperature and are the key drivers of the evolution of plant strategies. Trait-based demographic rates shape the contours of fitness landscapes, and these contours shift across environmental gradients.

6

Plant Functional Traits and the Multidimensional Phenotype

The global spectrum of plant form and function is thus, in a sense, a galactic plane within which we can position any plant—from star anise to sunflower—based on its traits.

—Sandra Díaz and colleagues (2016, p. 171), "The global spectrum of plant form and function," in *Nature*

I freely confess that some of these assumptions are likely to inspire intellectual hemorrhages in the trained botanist, but the multicellular morphospace is computationally intractable without them.

—Karl Niklas (1997, p. 236), *The Evolutionary Biology of Plants*

6.1 Embracing a blurred vision to avoid intellectual hemorrhages

Some of you may be wondering why, in a book about plant strategies, it has taken so long to arrive at a chapter devoted to functional traits, which are (after all) the building blocks of strategies. The answer to your query is that we can only call a trait "strategic" if it consistently demonstrates positive demographic outcomes that permit the persistence of a plant in a given environment. Considerable effort has been spent by comparative functional ecologists on measuring traits and defining the boundaries of trait spaces (Pugnaire and Valladares, 2007), but if no evidence is presented that traits affect demography, then we are simply defining what phenotypes exist, rather than demonstrating that traits are adaptive (Ackerly et al., 2000; Ackerly, 2003; Ackerly and Monson, 2003). The quantitative link between traits and fitness is made explicit in Chapters 8 and 9. The objective of this chapter is to define the dimensionality of the whole plant phenotype by focusing on the traits that matter most.

Given the diversity of plant form and function and the complexity of cellular differentiation that occurs within plant organs, it might seem an impossible task to reduce the dimensionality of plant traits down to a short list of orthogonal dimensions. This will no

Plant Strategies. Daniel C. Laughlin, Oxford University Press. © Daniel C. Laughlin (2023). DOI: 10.1093/oso/9780192867940.003.0006

doubt "mock the histological, anatomical, morphological, and physiological complexity observed among many kinds of plants" and will "inspire intellectual hemorrhages" among botanists in particular (Niklas, 1997, p. 236). When we marvel at the complexity of organelles, cells, and tissues at microscopic scales it can be difficult to see how organ-level traits could be useful at all. But recall that our goal is generality. Robert MacArthur (1972, p. 1) declared that "to do science is to search for repeated patterns, not simply to accumulate facts." Sometimes, to seek generality in science we need to view the world using MacArthur's "blurred vision" (MacArthur, 1968; Grime, 2001), and ignore the finer details to focus instead on the broader picture. This chapter includes just enough anatomy and physiology to gain an intuitive understanding of what plant traits represent, and what they gloss over to achieve generality. Many important anatomical and physiological details are left out, so I encourage the curious reader to consult other excellent textbooks on these topics (Larcher, 2003; Schulze et al., 2005; Evert, 2006; Lambers and Oliveira, 2019). I also mostly skip over the field and laboratory methodologies for how to measure traits—you should consult the plant trait handbooks for this information (Pérez-Harguindeguy et al., 2013; Klimešová et al., 2019; Freschet et al., 2021).

6.2   Modular indeterminate growth permits developmental flexibility

We begin by reviewing the basic features of the plant body, then evaluate the essential adaptations that evolved to inhabit terrestrial landscapes. Finally, we consider the form and function of roots, clonal storage organs, stems, leaves, and seeds. This chapter concludes with a synthesis of plant trait dimensionality.

Figure 6.1 illustrates the basic architecture of a typical dicot. One of the most important and distinguishing features of vascular plants is their modular construction (Harper, 1977; White, 1979; Klimešová et al., 2019). You and I are not modular organisms—we are unitary organisms. I would die if a large tree branch fell on my head. I couldn't just resprout a new head. In contrast, the tree that just lost its limb would simply shake it off as a mere flesh wound and would compartmentalize the wound like a master surgeon (Shigo, 1986) and would fill in the vacated space with new modular growth. This feature makes plants extremely flexible and able to adjust their growth in response to local environmental stimuli. It also makes them rather difficult to kill once they are established. Indeed, survival rates of mature trees can approach 99.9%.

Modules are the basic architectural building blocks of a plant (Figures 6.1 and 6.2). There exist numerous definitions of what constitutes a module (Vuorisalo and Mutikainen, 1999), but for our purposes this book defines a module as the phytomere: a unit of construction produced by apical meristems that consist of a node, an internode, a leaf, and an axillary bud (Raven et al., 2004; Klimešová et al., 2019). This basic modular unit can be repeated in a seemingly infinite variety of ways and can be arranged hierarchically within a plant. A collection of aboveground modules is unified through their connection to a single rooting system, and multiple aboveground shoots can be connected belowground through clonal growth organs. All modules need to be connected to the belowground root system through the vasculature to obtain resources from

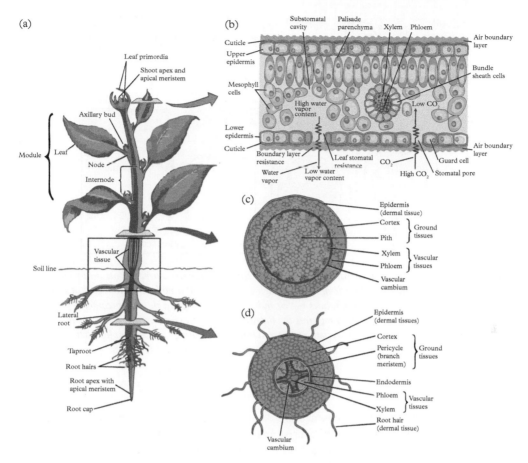

Figure 6.1 *Schematic representation of a typical dicot plant body and diagrammatic cross sections of (b) the leaf, (c) the stem, and (d) the root.*
Adapted from Taiz and Zeiger (2006) with permission.

belowground. The rooting unit is an important concept in plant architecture and differs between non-clonal and clonal species. Some modules can be connected to roots directly and form their own rooting unit, such as a stoloniferous herb. More often, a plant exhibits a shoot that is composed of several modules. Trees are rooted units with shoots that are composed of a large hierarchical body of repeating modules. The architecture of meristems is central to understanding the traits that promote growth and survival.

Meristems are regions of undifferentiated cells that give rise to all the tissues and organs that we see in plants (Figure 6.1). Meristems are formed by stem cells—undifferentiated cells defined by their abilities for self-renewal and for generating specialized cells (Munné-Bosch, 2007). All plant organs derive from these pluripotent stem cells in the apical meristem, known as the "plant's fountain of youth" (Bäurle and Laux, 2003). Aristotle observed that "plants are always being reborn; that is why they

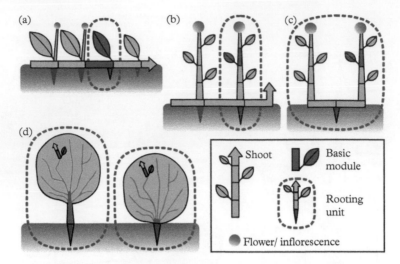

Figure 6.2 *Illustration of the hierarchical nature of plant modular structure. Shoots and rooting units are composed of aggregations of basic modules, which each consist of a node, an internode, a leaf, and an axillary bud. These vegetative modules repeat indeterminately until a module becomes reproductive, the development of an inflorescence programs the branch to end determinately. Clusters of basic modules that are connected to the same root system create a hierarchical rooting unit; examples include a (a) creeping herb, (b) clonal herb, (c) nonclonal herb, and (d) nonclonal woody plant.*
Adapted from Klimešová et al. (2019) with permission.

last so long. For some branches are always new, while others grow old" (quoted by Arber, 1950, p. 19). Shoot apical meristems are the source of all aboveground organs, including leaves and inflorescences, and root apical meristems are located at the tips of the roots. Shoot and root apical meristems that propagate vegetatively exhibit indeterminate growth, like a freely branching tree. Growth of axillary buds is suppressed by apical dominance, but this can be eliminated by the destruction of an apical meristem. As a point of contrast to indeterminate growth, soybeans are an example of a plant that exhibits determinant growth, where its height is predetermined. Vegetative modules are indeterminate, but flower-bearing modules are determinant. Apical and axillary meristems extend the length of shoots and roots, whereas lateral meristems extend the width of plants. Lateral meristems are only found in woody plants that exhibit secondary thickening of the xylem. The vascular cambium gives rise to new xylem and phloem cells and the cork cambium gives rise to the outer bark (i.e., phellem).

6.3 Land colonization spurred structural innovations

Rather than focusing on all the many hundreds of characteristics that are possible to measure on plants, let us blur our vision and consider the most salient features of a plant:

leaves, stems, roots, buds, and seeds. The significance of these organs can be taken for granted, but without these organs, plants would never have colonized all corners of the terrestrial biosphere (Bowles et al., 2022). In his monograph on plant evolution, Karl Niklas (1997) emphasized several critical plant traits molded by natural selection over the course of evolutionary time, and argued that traits evolved repeatedly because the laws of physics and chemistry constrain which traits solve the fundamental problems faced by plants. He extended Sewall Wright's (1932) fitness landscape of genotypes to adaptive walks of plant phenotypes, where phenotypes evolve toward more efficient trait combinations. The optimal phenotype for the aquatic ancestor of land plants differs significantly from one built to survive life on land. To emerge from their aqueous habitat, plants had to deal with the difficulty of acquiring water as well as the dehydrating effects of a parched atmosphere. Given these seemingly insurmountable problems, why did plants leave the water at all? Something must have enticed them. Plants had two things to gain by transitioning between habitats: water attenuates the intensity of sunlight and absorbs the wavelengths that are most useful to plants, and both CO_2 and O_2 are more accessible near the surface of the water than in deep water. If the colonization of land was in the interest of more efficient access to light, it must have come as quite a shock to early land plants that growing tall became the principal driver of light limitation on land.

The earliest land plants had cylindrical stem-like axes that served two functions: photosynthesis and support (Figure 6.3). However, these plants faced a fundamental engineering constraint that prevented them from growing tall (Niklas, 1997). First, to maximize photosynthetic efficiency, the light harvesting organs should be located at the surface of the stem to increase light absorption and diffusivity of CO_2, so natural selection would have minimized the distance between the photosynthetic machinery and the atmosphere. Second, to maximize the mechanical strength of the stem, the best location for stiff building material is at, or just beneath, the surface. Hence, the problem: the same tissue cannot simultaneously photosynthesize and support the plant with maximum efficiency. How did plants solve this fundamental design problem?

Some plants avoided the problem by supporting their cells hydrostatically with water-filled tissues, which limited their distribution to wet habitats of short stature (Niklas, 1997). Mosses and some herbs are living examples of this design. Other taxa were more adventurous and sought to escape the limitations of hydrostatic design. The limitations on plant height were summarized by a formula derived by Leonhard Euler,[1] whose equation states that

$$H_{max} = C\left(\frac{E}{\rho}\right)^{1/3} D^{2/3} \tag{6.1}$$

where H_{max} is the maximum height, C is a constant, E is Young's elastic modulus, ρ is the bulk density of the stem, and D is stem diameter. For a given stem diameter, the

---

[1] Yes, incredibly, this is the same Euler who derived the equation for population growth under stable age distributions (Eq. 4.11), a Renaissance genius who contributed to both demography and functional ecology, and therefore this book owes Euler a tremendous debt.

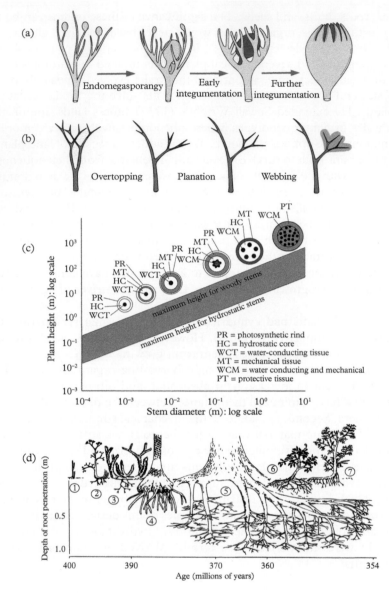

Figure 6.3 *The evolution of seeds, leaves, stems, and roots. (a) The theorized evolutionary development of the seed coat. (b) Theorized evolutionary development of macrophyll leaves. (c) Evolutionary development of stem anatomy with increasing stem diameter and the positive log-linear relationship between stem diameter and height predicted from Euler's Equation (Eq. 6.1). (d) Root systems and their depth of soil penetration in the Devonian (400–354 million years ago, axis not to scale). Representative taxa are as follows: (1) rhyniophytes (e.g.,* Aglaophyton *or* Horneophyton*), (2) trimerophytes (e.g.,* Psilophyton*), (3) early herbaceous lycopods (e.g.,* Asteroxylon *or* Drepanophycus*), (4) early tree lycopods (e.g.,* Lepidosigillaria *or* Cyclostigma*), (5) progymnosperm (e.g.,* Archaeopteris*), (6) early gymnosperms (e.g.,* Elkinsia *or* Moresnetia*), (7) zygopterid fern (e.g.,* Rhacophyton*).*

(a): Reproduced from Willis and McElwain (2014) with permission (Andrews 1961, Stewart and Rothwell 1993). (b): Reproduced from Willis and McElwain (2014) with permission. (c): Reproduced from Niklas (1997) with permission. (d): Reproduced from Algeo and Scheckler (1998) with permission.

maximum height of a stem depends on the stiffness and density of the tissue. Niklas demonstrated that plant height steadily increased throughout the Devonian period by finding alternative solutions to the design problem. Rather than making a compromised tissue that performed neither function well, plants were selected to separate each function into separate organs: leaves and stems.

As an apical meristem grows upward, each successive leaf develops above the previous leaf, leading to the problem of self-shading. But leaf arrangement did not evolve randomly. You may have heard of the celebrated Fibonacci series that converges on the golden mean: the irrational decimal fraction of 0.38197. Plants can minimize leaf overlap by orienting successive leaves at angles approaching 137.5°, which is the product of 360° and Fibonacci's golden mean. Because the golden mean is irrational (the number of decimal places never ends), if successive leaves are perfectly placed at this angle around the stem, no leaf will ever lie exactly above any other leaf *ad infinitum*! Niklas (1997) showed that the shape of leaves also matters: long and slender leaves cast less shade than short and round leaves, and the most efficient phenotype was comprised of long and linear leaves with divergence angles of 137.5°.

The evolution of secondary thickening of stem xylem was another key innovation. The earliest land plants consisted of stems surrounded by a photosynthetic rind and were limited to hydrostatic support mechanisms in wet habitats. A series of modifications ensued involving the development of specialized water conducting tissues. The crescendo in this sequence was the innovation of stems with an external layer of bark and an internal core of water conducting wood. This innovation marked the appearance of the first lateral meristems: the vascular cambium and the cork cambium. The cork cambium produces a porous outer layer that serves to protect the vascular cambium and allow for passive diffusion of O_2 through it. The deposition of wood by the vascular cambium satisfied the limitations of vertical growth. According to Niklas (1997, p. 284), "the great beauty in the design of woody stems lies in its ability to deal with Fick's law for passive diffusion as well as the mechanical requirements for vertical growth".

The evolution of seeds also played a pivotal role in the radiation of plants onto land by releasing plant reproduction from its dependency on water (Linkies et al., 2010). Prior to the seed, water was required for the growth and survival of the megagametophyte, fertilization of the egg, and the development of the sporophyte embryo. Embryos protected within a hard seed coat are better able to tolerate unfavorable seasons and disperse farther. A recent fossil discovery has uncovered the likely ancestor of the closed maternal seed coat (Shi et al., 2021).

The evolution of roots was equally critical, for no land plant can survive without access to water and mineral nutrients. Less is known about the evolution of root structures such as root caps and the development of lateral roots. According to Kenrick and Strullu-Derrien (2014, pp. 578–579):

> Roots evolved in a piecemeal fashion and independently in several major clades of plants, rapidly acquiring and extending functionality and complexity. Specific aspects of root evolution that are still poorly understood but amenable to investigation include cell patterning in root meristems, the development of lateral roots, and the evolution of root caps and root hairs.

The earliest root-like organs to evolve were simple rhizoid-bearing stems, but over 40 million years they developed into complex multicellular organs that became adept at anchoring the plant to substrates and acquiring mineral nutrients. Symbiosis with mycorrhizal fungi may have facilitated the invasion of land, as evidenced by well-preserved fossils from the Rhiny chert (Kenrick and Strullu-Derrien, 2014; Hetherington and Dolan, 2018; Freschet et al., 2021). The evolution of land plant morphology has quite naturally focused our attention on roots, stems, leaves, and seeds. We begin belowground, where a large body of meticulous field and glasshouse work has discovered a remarkable array of strategy dimensions in roots and belowground storage organs, and we then move our attention aboveground to focus on stems, leaves, and seeds.[2]

6.4   Roots acquire and transport resources, support the canopy, and anchor the plant

Roots anchor plants in place, provide support for growing stems, acquire water and mineral nutrients from the soil, and transport water and minerals to aboveground tissues. Roots perform many complex functions that make life on land possible. Early botanists such as Theophrastus took great interest in roots because of their high medicinal value (recall the decline in *Echinacea angustifolia* by herbalist collectors that value its roots, discussed in Chapter 4). For many centuries the roots of plants were thought to correspond to the mouths of animals because that is where food is absorbed, conjuring comparisons of plants to animals standing on their heads (Arber, 1950). Root trait importance, however, is inversely proportional to the ease at which they can be studied. It is very easy to ignore the "hidden half" of plants (Eshel and Beeckman, 2013), but our mission to define the multidimensional plant phenotype would fail if we did.

The gravitropic root apical meristem seeks out resources belowground. Starting from the lowest point and moving longitudinally upward, the root tip is comprised of the root cap, the meristematic zone (a region of cell division), the elongation zone, and the maturation zone (Figure 6.4). Starting from the outside of the root and moving cross sectionally inward, the root is comprised of the epidermis, the cortex, the endodermis, the pericycle, and the stele, which is where the vascular tissues are found. Lateral roots arise from the pericycle. A region of cells known as the quiescent center divide at an extremely slow rate and release dividing cells called initials at just the right pace to maintain the root's shape. However, its function is not altogether clear, and one hypothesis is that growth regulators are synthesized in the quiescent center.[3] The anatomy of roots can differ strongly between dicots and monocots, which can have strong implications on bulk root tissue densities.

---

[2] Technically, a seed is not an organ because a whole plant is contained within the seed coat. Some botany textbooks state that plants have three organs: roots, stems, and leaves. I assume reproductive structures are treated as modified leaves in these cases. While the diversification of flowering morphology clearly has significant effects on fitness, they cannot be measured on all seed-bearing land plants.

[3] May we all find our quiescent center.

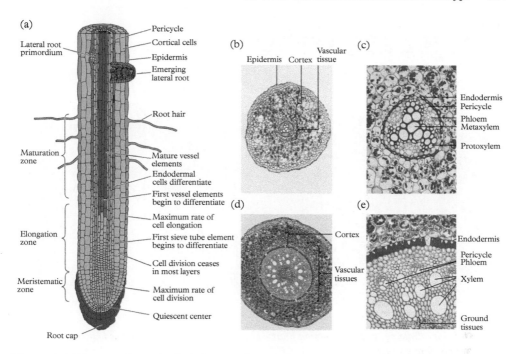

Figure 6.4 *Diagram of root tip. (b-c) Cross sections of monocot root and (d-e) dicot roots.*
Image adapted from Taiz and Zeiger (2006) with permission. (b–e): Reproduced from Mauseth (2009) with permission.

These anatomical basics give rise to a wide variety of architectural forms, but most root systems can be classified into two types: taproot and fibrous root systems (Figure 6.5). Tap roots represent the first root to emerge from a dicot seed that typically forms the central axis of the root system (Freschet et al., 2021). Taproot systems can send out lateral roots that develop sinker roots to explore deeper soil layers away from the central axis. In contrast, fibrous roots are synonymous with adventitious roots that arise from non-root tissue at the base of the shoot and is common in monocots. Roots of woody plants are typically tap root systems that develop secondary thickening of the xylem and will contain a lateral vascular meristem that controls root thickening. I encourage readers to explore the online open-access digitized drawings of root systems by Kutschera and Lichtenegger (2002).

Plants vary in their ability to extend roots into deep soil layers (Stone and Kalisz, 1991; Pierret et al., 2016). Root depth is more important for water acquisition than for nutrient acquisition. Rooting depth varies with size and age of the individual, so the appropriate trait for comparison at the species-level is maximum rooting depth (Schenk and Jackson, 2002; Tumber-Dávila et al., 2022). A global analysis found that the average rooting depth was 3.64 m for trees, 2.36 m for shrubs, 1.02 m for forbs, and 1.14 m for grasses (Canadell et al., 1996; Jackson et al., 1996; Tumber-Dávila et al., 2022). Some of

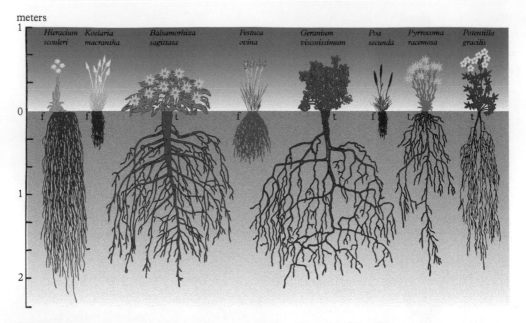

Figure 6.5 *Line drawings of eight common prairie plants illustrating the difference between taproot (t) and fibrous (f) root systems.*
Adapted into color from Weaver's (1919) original drawings with permission.

the most well-known deep-rooted species are short dryland trees and shrubs that avoid drought by tapping into deep water reserves even when their shallow-rooted neighbors are parched. For example, a mesquite (*Prosopis juliflora*) in Central America reportedly grows roots as deep as 53 meters, and shepherd's tree (*Boscia albitrunca*) in the Kalahari Desert has been shown to grow roots down to 68 meters! Only species with tap roots can extend roots that deep. Grasses and other monocots with fibrous root systems are inherently limited to shallow soil layers. Rooting depth is a notoriously difficult trait to measure. It requires either painstaking excavations that only paleontologists and anthropologists are qualified to do, or heavy equipment to dig very big pits in a short amount of time. Physiologists and hydrologists have been experimenting with using stable isotopes to determine whether plants are accessing water from deep or shallow soil layers (Beyer et al., 2018). Rooting depth is shallowest in succulents, annual herbs, and perennial grasses, and deepest in deciduous broadleaf and evergreen broadleaf trees (Figure 6.6). The strongest factor controlling rooting depth is the depth of the water table, which limits rooting depth to avoid anoxic conditions but also draws roots downward to tap into the capillary rise (Fan et al., 2017).

Roots were traditionally classified into size classes by their diameter: coarse roots were > 2.0 mm and fine roots were ≤ 2.0 mm. However, this arbitrary distinction proved insufficient because it does not distinguish between the diverse range of sizes

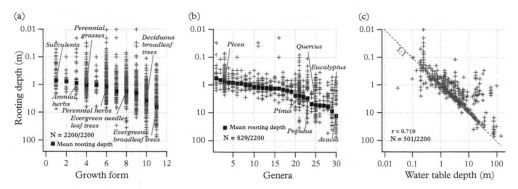

Figure 6.6 *Maximum rooting depth (log-scale) as a function of (a) growth form (selected growth forms are indicated in green text), (b) genera (selected genera are highlighted in green text), and (c) water table depth (log-scale).* r = *Pearson correlation coefficient and* N = *sample size.*
Reproduced from Fan et al. (2017) with permission.

of absorptive and transport roots across species. Absorptive roots have little to no secondary development and so have an intact cortex, an absence of cork periderm, and low suberization. Transport roots have developed secondary thickening with high stele-to-diameter ratios and a developed system of xylem conduits (McCormack et al., 2015). The problem with the arbitrary distinction is that many thin-rooted species develop roots ≤ 2.0 mm that transport water and nutrients, not just acquire them. To solve this problem, root trait ecologists proposed a functional classification and an order-based classification system (Figure 6.7). First-order roots are the most distal roots and would be most likely to be involved with acquisition. Most root experts acknowledge that this classification is not an exact science in practice, and so many agree that fine root measurements should be made on first-, second-, and third-order roots that exhibit no cork periderm (no suberization) to maximize the likelihood that they are involved in absorption and uptake.

Mass flow and diffusion are the dominant mechanisms by which plants acquire mineral nutrients without assistance from fungal partners. Mass flow is the movement of nutrients into the plant driven by water vapor loss through the transpiring leaves, a process that draws water up through the soil–plant–atmosphere continuum from high to low water potentials. Diffusion is the movement of nutrients to the root surface along a high-to-low concentration gradient. Root epidermis layers contain ion transport systems that can pump ions into the soil to change the diffusion gradients to acquire mineral nutrients. Nutrients can enter the root through a variety of mechanisms, including diffusion, facilitated diffusion, gated channel confirmation, symporter, antiporter, and active transport (Figure 6.8). However, we know little about how these microscale processes scale up to root segments or whole root systems (Griffiths and York, 2020). Determining whether anatomical or morphological root traits are associated with the prevalence of different ion uptake mechanisms is an important area of research.

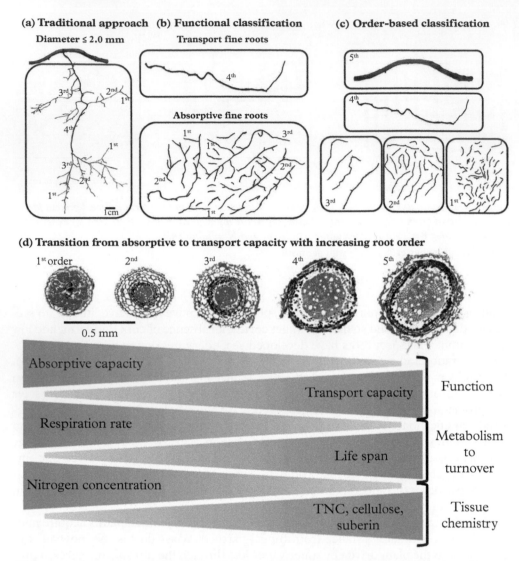

(a) Traditional approach   **(b) Functional classification**   **(c) Order-based classification**

Diameter ≤ 2.0 mm

Transport fine roots

Absorptive fine roots

(d) Transition from absorptive to transport capacity with increasing root order

Figure 6.7 *Comparison of fine root classification systems and changes in primary function with increasing root order. (a) Intact root branch of* Liriodendron tulipifera *with all fine roots traditionally defined as* ≤ 2.0 mm. *(b) The same branch is then separated into transport and absorptive fine roots based on a functional classification. (c) Again, the same branch is separated into root orders. (d) Root cross-sections of* Acer plantanoides *illustrate a common pattern of increasing diameter and development of secondary vascular tissue with increasing root order from left to right. Green triangles represent hypothesized changes in form and function with increasing root order. TNC = total nonstructural carbohydrates.*

Reproduced from McCormack et al. (2015) with permission.

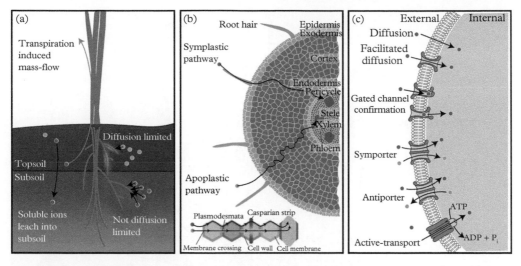

Figure 6.8 *Downscaling from the soil environment to ion uptake mechanisms. (a) Mass flow, diffusion, and interception are the main pathways of nutrient uptake. (b) Ions travel into the root through the symplastic pathway within cells or the apoplastic pathway around cells to reach the xylem. (c) Nutrient ions enter the root surface through a variety of mechanisms. We know little about how functional traits of roots may correlate with these different ion uptake mechanisms.*
Reproduced from Griffiths and York (2020) with permission.

Diffusion and mass flow are not the only mechanisms of nutrient acquisition. More than 80% of vascular plants associate with arbuscular mycorrhizal fungi (Tedersoo et al., 2020), and only a fraction of species are truly non-mycorrhizal. The rest associate with endomycorrhiza, ericoid mycorrhiza, arbutoid mycorrhiza, or orchidoid mycorrhiza (Figure 6.9) (Selosse and Le Tacon, 1998). In a complex relationship that varies from symbiotic to parasitic depending on the environmental context (see Gibert et al., (2019) for frequency distributions of positive and negative outcomes for various types of mutualisms), plants exchange carbon for nutrients with the fungi. Emphasis has been placed on mycorrhizal exchange for phosphorus due to the steep diffusion gradients that result from phosphorus depletion, but mycorrhiza are increasingly viewed as being important for nitrogen and water uptake as well. Many other plant species engage in symbiosis with nitrogen-fixing bacteria (including *Rhizobia* spp. and *Frankia* spp.) harbored in root nodules and thrive in nutrient-poor soil where even mycorrhizal species struggle to acquire enough nutrients.

Mycorrhiza have interacted with plant roots since the dawn of land plants. Mycorrhizal colonization is greatest in large diameter roots likely because fungi dwell in the cortex rather than the stele (Comas et al., 2014; Kong et al., 2014; Ma et al., 2018;

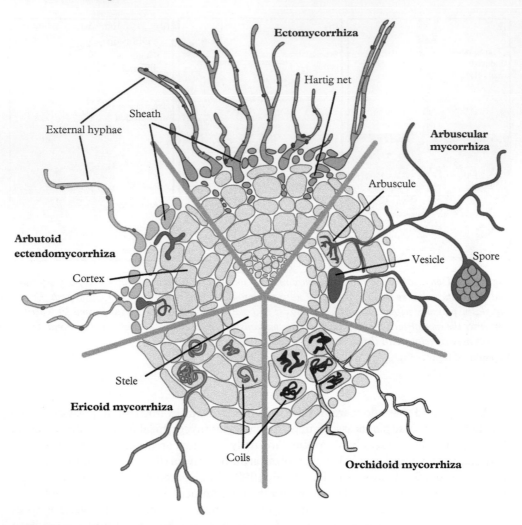

Figure 6.9 *Illustration of five different mycorrhizal association types and associated anatomy.*
This image was reproduced from Selosse and Le Tacon (1998) with permission, and adapted into English from an open-access Wikimedia file created by the Ukrainian bio project, licensed under the Creative Commons Attribution-Share Alike 4.0 International license.

Bergmann et al., 2020) and larger diameter roots contain proportionally more cortex than stele (Valverde-Barrantes et al., 2016). However, fine root diameter has become thinner over evolutionary time in the angiosperms (Figure 6.10), suggesting a tendency to become less dependent on mycorrhizal symbionts (Baylis, 1975; Comas et al., 2012; Ma et al., 2018). Bergmann and colleagues (2020) proposed that two morphological

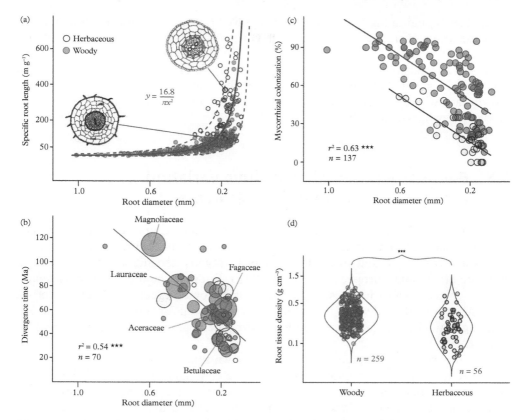

Figure 6.10 *Root diameter and specific root length (SRL) are inversely correlated (note that the x-axis is illustrated against convention in descending values to the right). (b) Evolutionary time of divergence of major taxonomic groups is positively correlated with root diameter. (c) Mycorrhizal colonization is greatest in large diameter roots. (d) Root tissue density is higher in woody than herbaceous plants.* Reproduced from Ma et al. (2018) with permission.

traits define a gradient in mycorrhizal collaboration: fine root diameter (mm) and specific root length (length of root per dry mass; m g^{-1}). Specific root length (SRL) is thought to reflect the potential rate of proliferation of roots into patches of resource-rich soil because thin-rooted species require less carbon per unit root length. Species with thin fine roots and high SRL represent a "do-it-yourself" strategy, whereas species with thick fine roots and low SRL "outsource" nutrient acquisition to mycorrhizal fungi.

Fine root nitrogen and phosphorus concentrations vary considerably across plant species and are important plant traits that are thought to be indicators of the metabolic activity of the roots because tissue nutrient concentrations are strongly correlated with respiration rates (Reich, 2014). Nutrient concentrations tend to vary inversely with root tissue density, indicating that denser roots are likely comprised of structural and vascular tissues compared to the nutrient rich cortical tissue. Root tissue density (RTD) is measured as the dry mass of the fine roots divided by the fresh fine root volume (g cm^{-3}). Bergmann and colleagues (2020) made a second proposition: that nutrient concentrations and RTD define a root conservation axis that is aligned with relative growth rate, where high RTD and low nutrient concentrations are associated with slow rates of growth (Kramer-Walter et al., 2016).

6.5   Belowground bud banks drive variation in vegetative reproductive strategies

Roots are just one component of belowground function. Clonal growth organs can be found belowground where buds sprout from roots and modified stems. Buds are dormant meristems found in a variety of plant organs depending on growth form and type of vegetative propagation (Figure 6.11). Buds can be found on roots, stems, stolons, epigeogenous rhizomes, hypogeogenous rhizomes, tubers, and bulbs (Herben and Klimešová, 2020). One of the grand challenges facing functional ecology is understanding how plant clonality and bud banks drive demographic outcomes (Klimešová et al., 2019).

The importance of bud location has been known since Raunkiaer (1934), but ecologists are devising new techniques to capture the continuous variation of clonal traits and bud bank traits (Figure 6.12). Clonal trait data was once limited to a simple binary designation of whether a plant was clonal or not because this information could be extracted from a local flora. However, there are several continuous traits that have strong potential to inform plant strategy theory. Plants can propagate themselves horizontally through lateral clonal spread and can sprout shoots from a reserve of dormant meristems throughout the body of the plant. Therefore, lateral spreading distance from the mother to the young ramet, multiplication rate (also known as offspring number), and persistence of connection are important clonal traits in herbs. Bud bank size and depth can also be measured on continuous scales and, given the central role that dormant meristems play in plant survival, bud bank traits should be important predictors of plant fitness in population models.

Clonal growth organs are widely spaced across the angiosperm phylogeny, suggesting that clonality arises frequently and independently across clades; however, stem-based clonal growth and root-based clonal growth are evolutionarily unrelated types of clonal growth (Herben and Klimešová, 2020). Stoloniferous plants attain greater spreading distances and higher offspring number, but reduced bud bank depth and connection persistence (Figure 6.12). Epigeogenous rhizomatous plants have large bud bank sizes. Hypogeogenous rhizomatous plants exhibit larger bud bank size, bud bank depth, spreading distance, and offspring number. Bud-bearing roots that allow for resprouting slightly increase bud bank depth. More attention must be given to clonal growth traits to gain a deeper understanding of perennial plant survival.

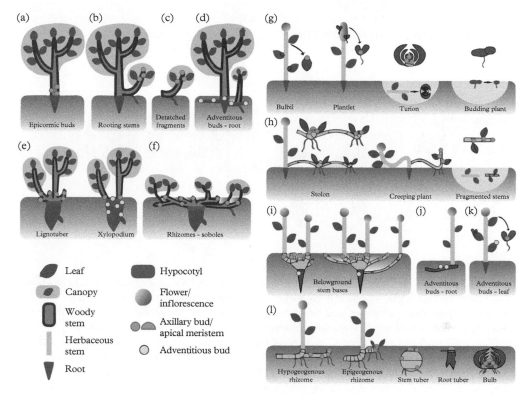

Figure 6.11 *Stylized illustrations of the large variety of clonal growth organs and the locations of the buds on each type: (a–f) woody plants, (g–l) herbaceous plants.*
Reproduced from Klimešová et al. (2019) with permission.

6.6 Stems support the canopy, protect the vascular system, and drive hydraulic function

Early land plants were photosynthetic stems, but they couldn't grow taller than a meter or so before buckling due to failure of the hydrostatic forces. The development of lateral meristems—the vascular cambium and the cork cambium—permitted land plants to grow much taller because they were no longer dependent on hydrostatic forces for support (Niklas, 1997). Early land plants separated the functions of support and photosynthesis into stems and leaves. But stems serve multiple functions. Not only do they support large, heavy canopies of leaves and branches, but also they transport and store water, minerals, and carbohydrates. It is impossible for stem tissue to optimize each function simultaneously. The most efficient design for rapid water transport would be many wide, thin-walled conduits, but this design can lead to hydraulic failure in a drought and mechanical failure in a wind storm before the plant even reaches reproductive maturity

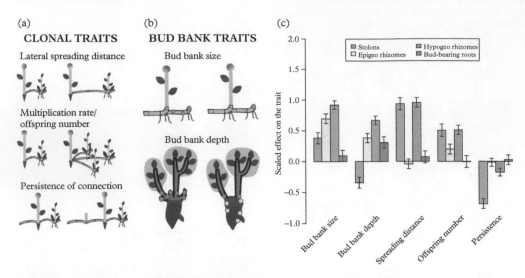

Figure 6.12 *(a–b) Illustration of five continuous traits that capture the diversity of form and function of clonality and bud banks. (c) Effects of possessing clonal growth organs (stolons, epigeogenous rhizomes, hypogeogenous rhizomes, and bud-bearing roots) on expression of continuous clonal traits (bud bank size, bud bank depth, spreading distance (i.e., lateral spread), offspring number (i.e., multiplication rate), and persistence of connection). Bars less than zero indicate negative effects of the clonal growth organs on the traits, and vice versa. For example, plants with stolons tend not to possess deep bud banks whereas plants with hypogeogenous rhizomes tend to possess deep bud banks.*
(a–b): Adapted from Klimešová et al. (2016) with permission. (c): Reproduced from Herben and Klimešová (2020) with permission.

(Baas et al., 2004). Chave and colleagues (2009) discuss several aspects of a proposed wood economics spectrum, such as resistance to decay, storage capacity, and mechanical strength, each of which are important problems faced by the plant stem.

Herbaceous plants generally do not produce secondary thickening of the xylem and so generally rely on hydrostatic forces for support. Starting from the outside and moving inward, a typical herbaceous stem is composed of an epidermis, a cortex, and a pith in the core. A ring of vascular bundles separates cortex from pith and contains the vascular cambium, xylem, and phloem (Figure 6.13a).

Woody stems start out like herbaceous stems early in development, but differences appear as soon as the vascular cambium and cork cambium develop. Starting from the outside and moving inward, the woody stem is composed of the cork (i.e., outer bark), the cork cambium that produced the cork, and the phloem (i.e., inner bark), the vascular cambium, the xylem, and the pith (Figure 6.13b). Woody plants grow wider at the same time they grow taller by adding new layers of xylem internal to the vascular cambium. Xylem anatomy differs among gymnosperms, dicots, and monocots, which can cause complications when we seek a single trait to compare across all species (Carlquist, 1975). Tracheids are the main conducting cells in gymnosperms and ferns and are narrow in

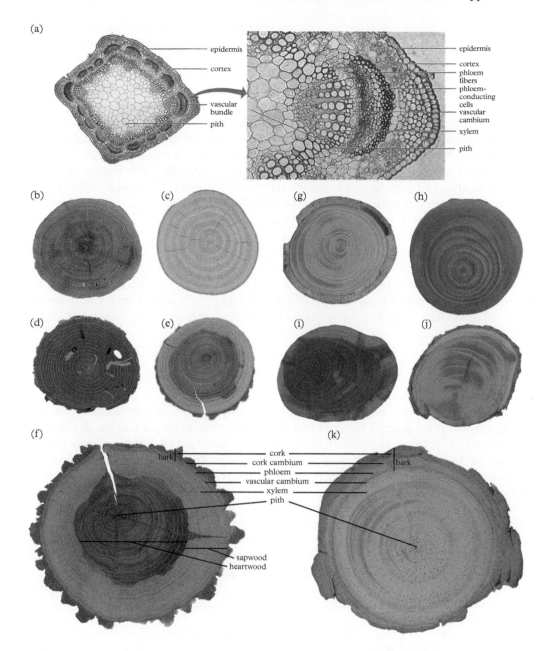

Figure 6.13 *(a) Stem cross section of alfalfa (*Medicago sativa*) with anatomy that is typical of an herbaceous dicot. Stem-specific density, reported here as stem-specific gravity, generally increases from top to bottom. Angiosperms: (b)* Populus deltoides, *6.7 cm diameter, density = 0.36 g cm⁻³, (c)* Platanus occidentalis, *6.9 cm diameter, density = 0.46 g cm⁻³, (d)* Robinia pseudoacacia, *11.7 cm diameter, 0.66 g cm⁻³, (e)* Maclura pomifera, *8.5 cm diameter, 0.76 g cm⁻³, (f)* Quercus marilandica,

continued

diameter and contain few large pits that serve as passageways through which water can flow. Vessels are the conducting cells of angiosperms contained within a fibrous matrix and can be much longer than tracheids and contain smaller, more numerous pits. When water is in good supply, there is a continuous strand of water molecules that is pulled by cohesion–tension from the roots, through the xylem, to the stomata in the leaves where the water evaporates into the atmosphere. However, when water is in short supply, the negative tension in the xylem becomes even more negative causing air bubbles to seed into the water column, called an embolism, which causes cavitation of the water column and leads to hydraulic failure. Amazingly, it is still not clear how embolisms are repaired, but they have been observed so we know they happen regularly under negative tension.

Vulnerability to embolism can vary strongly across species (Choat et al., 2012). One common way to compare vulnerability to embolism is to quantify the xylem pressure potential (MPa) at which 50% loss of conductivity occurs (P_{50}). There is interest in other quantities, such as P_{12} and P_{88} (Mrad et al., 2021) and the inflection points on the curves, to represent the pressure at which embolisms begin and when they cause death; however, far more data is available on P_{50} globally so I emphasize P_{50} here. These quantities are measured on branches that are subjected to increasingly negative tensions. The relationship between percent loss of conductivity (*PLC*) and xylem pressure potential are fitted with the following sigmoidal equation (Pammenter and Van der Willigen, 1998):

$$PLC = \frac{100}{\left(1 + exp\left(\frac{S}{25}\left(P_i - P_{50}\right)\right)\right)} \tag{6.2}$$

where S (% MPa^{-1}) is the slope of the vulnerability curve at the inflection point. My colleagues and I measured this data and fit these curves on native tree species in New Zealand (Laughlin et al., 2020a) and found a broad range of xylem vulnerabilities (Figure 6.14). To find P_{50} on the graph, simply locate the xylem pressure at which the curve intersects the dotted line representing 50% loss of conductance. *Schefflera digitata* ($P_{50} = -1.69$ MPa) and *Fuchsia excorticata* ($P_{50} = -1.71$ MPa) were the most vulnerable to cavitation, whereas *Prumnopitys ferruginea* ($P_{50} = -7.8$ MPa) and *Phyllocladus alpinus* ($P_{50} = -6.9$ MPa) were the most resistant.

The vasculature of plants can be thought of as an elaborate plumbing system. Species with wide conduits and thin walls exhibit fast rates of water flow through their pipes but lack resistance to embolism. Species with narrow conduits and thick walls exhibit

---

Figure 6.13 (*continued*)
21.6 cm diameter, 0.59 g cm^{-3}. Gymnosperms: (g) Calocedrus decurrens, 3.1 cm diameter, density =
0.35 g cm^{-3}, (h) Sequoia sempervirens, 1.9 cm diameter, density = 0.36, (i) Juniperus virginiana,
7.1 cm diameter, density = 0.44 g cm^{-3}, (j) Pinus edulis, 10.0 cm diameter, density = 0.50 g cm^{-3},
(k) Pinus elliottii, 4.5 cm diameter, density = 0.54 g cm^{-3}.

(a): Reproduced from Bidlack and Jansky (2011) with permission. (b–k) Woody stem cross sections of (b–f) angiosperms and (g–k) gymnosperms, provided by and reproduced here with permission from Joe Buck (https://www.oldtrees.co).

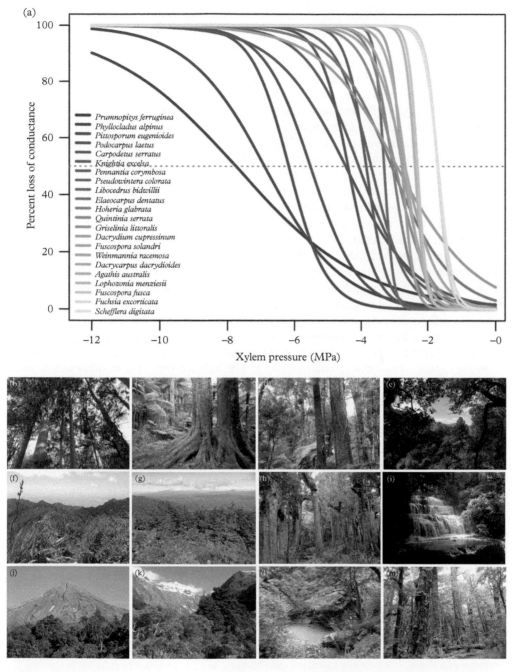

Figure 6.14 *(a) Xylem vulnerability curves measured on indigenous tree species of Aotearoa New Zealand. (b)* Agathis australis *(kauri) on the Manginangina kauri walk, Puketū Forest. (c) Pair of* Dacrycarpus dacrydioides *(kahikatea) at Waingaro Landing. (d)* Dacrydium cupressinum *(rimu) on Mt Pirongia. (e) Mixed stand including* Elaeocarpus dentatus *(hinau) and* Prumnopitys ferruginea *(miro) at Whirinaki Te Pua-a-Tāne Conservation Park managed by Ngāti Whare iwi. (f)* Knightia excelsa *(rewarewa) and* Quintinia serrata *(tāwheowheo) on Mt Karioi. (g)* Griselinia littoralis *(kapuka),* Fuscospora solandri *var.* cliffortiodes *(mountain beech),* Phyllocladus alpinus
continued

low flow rates through their pipes but greater resistance to embolism. These observations have led physiologists to predict the existence of a trade-off between hydraulic efficiency and safety. Gleason and colleagues (2016) define hydraulic efficiency as "the rate of water transport through a given area and length of sapwood across a pressure gradient—the xylem-specific hydraulic conductivity (K_S)." Hydraulic safety is defined as P_{50}—with the caveat that 50% loss of conductivity is an arbitrary percentage for standardization. Gleason and colleagues (2016) also analyzed the largest compilation of these two traits across 335 angiosperms and 89 gymnosperms and found only a weak negative correlation. No species had high safety and efficiency, supporting the idea of the trade-off, but many species had low safety and low efficiency, a result that remains difficult to understand. Despite the weak trade-off, resistance to embolism remains one of the best predictors of drought resistance in woody plants. K_S and P_{50} are difficult to measure without specialized equipment, but there is a reasonably predictable relationship between wood density and P_{50}, although the equations differ for angiosperms and gymnosperms (Hacke et al., 2001; Laughlin et al., 2020a).

Stem specific density (SSD) is a relatively easy-to-measure property that can be used to compare species investment in stem strength and rigidity (Chave et al., 2009). It can be measured on both woody plants and herbs, though less data is currently available on herbaceous species at global scales. Wood density—the quotient of wood dry mass per unit fresh volume (g cm^{-3})—is a composite trait and an important component of Euler's equation for limitations on tree height (Eq. 6.1). Wood density ranges between 0.1 g cm^{-3} in the fast-growing balsa tree (*Ochroma pyramidale*) to 1.2 g cm^{-3} in the slow-growing Lignum vitae tree (*Guaiacum* spp.). The lightweight balsa wood is well known for its use in model airplanes, and the dense lignum vitae was sought out to make bails that were sturdy enough to maintain the integrity of cricket wickets in gusty weather. Species with dense wood have slower growth rates because they allocate more carbon per unit volume to the stem.

Bark thickness is another important stem trait in woody plants. The outer bark is produced by the cork cambium and gets thicker as the diameter of the tree increases in size with age. Species vary greatly in the thickness and density of the outer bark (Laughlin et al., 2011a; Lawes et al., 2013; Richardson et al., 2015). The cork is breathable because gases need to diffuse into the stem from the atmosphere to feed the respiratory demands

---

Figure 6.14 *(continued)*
(mountain toatoa) on the Urchin rig Track, Kaimanawa Forest Park. (h) Libocedrus bidwillii *(kaikawaka) on Ngatoro Loop Track, Mount Egmont National Park, Taranaki. (i)* Carpodetus serrata *(putaputāwētā),* Pennantia corymbosa *(kaikomako),* Pittosporum eugenioides *(tarata), and* Schefflera digitata *(patē) at Purakaunui Falls in the Catlins. (j)* Weinmannia racemosa *(kamahi) and* Podocarpus laetus *(mountain tōtara) on Taranaki. (k)* Fuchsia excorticata *(kotukutuku) and* Lophozonia menziesii *(silver beech) at Lake Marian, Fiordland National Park. (l)* Hoheria glabrata *(mountain lacebark),* Lophozonia menziesii *(silver beech), and* Pseudowintera colorata *(mountain horopito) on Rob Roy Glacier Track, Mt Aspiring National Park. (m)* Fuscospora fusca *(red beech) on Lake Gunn Nature Walk, Fiordland National Park.*

(a): Reproduced using original data from Laughlin et al. (2020a). (b–m): All photos taken by the author in New Zealand and are freely available from the Global Vegetation Project (www.gveg.wyobiodiversity.org).

of the vascular cambium (Niklas, 1997). The presence of small openings called lenticels enhance gas exchange. Bark thickness is an important innovation that protects species from fire by insulating the vascular cambium from heat-induced damage (Vines, 1968; Uhl and Kauffman, 1990; Pausas, 2015).

6.7 Leaf traits respond to drought, light, and temperature

Leaves evolved when short photosynthetic stems were no longer the most competitive phenotype. Sturdy, non-photosynthetic stems were the evolutionary solution to attain height while providing structural support and protection of the vascular cambium. The function of photosynthesis was given to leaves comprised of mesophyll cells connected by a network of veins and protected by a layer of epidermis. Leaves have one primary job: to harvest light energy to synthesize sugar from CO_2 that enters through their stomata. Leaves are enormously variable in size, shape, thickness, and density. Starting from the top and moving down, the leaf is composed of a cuticle, an epidermis with stomata, layers of palisade parenchyma in the mesophyll, vascular tissue of xylem and phloem surrounded by bundle sheath cells, and a lower layer of epidermis with stomata (Figure 6.1). The arrangements of these basic features vary among conifers and angiosperms and among dicots and monocots.

The most conspicuous macroscopic trait of a leaf is its size. Leaf size varies by over 100,000 times among plant species globally and early plant geographers were keen to point out the prevalence of small leaves in arid shrublands compared to massive leaves in humid forests (Schimper, 1903). *Wolffia* spp. (duckweeds) produce leaves that are the size of a pin, whereas *Victoria amazonica* produces leaves up to three meters in diameter. The key to understanding the importance of leaf size is contained in the physics of boundary layers. Invisible boundary layers are thin zones of calm air that envelop every leaf. Larger leaves have thicker boundary layers (Parkhurst and Loucks, 1972), which reduce the transfer of heat and gases between the leaf and surrounding atmosphere. The prevalence of small leaves in dry environments reflects the vulnerability of large leaves to overheating when water limitation prevents cooling by transpiration. Large leaves are also more frost-prone in sites where water is not limiting (Lusk et al., 2018). Among long-lived evergreen tree species, leaves tend to increase in size from cold to warm climates; however, this pattern does not hold among deciduous species (Lusk et al., 2018). Large-leaved deciduous species can extend further into cold climates because they shed their leaves in the autumn. Leaf size is controlled by an interaction between precipitation and temperature, where leaf size increases strongly with increasing temperature in wet climates, but slightly decreases with increasing temperature in dry climates (Wright et al., 2017).

Two species can display leaves with identical leaf areas yet photosynthesize at vastly different rates. The thickness of the leaf and the density of the tissue drive this variation. Some leaves are pliable and floppy, others are dense and rigid. To standardize comparisons, we divide leaf size (m^2) by its dry mass (g) to compute the specific leaf area (SLA; m^2 g^{-1}). The inverse of SLA is leaf mass per area (LMA; g m^{-2}), which varies over three orders of magnitude globally (Poorter et al., 2009). SLA is often used in growth equations, but LMA is often used when decomposing it into its fundamental base quantities (Poorter et al., 2009). Both traits are used in this book, depending on the context. LMA is lowest in aquatic plants, ferns, herbs, and deciduous shrubs and trees, and highest in evergreen shrubs and trees and succulents. However, there is considerable variation within each life form as well (Figure 6.15). LMA is on average higher in gymnosperms than angiosperms (Flores et al., 2014). LMA represents a trade-off between leaf construction cost and rate of photosynthesis and respiration. While there are debates regarding the underlying cause of this trade-off (Shipley et al., 2006; Blonder et al., 2011; Sack et al., 2013; Mason et al., 2016), high LMA leaves tend to be composed of more sclerenchyma tissue, higher lignin content, and thicker cell walls, whereas low LMA leaves have a higher proportion of mesophyll with higher protein content and thinner cell walls. Some consider LMA as simply a convenient approximation to anatomical detail, but it is also an important measure of the cost of constructing light-capturing leaf area.

Leaves are sugar factories. Plant ecologists have long recognized that plant resource acquisition and conservation can be understood in terms of human economic systems (Bloom et al., 1985; Givnish 1986). Just as businesses need to decide whether to save their capital or to invest it in new infrastructure, plants can allocate their fixed carbon in different ways. Factories make products, those products are sold to gain capital, and that capital can either be spent on more factories to make more products or invested in the current factories to maintain their long-term viability. Likewise, leaves fix carbon, and that carbon can be spent in different ways. The carbon can either be used to build lots of inexpensive leaves that fix carbon rapidly which generates a fast rate of return on the investment, or it can be invested to build a few expensive leaves that fix carbon slowly which generates a slow rate of return on the investment. The leaves generating fast rates of return live shorter lifespans than the leaves that generate slow rates of return (Grime, 1979; Chapin, 1980). Global scale analyses confirmed the singular dimensionality of leaf economics traits and captured the attention of ecologists around the world (Reich et al., 1997; Wright et al., 2004), yet the strength of these leaf trait correlations can sometimes be weak within communities (Funk and Cornwell, 2013).

The leaf economics spectrum describes a fundamental trade-off between the rate of carbon fixation and the cost of leaf tissue construction (Reich et al., 1997). The strong correlation among a variety of leaf traits represent this important trade-off, including SLA, the maximum rate of photosynthesis per unit leaf mass (A_{mass}; nmol g^{-1} s^{-1}), leaf lifespan (LL; months), and leaf nitrogen concentration (N_{mass}; %). The key trade-off between the potential rate of carbon return versus the duration of that return is exemplified by the negative relationship between photosynthetic rate on a mass basis and leaf lifespan—a leaf that photosynthesizes rapidly can do so for a short amount of time, whereas a leaf that photosynthesizes slowly can do so for a long duration (Figure 6.16). This trade-off is consistent with cost–benefit models of optimization of leaf carbon gain (Kikuzawa and Ackerly, 1999) and leaf amortization theory (Orians and Solbrig, 1977).

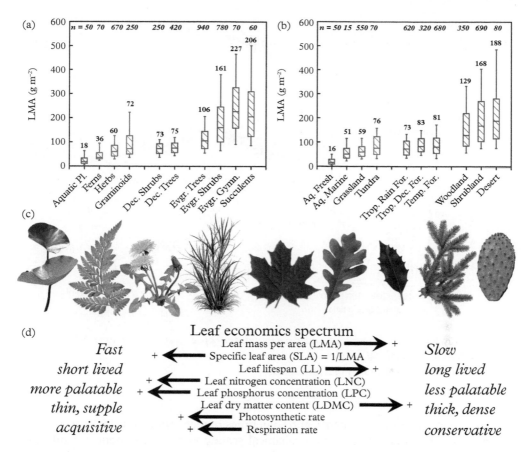

Figure 6.15 *Distributions of leaf mass per area (LMA) among (a) functional groups and (b) habitats. (c) Arrangement of leaves that represent the different functional groups in (a): aquatic plant—*Nelumbo nucifera*; fern—unknown sp.; herb—*Taraxacum officinale*; graminoid—unknown sp.; deciduous trees—*Platanus *sp. and *Quercus *sp.; evergreen shrub—*Ilex aquifolium*; evergreen gymnosperm—*Picea *sp.; succulent—*Opuntia *sp. (d) The central leaf economic traits arranged to describe their alignment with the fast–slow spectrum.* Reproduced from Poorter et al. (2009) with permission.

Leaf economic traits capture the gradient between acquisitive and conservative resource use strategies (Chapin, 1980). Inexpensive leaves are expendable, and are generally preferred by herbivores because they have less lignified cell walls and higher protein content (Mason and Donovan, 2015). It is not such a big deal to lose a few leaves to herbivores when you can rapidly grow more cheap leaves. However, it can be tragic for a plant to invest in a few expensive leaves only to have them ripped off and eaten by an herbivore (Grime, 1979). The leaves of conservative species are more strongly defended against herbivores owing to a higher concentration of lignin, thicker cell walls,

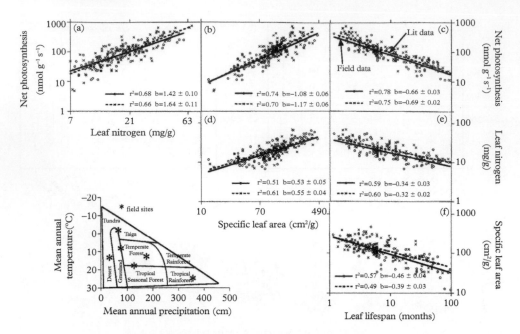

Figure 6.16 *Relationships among mass-based photosynthetic capacity, specific leaf area (SLA), leaf nitrogen concentration, and leaf lifespan.*
Reproduced from Reich et al. (1997) with permission.

lower nitrogen and protein contents, and sometimes higher concentrations of secondary metabolites. The discovery of a biomechanical scaling law between petiole width and leaf mass (Royer et al., 2007) has enabled paleobotanists to reconstruct the LMA of fossil floras to infer changes in functional strategies in ancient ecosystems across critical geological events (Blonder et al., 2014; Butrim and Royer, 2020). Rates of types of herbivory can also be extracted from scars on leaf fossils (Currano et al., 2021). Currano and colleagues (2008) used these techniques to demonstrate an increase in insect herbivory during a warming event called the Paleocene–Eocene thermal maximum. Roots and stems are not well preserved in the fossil record, but the functional traits of leaves are providing powerful insights in paleobotany, including reconstructing past climate (Peppe et al., 2011; Peppe et al., 2018).

Leaves face a fundamental constraint that limits their rate of photosynthesis. Every time leaves open their stomata to allow CO_2 to diffuse into the mesophyll, water vapor is released into the atmosphere from the inside of the leaf, that is, plants must lose water to gain carbon. When water becomes limiting, leaves struggle to maintain turgor pressure. When leaves lose turgor, they close their stomates and shut down photosynthesis. Plants vary in their ability to function at low water potentials (Bartlett et al., 2012b). Turgor loss point (TLP) is the leaf water potential at turgor loss, that is, when the leaf ceases to function or wilts. Pressure–volume curves describe how leaf water potential changes as a function of leaf water content. Anisohydric species can maintain leaf function at lower water potentials and so generally exhibit more negative TLPs. Isohydric species close their stomates at the earliest sensation of drought and exhibit less negative TLPs. Advances in methodologies now allow TLP to be estimated by measuring the leaf osmotic potential, that is, the water potential at full turgor, as it is a good predictor of TLP (Figure 6.17). Prior to this advance, TLP was constrained to physiology laboratories and measured on only a few species, but now field ecologists can harvest leaves from a large number of species in the field, rehydrate the leaves in the lab, freeze them, then measure their osmolality at full turgor using an osmometer (Bartlett et al., 2012a; Bartlett et al., 2012b; Zhu et al., 2018; Griffin-Nolan et al., 2019; Blumenthal et al., 2020). TLP is strongly correlated with leaf dry matter content (LDMC, g g^{-1}), which is quantified as leaf dry mass divided by leaf fresh mass at full turgor (Garnier et al., 2001). However, TLP is not as strongly correlated with SLA (Blumenthal et al., 2020). LDMC may be a good surrogate for leaf TLP because it predicted survival of shortgrass prairie graminoids in dry years as good as TLP, if not better (Stears et al., 2022b). Additional analyses are required to determine the global correlation between TLP and leaf economic traits such as SLA.

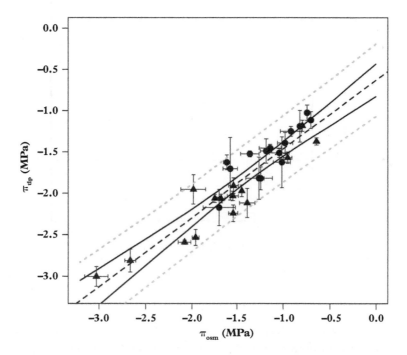

Figure 6.17 *Measurements of osmotic potential (π_{osm}, leaf water potential at full turgor measured using an osmometer) are good predictors of turgor loss point (π_{tlp}, leaf water potential at turgor loss measured using pressure–volume curve analysis) for species of a wide range of leaf structure and drought tolerance.* Reproduced from Bartlett et al. (2012a) with permission.

If water and nutrients are not limiting, then the vegetation develops into a dense closed canopy, causing light to become limiting. The photosynthetic response of a leaf to increasing light availability is quantified using light response curves (Figure 6.18). Light compensation points (LCP) are the light intensity at which net photosynthesis equals zero, that is, where rate of photosynthesis equals rate of respiration. Light saturation points (LSP) are the light intensity at which the rate of photosynthesis is first maximized (A_{max}). Light limitation should select for species with lower LCPs because they can continue to maintain positive carbon balance at lower light levels. Comparisons of light-demanding and shade-tolerant species do suggest that LCPs are lower for shade-tolerant plants, but the differences are small (Craine and Reich, 2005). This prompted researchers to consider carbon budgets at the whole-plant level, given the high respiratory demands of stems. Few studies have measured LCP at the whole-plant level (Baltzer and Thomas, 2007a; Lusk and Jorgensen,

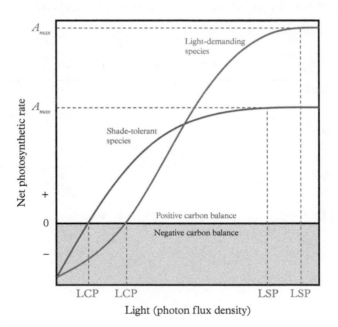

Figure 6.18 *Theoretical light response curve illustrating how net photosynthetic rate is a positive yet saturating function of light availability. The light compensation point (LCP) is the light intensity at which net photosynthetic rate equals zero. The light saturation point (LSP) is the light intensity at which the rate of photosynthesis is maximized and becomes saturating (A_{max}). Shade-tolerant species exhibit lower LCP and LSP than light-demanding species, which sets up a trade-off between survival in shade and growth rate in high light conditions. Net growth can also be quantified for whole plants, which is likely the more ecologically relevant scale.*

2013), but one study demonstrated that species with low whole-plant LCP exhibited lower mortality rates in the shade but slower growth rates in full light, suggesting that this trait is a major driver of the growth-survival trade-off in forests (Lusk and Jorgensen, 2013).

The most abundant protein on Earth is Ribulose-1,5-bisphosphate carboxylase-oxygenase (or Rubisco for short), and this protein is the engine that drives photosynthesis in the leaf (Raven, 2013). This stunning fact alone demonstrates the pivotal role of plants in the biosphere. Rubisco has a problem, however: at high temperatures, Rubisco can have a lower affinity for CO_2 and can instead take up O_2 in a process called photorespiration, which is wasteful. Typically, when a leaf opens its stomata, CO_2 diffuses into the leaf and O_2 and water vapor diffuse out. Under these conditions photorespiration is minimized. However, when a plant closes its stomata to reduce water loss by evaporation when it is hot, O_2 from photosynthesis builds up inside the leaf and photorespiration increases due to the higher ratio of O_2 to CO_2. At mild temperatures, Rubisco's affinity for CO_2 is about 80 times higher than its affinity for O_2, but at high temperatures Rubisco's affinity for O_2 increases (Brooks and Farquhar, 1985). The C_4 photosynthetic pathway evolved independently dozens of times as a solution to this problem. In C_4 plants, the light and light-independent reactions occur in separate cells and another enzyme, PEP carboxylase, maintains high rates of CO_2 uptake even at high temperatures. This gives C_4 plants greater water-use efficiency and allows them to grow in the late summer heat. C_3 plants are often called cool-season plants and grow best in late spring/early summer, whereas C_4 plants are warm-season plants and grow best in the long hot days of late summer. CAM photosynthesis uses the same biochemical machinery as C_4 plants except CAM plants temporally partition the light and light-independent reactions. CAM plants close their stomata during the heat of the day to prevent water loss and only open them at night to uptake CO_2. This strategy makes succulent CAM plants well adapted to hot and dry deserts. Table 6.1 lists the differences among C_3, C_4, and CAM plants.

6.8 Phenological variation permits the partitioning of time

C_3, C_4, and CAM species distributions vary across space, but they also exhibit phenological variation over time (Craine et al., 2012b). C_3 species are often called early-season species because they leaf out and flower in spring and early summer, whereas C_4 species are called late-season species because they produce most of their new growth in later summer and can flower late into the autumn months (Figure 6.19). Phenological traits are included in the very definition of functional traits (Violle et al., 2007; Garnier et al., 2016), yet phenological traits are less commonly included in large trait datasets and analyses. Post (2019) proposed that phenological traits are adaptive strategies that

Table 6.1 *Traits of plants that use C₃, C₄, and CAM photosynthesis.*

Trait	C$_3$	C$_4$	CAM
Leaf anatomy	Laminar meso-phyll, chloroplasts concentrated in mesophyll	Mesophyll arranged radially around bundle sheaths ("Kranz"-type anatomy)	Cells have large vacuoles
Chlorophyll a:b ratio	~ 3	~ 4	≤ 3
Primary CO$_2$ acceptor	RUBP	PEP	In light: RuBP In dark: PEP
First product of photosynthesis	C$_3$ acids (PGA)	C$_4$ acids (oxaloacetate, malate, aspartate)	In light: PGA In dark: malate
Carbon isotope ratio in photosynthates (δ^{13}C)	−20 to −40‰	−10 to −20‰	−10 to −35‰
Photorespiration	Yes	No	No
Net photosynthetic capacity	Slight to high	High to very high	In light: slight In dark: medium
Light saturation of photosynthesis	At intermediate intensities	No saturation, even at highest intensities	At intermediate to high intensities
Dry matter production	Medium	High	Low
Optimum temperatures	5–35°C	30–45°C	10–25°C (at night)
Seasonality	Cool-season	Warm-season	

Reproduced from Larcher (2003) with permission.

allow organisms to capture and partition the resource of time. Early-season species may deplete water that is available in spring, but this won't influence water supply in late summer when late-season species are active.

It is easy to define phenological traits, such as date of first flowering, within a given climate zone and seasonality, but it has proven much more difficult to define phenological traits that are sensible to compare across all climate zones worldwide, and it is an important unsolved problem. Phenological traits such as pollination time, flowering time, bud burst, leaf out time, and peak biomass production differ across species and have proven to be sensitive indicators of climate change (Panchen et al., 2014; Wolkovich and Ettinger, 2014; Piao et al., 2019). Flowering time among trees has been shown to vary strongly along climatic gradients where early pollination occurs in warmer elevations with longer growing seasons (Laughlin et al., 2011a). Change in climate is likely to lead

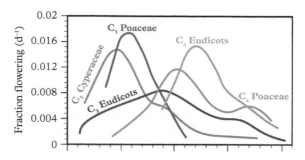

Figure 6.19 *First flowering dates expressed as a fraction of all recorded species across five sets of North American prairie species grouped by photosynthetic pathway and phylogeny.*
Reproduced from Craine et al. (2012b) with permission.

to phenological reassembly, where species-specific shifts in phenology could ultimately lead to mismatches between flowering time and pollinator activity. A study conducted on Mt. Rainier demonstrated that timing and duration of flowering was sensitive to timing of snowmelt and temperature, and the authors observed a reassembly of phenology in a year that represented projected climate conditions in the 2080s (Theobald et al., 2017).

Phenological traits relate to other key trait dimensions such as leaf economics and height. In one study, flowering time and duration appeared to be strongly correlated with maximum height and unrelated to leaf economics, where taller species flowered later in the season for shorter durations (Laughlin et al., 2010). Using this trait data, Laughlin and colleagues (2011b) found that a century of increasing tree densities in Arizona ponderosa pine forests was associated with a shift toward shorter and earlier-flowering C_3 species in the understory. Flowering time was also positively related to reproductive height across annuals and perennials in a Mediterranean flora in the southern France (Segrestin et al., 2020). Many studies have linked functional traits to shifts in phenology in response to climate change (König et al., 2018), suggesting that as more data are assessed we may find that many phenological traits will be correlated with commonly measured traits. Invasive plants tend to be fast-growing, early-flowering species, which has implications for how invasions will advance under climate change (Wolkovich and Cleland, 2014).

6.9 Seeds enable reproduction on dry land

We have worked our way from the roots through the shoot and have now arrived at the reproductive structures. Floral traits are just beginning to be fully integrated into discussions of plant strategies and are clearly important for rates of reproduction (Maglianesi et al., 2014). Like leaves, flowers have production and maintenance costs, but the paybacks are in terms of seed production and eventual recruitment (Roddy et al., 2021). Floral morphology can have strong effects on reproduction, but these traits have been difficult to generalize across species because different plant families exhibit different sets of floral parts. For evidence, simply compare the diversity of floral traits among the Asteraceae, Poaceae, and Orchidaceae. Recently, Paterno and colleagues (2020, p. 10921) showed that "heavier angiosperm flowers tend to be male-biased and invest strongly in petals to promote pollen export, while lighter flowers tend to be female-biased and invest more in sepals to insure their own seed set." Flower biomass may prove to be a useful trait for comparing reproductive allocation among species. Much more work has been conducted on seeds.

The innovation of seeds revolutionized plant reproduction because fertilization was no longer limited to wet environments, and it enabled the young plant to travel in a dormant state to find more favorable opportunities for colonization. A spore-dominated world gave way to a seed-dominated world after the radiation of seed plants into the terrestrial biosphere. Seeds are not organs: they are young dormant plants protected by a hard outer coat. It is challenging to be accurate with terminology, as evidenced by the ribbing I get from my systematist colleagues when I carelessly throw around the word "seed" in the field. What many might casually call a seed are actually fruits. Grass seeds are fruits, and sometimes fruits are the dispersal unit. For the purposes of this book, we are primarily interested in the diaspore, which is the plant dispersal unit consisting of a seed or spore plus tissues that aid in dispersal. Seed morphology varies across plant phylogenies (Figure 6.20). All seeds have a seed coat (testa), but the energy for the developing seedling may be contained within the embryo, endosperm, or perisperm. The mass of this energy reserve contributes to the great variation in size among seeds.

Seed size variation is enormous, varying by at least 11 orders of magnitude (Moles et al., 2005b). Small-seeded species produce more seeds per unit canopy area per unit time than large-seeded species (Figure 6.21). Hence, there is a general trade-off between seed size and output, aligning with general life history theory (Stearns, 1992). Note that this linear trade-off is not strong in fleshy fruited tropical species (Grubb, 1998b). In fact, small plants can only make small seeds due to biophysical constraints, but large plants produce a wide variety of seed sizes, which suggests that Charnov's (1993) idea that size of infant animals are inevitably coupled with adult size is less relevant to plants (Grubb et al., 2005). Large-seeded species invest more into energy reserves than small-seeded species, which contributes to greater seedling establishment rates in resource-limited environments (Jakobsson and Eriksson, 2000; Westoby et al., 2002). However, Moles and Westoby (2006, p. 91) conducted a meta-analysis of seed mass and demographic data and discovered that the theory of a

> trade–off between producing a few large offspring, each with high survival probability, versus producing many small offspring each with a lower chance of successfully establishing was incomplete. It seems more likely that seed size evolves as part of a spectrum of life history traits, including plant size, plant longevity, juvenile survival rate and time to reproduction.

(a)

(b)

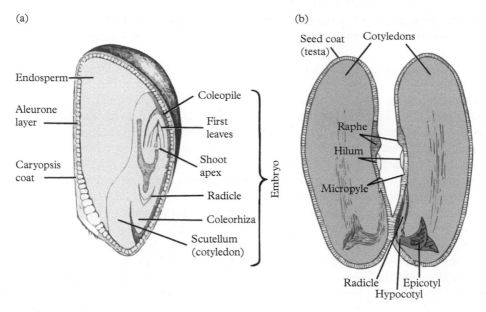

Figure 6.20 *Seed anatomy for (a) a monocot and (b) dicot.*
Adapted from Rost et al. (1984) with permission.

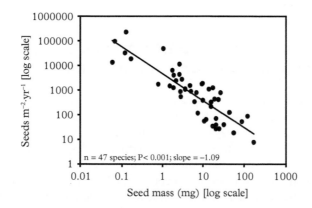

Figure 6.21 *Relationship between seed mass and seed output (number) per square meter of canopy per year among plant species measured in Ku-ring-gai Chase National Park, Sydney, Australia.*
Reproduced from Henery and Westoby (2001) with permission.

On the basis of a positive relationship between plant height and lifespan, you can expect a positive relationship between seed mass and lifetime reproductive output (Moles and Leishman, 2008), and the trade-off between seed size and output may only hold for a fixed amount of energy devoted to seed production. Small seeds can be borne on plants of any size, but small plants cannot support large seeds (Grubb, 1998b). This suggests that strategies appear when viewing traits at the whole-plant level. When evaluating the dimensionality of plant traits synthetically, seed size tends to be positively correlated with plant stature (recall Figure 2.15).

The diversity of seed sizes within any plant community is considerable and is proposed to reflect a competition–colonization trade-off that maintains coexistence in species-rich communities (Tilman, 1994). However, competitive asymmetry must be quite strong to explain coexistence (Levine and Rees, 2002), so additional forms of trait-based niche differentiation are likely needed to maintain coexistence in the long term (Coomes and Grubb, 2003). The evolution of seed mass strategies within communities is discussed further in Chapter 9 (Rees and Westoby, 1997).

Vegetative plants are sedentary, but they can travel considerable distances as seeds. Small seeds can travel far distances with the help of the wind, but large-seeded palm trees can disperse their seeds massive distances with the help of water. The dispersal agent is a major determinant of dispersal distance (Van der Pijl, 1982). Fruits adorned with wings can travel farther in the wind, but generally speaking, light seeds with low terminal velocities will travel farthest in the wind. Fruits encased in fleshy pericarp will attract birds and mammals to be eaten and hopefully dispersed away from the mother plant. Dispersal distance for wind-dispersed seeds can be modeled as a function of releasing height, wind speed, and descent velocity (Greene and Johnson, 1989). A meta-analysis demonstrated that plant height was a far stronger predictor of dispersal distance than seed mass, but after accounting for height, smaller seeds did disperse farther than larger seeds (Thomson et al., 2011).

6.10   Five plant strategy spectrums explain global variation in plant form and function

So far, in this chapter we explored global variation in roots, clonal growth organs and bud banks, stems, leaves, and seeds, but each of these categories of traits can be measured in a variety of different ways. The anatomical diversity of cell and tissue structure within these organs deserves study by its own right, but this approach is not tractable on our quest to ascertain the dominant dimensions of plant form and function at a global scale. At this stage in the history of plant ecology, we have focused on comparing general morphological differences such as LMA in leaves, tissue density in stems, and mass of seeds. These traits may gloss over the cellular details, but they represent key trade-offs in plant function. The good news is that we can now compare thousands of species worldwide. The global quantification of plant traits flourished when plant strategies began to be viewed as continuous spectrums (Parsons, 1968; Grime and Hunt, 1975; Westoby, 1998). In fact, we can now compare LMA across 16,460 species using the TRY plant trait database (Kattge et al., 2020)! LMA is a popular trait to measure because it is an indicator of the economic strategy of the plant and because it is a joy to measure; it is tangible. But one inherent challenge of this global research program is the operational trade-off between measuring easy traits on many species versus measuring more difficult traits on just a few (Garnier et al., 2016). This trade-off inherently biases our understanding of the primary strategy axes.

There are traits that are just as informative but far scarcer in the plant trait databases. Xylem vulnerability to embolism is the key trait for measuring drought resistance in woody plants, but this trait requires specialized knowledge and equipment, and so it has only been measured on fewer than 2,000 species globally (Choat et al., 2012). This trade-off generated by the ease of trait measurement constrains comparative plant ecology and limits global analyses to traits that have been measured on the most species. As we explore the primary axes of functional variation in plants, we need to be cognizant of the fact that some traits could be just as, or even more important, but we haven't measured them on enough species yet to know.

Global scale compilations of plant traits made it possible to start estimating the dimensionality of the plant phenotype (Kattge et al., 2011; Kattge et al., 2020). Recall from Chapter 3, the intrinsic dimensionality of plant traits represents the number of independent axes of functional variation among plants and is therefore a fundamental quantity in comparative plant ecology (Laughlin, 2014a; Mouillot et al., 2021). The largest global-scale analysis of the dimensionality of plant traits evaluated the covariation among six aboveground functional traits, including adult plant height, stem specific density, leaf size, leaf mass per area, leaf nitrogen concentration, and diaspore mass (Díaz et al., 2016). Recall how important these traits were in the discussion of land plant evolution. Figure 6.22 illustrates the distributions of these variables, and several more. Díaz and colleagues' (2016) landmark paper was one of the first major achievements of the founding members of the TRY plant trait database, and it included traits measured on no less than 46,085 vascular plant species!

They subjected the trait data to a dimensionality reduction technique called principal components analysis (PCA), which uses eigenanalysis to summarize the correlation structure of a multivariate dataset and can be used to determine how many independent dimensions of variation exist in the data (Johnson and Wichern, 1992). For example, consider a dataset containing three variables. If two variables were strongly correlated with each other but both were uncorrelated with the third variable, then the dataset would be two-dimensional, not three-dimensional. It is important to recognize that the dimensionality of traits in such analyses is a feature of the dataset (our finite sample), not necessarily the population of interest (all plant phenotypes) because trait data is limited.

Díaz et al., (2016) analyzed six aboveground traits and concluded that plants span a two-dimensional plane (Figure 6.23). The first dimension represented variation in plant size and included the following traits: adult height, leaf area, wood density, and seed mass. Given the theory that links body size to variation in life history in animals (West et al., 1997), it is not surprising that size is an important dimension for plants (Enquist et al., 1998). Body size has been shown to scale negatively with metabolic rates in organisms, suggesting it is a powerful integrator of multiple phenotypic traits and it has long been studied by life history theorists (Enquist et al., 1999). The concept of plant size is a bit slippery because it is notoriously difficult to measure given that there are many dimensions of size. Stem girth, total height, canopy diameter, and total biomass are all defensible measures of plant size, so which one is best? Plant vegetative height at reproductive maturity is an important trait that provides information about life history strategies, but maximum height is a more commonly measured trait. Maximum height is

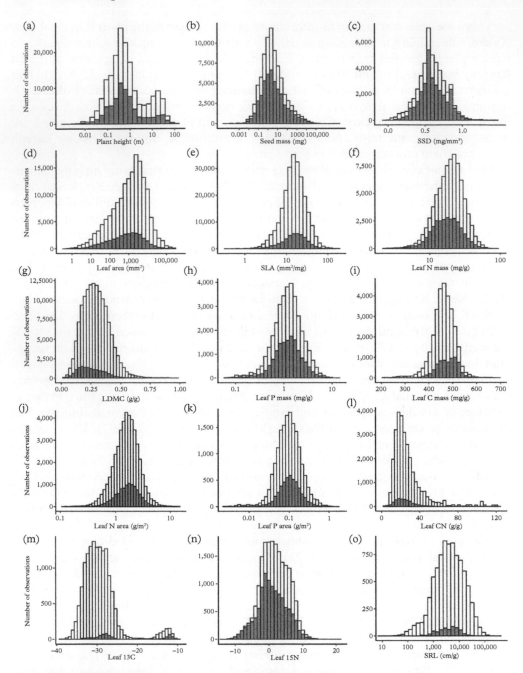

Figure 6.22 *Frequency distributions and range of variation of the most commonly measured continuous plant functional traits. The dark bars represent the number of observations in 2011, the white bars represent the number of observations in 2020. Note that the total number of observations per trait decreases from top to bottom, and root traits are far under-sampled compared to leaves.*

SSD = stem specific density, SLA = specific leaf area, N = nitrogen, LDMC = leaf dry matter content, P = phosphorus, C = carbon, CN = carbon-to-nitrogen ratio, 13C = isotopic composition δ^{13}C, 15N = isotopic composition δ^{15}N, SRL = specific root length.
Reproduced from Kattge et al. (2020) with permission.

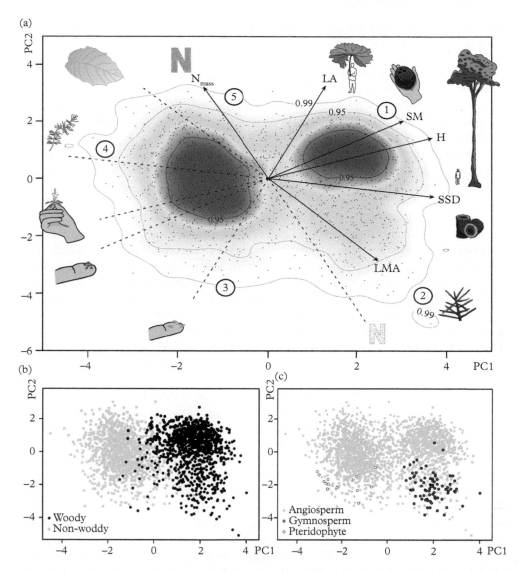

Figure 6.23 *Two of many global spectrums of plant form and function. The first dimension represents variation in plant size and the second dimension represents variation in leaf economics. Note the strong differences between (b) woody and non-woody species and between (c) angiosperms and gymnosperms.* Reproduced from Díaz et al. (2016) with permission.

clearly important given the asymmetric competition for light (Keddy and Shipley, 1989; Westoby, 1998), but maximum height at maturity, the speed at which maximum height is attained, and the length of time a species maintains its maximum height all have costs and benefits in their measurement (Westoby et al., 2002). Seeds vary in their ability to disperse away from the parent plant, successfully germinate, and become established seedlings (Grubb, 1977), and the large variety of seed sizes and shapes is indicative of the range of regeneration strategies in plants. Seed mass and seed shape can sometimes, although not always, influence persistence in the seed bank (Thompson et al., 1993; Moles et al., 2000). Stems provide structural support in the gravity-laden terrestrial environment, they transport water, nutrients, and sugars, and they can be important for defense and storage. Stem density (i.e., specific gravity) is an important property of plant stems that can represent a trade-off between the efficiency of hydraulic conductivity and resistance to drought—or freezing-induced cavitation (Hacke et al., 2001; Baas et al., 2004). Stem density also reflects a trade-off between growth rate and survival (Chave et al., 2009; Wright et al., 2010), but note that wood density and maximum height are not typically correlated among tropical trees (Rüger et al., 2018). There is clearly a lot going on in this first dimension.

The second dimension represented variation along the leaf economics spectrum. Leaf nitrogen concentration and SLA described variation along this axis. The leaf economics spectrum has perhaps become the most well-known dimension of plant function, describing a trade-off between leaf lifespan and rate of carbon acquisition (Reich et al., 1997; Reich et al., 1999; Wright et al., 2004), and is the clearest indicator of a plants' location along the continuum from conservative to acquisitive phenotypes. One important feature of this analysis also highlighted two hotspots within the plane, representing a contrast between herbaceous and woody plants. Theophrastus smiled when he read the news in *Nature* that week.[4] These two principal dimensions have also been shown to be explained by climate and soil at the global scale: the size axis is mostly explained by climatic variation along a latitudinal gradient, and the economics axis is explained by climate-soil interactions (Joswig et al., 2022).

Díaz and colleagues' (2016) impressive analysis set a high bar for the number of species that should be compared in functional ecology. However, it was also slightly disappointing to those of us who have long considered roots to be rather fundamental to plants. Roots were left out for good reason: so few data on root traits existed at that time, so it was not possible to integrate roots into the analysis and retain a large number of species. Tests for the dimensionality of plant traits face a fundamental challenge: increasing the number of traits always reduces the number of species on which all traits have been measured. Gaps in the dataset arise with increasing numbers of species. It is popular to fill these gaps in the data using algorithms that predict the traits given trait correlations, environmental relationships, phylogenetic constraints, or all three sources of information (Schrodt et al., 2015). However, I always encourage my students to measure traits on all the species in their study system and not to rely on global databases for three reasons: to eliminate the gaps, to increase the accuracy, and to increase the global distribution of traits!

Reich (2014) boldly hypothesized that economic trade-offs should hold across the entire plant body—that is, rates of physiological processes in stems and roots should also trade-off with construction costs of stem and root tissue. Early tests of Reich's whole-plant economic spectrum hypothesis were promising (Freschet et al., 2010), but

---

[4] Recall from Chapter 2 that the Greek philosopher Theophrastus is thought to have proposed one of the first plant life form classifications: trees, shrubs, sub-shrubs, and herbs (i.e., literally "grasses").

others started to indicate that root tissue density and nitrogen concentrations were correlated with leaf economic traits but that root diameter and specific root length were independent of the leaf economics spectrum (Kramer-Walter et al., 2016). The story has become even more complex over time. Leaf economic traits align along a single axis because they primarily have one important function—to acquire light and fix carbon. Roots, on the other hand, have multiple functions, including storage, anchorage, transport, and support, so it stands to reason that the traits of these organs will exhibit higher dimensionality (Laughlin, 2014a). Root trait ecologists convened a workshop in Leipzig to test this result on a global dataset, and more importantly, to provide a reasonable explanation for it. They compiled a global root trait database (Iverscn ct al., 2017; Guerrero-Ramírez et al., 2021), and focused their attention on the four most commonly measured fine root traits: root diameter, specific root length, root tissue density, and root nitrogen concentration.

Fine root traits span at least two dimensions (Figure 6.24) (Bergmann et al., 2020). The first dimension has been called the "collaboration axis", which is defined by a trade-off between specific root length and root diameter that evolved in concert with symbiosis with mycorrhizal fungi. Thick-rooted species "outsource" their soil resource acquisition,

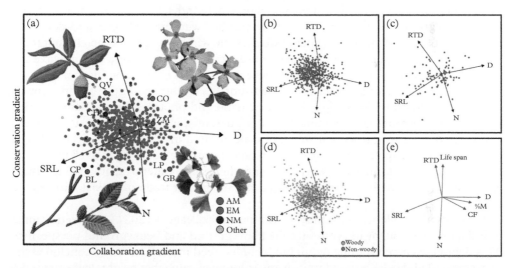

Figure 6.24 *The four most commonly measured fine root traits span two dimensions. The first dimension represents a collaboration axis (a contrast in root diameter [D] and specific root length [SRL]) and the second dimension represents a conservation axis (a contrast in root tissue density [RTD] and root nitrogen concentration [N]). Tree species are illustrated that represent different corners of the trait space (clockwise from upper left):* Quercus virginiana, Cornus officinalis, Ginkgo biloba, *and* Betula lenta. *The two-dimensional model holds within (b) species that host AM fungi, and (c) species that host EM fungi. (d) Non-woody species tend toward the high SRL and high RTD corner but are also represented throughout the space. (e) Root lifespan is most correlated with RTD, and cortex fraction (CF) and % mycorrhizal colonization (%M) are most strongly correlated with root diameter.*

AM = arbuscular mycorrhizal, EM = endomycorrhizal, NM = nonmycorrhizal.
Reproduced from Bergmann et al. (2020) with permission.

whereas "do-it-yourself" species are exemplified by those with the thinnest roots that can explore the soil and acquire resources by themselves. Plants species that form cluster roots (also called proteoid roots, given their prevalence in the Proteaceae) are extreme examples of plants that mine limiting nutrients such as phosphorus by proliferating a high density of extremely thin roots (Laliberté et al., 2015). Root traits span a second independent dimension called the "conservation axis", where conservative species invest in high root tissue density (fine root mass per unit volume), and acquisitive species construct more metabolically active tissue with low root tissue density and high root nitrogen concentration (Kramer-Walter et al., 2016; Weemstra et al., 2016; Bergmann et al., 2020).

The critical question that follows from Bergmann and colleagues' (2020) proposal immediately becomes how the root economic space relates to the aboveground trait dimensions. When we combine aboveground and belowground traits, how many independent dimensions remain? One recent synthesis did just that and indicated that at least four dimensions remain (Weigelt et al., 2021): the leaf and root economics spectrum (PC1), the belowground collaboration gradient (PC2), plant height aboveground (PC3), and another belowground dimension of rooting depth (PC4) (Figure 6.25). Previous studies (Schenk and Jackson, 2002; Garnier et al., 2016) have suggested that rooting depth was correlated with plant height, where forbs, grasses, and trees that were taller aboveground also extended roots deeper belowground. Weigelt and colleagues (2021) hypothesized that rooting depth would scale with aboveground height, but found that these two traits aligned on independent dimensions. There were two possible reasons for this. First, greater height can achieve more access to light, but deeper roots can be important for water acquisition, nutrient uptake, and anchorage. Second, growing tall increases transpiration-induced water demand and photosynthesis-related nutrient demand, but growing deeper roots does not guarantee improved access to water if it is not limiting or to nutrients if they are concentrated in shallow soil layers (Weigelt et al., 2021). Therefore, rooting depth should be unrelated to light availability, and rooting depth should be shallow in sites with either high water availability, high nutrient availability, or both. It is still unclear what other trait trade-offs exist with rooting depth.

In yet another independent study, Carmona and colleagues (2021) analyzed a similar trait dataset that sought to synthesize the aboveground and belowground perspectives. This dataset excluded rooting depth but included seed mass. The study concluded that aboveground and belowground traits were orthogonal, such that height, leaf economics, root collaboration, and root conservation were independent dimensions. Slightly different results are obtained when a similar analysis was conducted on a large dataset of 1467 tropical tree species. Vleminckx and colleagues (2021) found that the most important dimension was related to root diameter, the second and third dimension were related to leaf economics, and the fourth dimension was related to wood density. This study also emphasized the importance of bark thickness—a trait often disregarded in global datasets that include herbaceous plants. The importance of bark thickness in ecosystems that do not regularly burn is not well understood. One strong pattern, however, is that species with thick bark are often found in infertile soil (Jager et al., 2015; Richardson

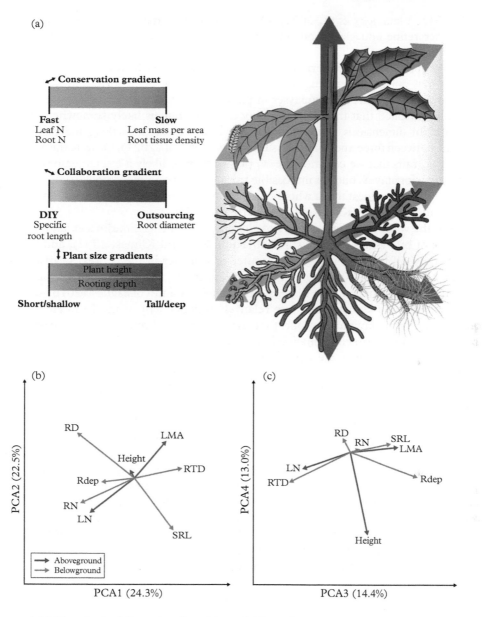

Figure 6.25 *Hypothesized dimensionality of the multidimensional plant phenotype. (b) Empirical dimensions of plant traits: the leaf and root economic spectrum (PC1), the root collaboration gradient (PC2), rooting depth (PC3), and plant height at maturity (PC4).*

RD = root diameter, SRL = specific root length, RN = root nitrogen concentration, RTD = root tissue density, LN = leaf nitrogen concentration, Rdep = rooting depth, LMA = leaf mass per area, Height = maximum height.
Reproduced from Weigelt et al. (2021) with permission.

et al., 2015; Vleminckx et al., 2021). Only as more trait data is made open access will we further refine and clarify our understanding of the dimensionality of these plant traits.

Are we missing other important traits? I am sure we are! It would be presumptuous to think that our understanding of the dimensionality of plant form and function will remain stable. Previous meta-analyses of plant trait matrices and models of community assembly indicated that the dimensionality of plant traits is likely somewhere between four to eight dimensions (Laughlin, 2014a). A more recent analysis indicated that it might be between three and six dimensions (Mouillot et al., 2021). Despite the enormous number of traits that we can measure on plants, there is likely a tractable upper limit to their dimensionality. While it is impossible to know the true dimensionality without larger global datasets, I suggest that a whole class of traits have still been ignored: clonal growth organs and bud banks (Klimešová and Herben, 2015).

Given the widespread importance of clonality in plants (Klimešová et al., 2018), the relevance of bud banks to plant survival (Klimešová and Klimeš, 2007), and the well-known trade-off between being a resprouter or a seeder in the context of disturbance regimes (Clarke et al., 2013; Pausas and Keeley, 2014), I propose to include a seeder–resprouter spectrum as a fifth dimension. Klimešová and colleagues (2016) indicated that bud bank and clonal traits were relatively independent of the Leaf–Height–Seed traits, providing preliminary support for this prediction. Admittedly, there are fewer data on this axis, and our best information is concentrated in Europe. Thankfully, researchers in tropical grassy biomes are contributing rapidly to this base of knowledge because resprouting is a critical adaptation to drought and fire in the savannas of South America and Africa (Fidelis et al., 2014; Filartiga et al., 2017; Nolan et al., 2021; Pilon et al., 2021).

Five strategy spectrums, or six or seven, will not explain all the variation in plant traits worldwide. The only way to capture all the variation in traits is to include all traits in our models. But we seek a lower dimensionality that captures the most important variation. For those of us that prefer to dwell in the unique corners of trait space and revel in the unique exceptions, this approach will be unsatisfying. But to those of us that seek generality—that blur our vision and let go of the rich details of natural history to discover something larger that unifies our understanding of plant ecology and evolution—then determining the dimensionality of traits will continue to be a central goal. What is most important, is that traits need to be chosen carefully and included using substantive arguments. Figure 6.26 summarizes the fundamental trade-offs that each functional trait is thought to reflect.

6.11   Do functional traits determine life history?

Let us reflect on how the multidimensional phenotype determines the life history of the plant. Recall from Chapter 4 that vital rate elasticities quantify the importance of a vital rate to population growth rate (Silvertown et al., 1993). Adler and colleagues (2014) analyzed vital rate elasticities across a large number of plant species that occurred

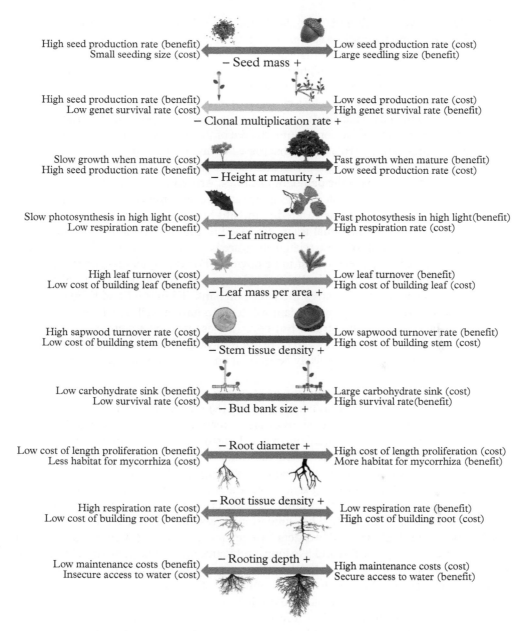

Figure 6.26 *Costs and benefits for each end of key functional trait spectrums. Each trait represents fundamental trade-offs faced by plants seeking to grow, survive, and reproduce on earth. See discussion of each of these proposed trade-offs throughout this chapter.*

in both the COMPADRE and TRY databases and asked whether the importance of survival, growth, and fecundity to population growth rate could be explained by variation in functional traits. They showed that species with larger seed mass had higher survival elasticities, and that species with higher SLA and leaf nitrogen concentration had higher fecundity elasticities. These results demonstrated that functional traits do underlie life history differences, including the fast–slow continuum.

According to the largest analysis to date, the three most important spectrums of life history traits are the fast-slow continuum, degree of iteroparity, and (possibly) lifespan (Salguero-Gómez et al., 2016). Recall that these life history traits are computed directly from life tables and projection matrices, and the life tables and projection matrices are built directly from stage-specific demographic rates. Arnold (1983), and later Violle and colleagues (2007), proposed that demographic rates are determined by phenotypic traits. Therefore, it should be possible to predict the life history strategy of a species based on the combinations of functional traits expressed in the multidimensional phenotype (Figure 6.27). The details of these relationships are best worked out with empirical data, but some relationships appear at first to be obvious. For example, lifespan is positively related to plant height (Moles et al., 2009), but lifespan is also enhanced by possession of well-developed perennating organs and a large bud bank (Klimešová and Klimeš, 2007). Likewise, a short-lived fast-growing plant will tend to have a well-developed seed bank and exhibit fast leaf economic traits. But beyond these general statements, it becomes clear that each life history strategy can be realized through many different combinations of functional traits, and these are likely dependent on the environment in which the plant is growing.

A different set of trade-offs are important when the focus is on tree demographic rates in tropical forests. Rüger and colleagues (2018) analyzed long-term demographic data and proposed that two dimensions of life history traits describe neotropical forest dynamics driven by gaps created by tree fall disturbances and that these life history dimensions are explained by functional traits (Figure 6.28). The first dimension recognizes the broad differences between fast-growing species that dominate early successional stages, and slow-growing shade-tolerant species that reach dominance at later successional stages. However, the authors also describe a second dimension called the "stature-recruitment" trade-off. This dimension represents a distinction between "long-lived pioneers" and "short-lived breeders." Rüger and colleagues (2020, p. 165) say that long-lived pioneers "grow fast and live long, and hence attain a large stature, but exhibit low recruitment," whereas short-lived breeders "grow and survive poorly, and hence remain short-statured, but produce large numbers of offspring." Their key insight is that both demographic dimensions were needed to obtain accurate predictions of forest succession using the PPA model of forest dynamics (Purves et al., 2008). Like Noble and Slatyer (1980) before them, Rüger and colleagues (2020) use the information contained in the life history traits to predict basal area and compositional changes that occur in a Neotropical forest on Barro Colorado Island, Panama, driven by the gap dynamics of tree fall and the competition for light. Rüger et al., (2018) demonstrated that these two demographic dimensions can be explained by functional traits: wood density and LMA explained variation in a growth-survival trade-off, whereas height and seed mass explained variation in the stature-recruitment axis.

Functional strategies   Life history strategies

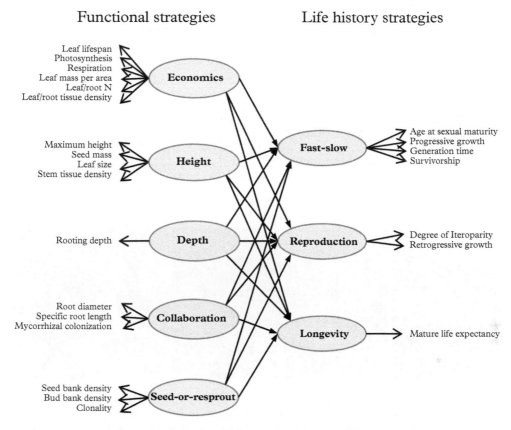

Figure 6.27 *Hypothesized directional effects of functional traits (Klimešová and Herben, 2015; Díaz et al., 2016; Bergmann et al., 2020; Weigelt et al., 2021) on life history traits (Salguero-Gómez et al., 2016). The leading dimensions of functional variation are represented as ovals, which point to the directly measurable variables.*

Kelly and colleagues (2021) set out to examine these relationships at a global scale and proposed a framework for understanding the drivers of life history strategies as interactions between functional traits and environments. They merged the TRY dataset of plant traits with the COMPADRE dataset of matrix population models to evaluate the framework with data. Despite the size of these global datasets, only 80 species could be included in their study. It is easy to see how taller plants can attain longer lifespans, but height interacted with climate in this relationship (Figure 6.29). In cold climates, short cushion plants can attain long lifespans and generation times even though they are extremely short, but the longest living plants and the longest generation times in warm climates are exhibited by the tallest species. Similar interactions occurred between the distribution of mortality across the lifespan and climate. High juvenile mortality was observed in large plants in hot and dry climates, but juvenile mortality in small plants were more common in cool and wet climates. These results suggest that phenotypic

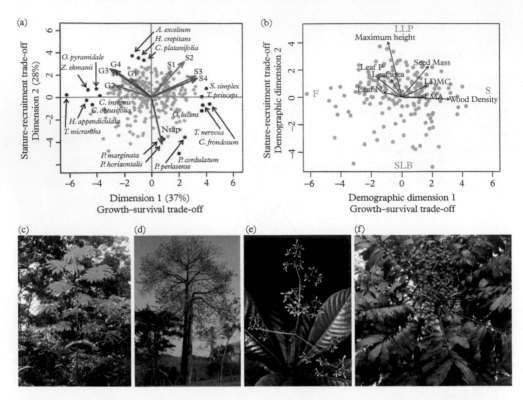

Figure 6.28 *Life history trade-offs for 282 trees in the tropical rain forest of Barro Colorado Island, Panama. The growth–survival trade-off (dimension 1) was orthogonal to a stature-recruitment trade-off (dimension 2), and together explained 65% of the variation in demographic traits. G1–G4 and S1–S4 represent growth rates and survival rates, respectively, in different layers of the canopy. (b) Correlation vectors of functional traits superimposed on the life history space indicate that wood density and LMA loaded strongly on the growth-survival axis and maximum height and seed mass loaded onto the stature-recruitment axis. The acronyms in green are the names of the strategies at the end of each axis. (c)* Cecropia obtusifolia *is a "fast" (F) species. (d)* Cavanillesia platanifolia *is a "long-lived pioneer" (LLP) species. (e)* Psychotria marginata *is a "short-lived breeder" (SLB) species. (f)* Talisia princeps *is a "slow" (S) species.*

LMA = leaf mass per area, LDMC = leaf dry matter content.
(a, b): Reproduced from Rüger et al. (2018), with permission. (c–f) Photos of species that represent each end of the demographic axes. (c): Photo provided by Rolando Pérez and Richard Condit with permission. (d): Photo provided by Rolando Pérez with permission. (e): Photo provided by Steven Paton with permission. (f): Photo provided by Rolando Pérez with permission. Permission to use these photos was provided by the Smithsonian Tropical Research Institute (https://stri.si.edu/).

effects on life history patterns depend on the environmental conditions. Other studies have shown that hydraulic traits predict the growth–survival trade-off in tropical forest species (Oliveira et al., 2021). It is important to continue to test these relationships as data on traits, demography, and environmental conditions are synthesized.

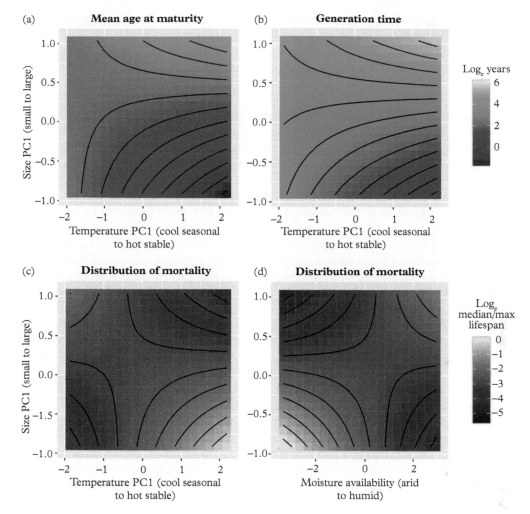

Figure 6.29 *Interactions between climate (x-axes) and functional traits (y-axes) explain some variation in life history traits (color and contour of surface).*
Reproduced from Kelly et al. (2021) with permission.

6.12 Phylogenetics enriches our understanding of trait spectrums

Traits are heritable properties that are passed onto future generations. Therefore, species that are close cousins on the evolutionary tree of life may share similar traits because of their relatedness, as well as because they are adapted to similar environments. This tendency for species to retain ancestral traits is called phylogenetic conservatism and is a persistent problem that must be addressed in comparative plant ecology. There are two perspectives on this problem.

Comparative plant ecology considers species to be evolutionary experiments in phenotypic variation to (1) explore patterns of correlations between traits to understand phenotypic constraints and trade-offs, and to (2) explore relationships between traits and environments to understand the adaptive value of traits. We have explored the former in this chapter, and we explore the latter in later chapters. One of the assumptions in statistics is that the sample units of our analysis must be independent replicates. However, evolution is a branching process and species that are close together on the evolutionary tree are more likely to share similar phenotypic traits because they are more closely related. This biological fact means that species are not strictly independent and if we treat them as independent we violate a statistical assumption (Felsenstein, 1985). Importantly, the residual error of the models are correlated with the phylogeny. The argument is that, for plant strategies to represent convergent evolutionary solutions to common environmental problems, they must be valid after accounting for evolutionary relatedness. Comparative methods have made great advances in the last few decades to improve our ability to account for evolutionary relatedness in the search for convergent solutions.

There are two equivalent approaches to account for evolutionary relationships in analyses of trait covariation: phylogenetic independent contrasts (PICs) and phylogenetic generalized least squares (PGLS) regression estimators (Blomberg et al., 2012; Revell and Harmon, 2022). This former involves calculating contrasts as the difference between trait values of the two species that branch from a common node and assigning trait values to the internal nodes of the tree (Felsenstein, 1985). This results in the loss of one degree of freedom in the analysis and is more difficult to interpret because you are no longer analyzing the N species in your dataset, but rather $N-1$ independent contrasts that represent common ancestors. PGLS regression estimators for phylogenetic analyses were developed more recently and account for relatedness by modeling the residuals of the regression model as a function of phylogenetic distances (Revell, 2012). This has the effect of accounting for evolutionary relatedness at the same time you test for trait–trait and trait–environment correlations, and it also has the advantage of preserving the species in your sample (rather than analyzing common ancestors).

Let us consider a simple and admittedly contrived example to make the point clear. I always find it instructive to consider ancient evolutionary divergences, such as between gymnosperms and angiosperms, or even more recent radiations, such as between monocots and dicots, when trying to understand the evolution of continuous traits. Consider a small dataset of six species, three of which are gymnosperm trees and three of which are angiosperm grasses. Imagine that we want to know whether traits are correlated across the species to see whether groups of species represent sets of correlated traits, and therefore convergent strategies. If we evaluated the relationship between height and SLA across these six species, what would we find?

To answer this question we need two sets of information: trait data and the phylogenetic tree for these species. The phylogenetic tree shows us that the conifers and grasses each form separate polytomies, that is, they belong to different clades (Figure 6.30). This might not matter if variation in traits is greater across than within clades. When we

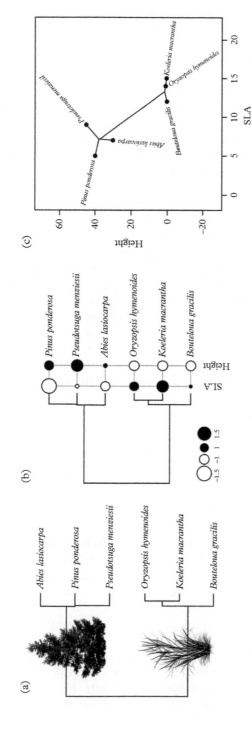

Figure 6.30 (a) Phylogenetic tree illustrating the evolutionary relationships among six species (three conifers at top and three grasses below). (b) Specific leaf area (SLA) and height for each species are illustrated as black and white circles of different standardized sizes, and the negative correlation among the two traits is clear. (c) Bivariate scatterplot of the two traits with the phylogenetic relationships of the species superimposed. Given that there is zero correlation between the traits within each of the two clades, the trait correlation vanishes after accounting for phylogenetic relatedness.

calculate a measure of phylogenetic signal to test the strength of phylogenetic conservatism in each trait (Cadotte and Davies, 2016; Swenson, 2019), we find that each trait exhibits significant phylogenetic signal. This means that traits are more similar within clades then between clades, that is, the traits are conserved within clades.

Phylogenetic analyses allow us to test whether the correlation between traits is affected by phylogenetic relatedness. The raw correlation between the two traits is significantly negative, which suggests that species with high SLA are short and species with low SLA are tall, but a statistician would raise their eyebrows and say that these data appear to belong to different populations. After incorporating a phylogenetic correlation structure into a phylogenetic generalized least squares regression model, the trait–trait relationship disappears. If the negative correlation existed within each clade, then the phylogenetic correlation would have remained significant. Some would argue that the negative correlation between SLA and height is driven entirely by an ancient divergence, and therefore does not tell us something important about modern relationships among traits.

This has prompted a strong movement within comparative ecology to always incorporate phylogenies to enrich analyses of trait–trait or trait–environment relationships (Weber and Agrawal, 2012). I enjoy doing so in my work investigating relationships among traits (Kramer-Walter et al., 2016; Bergmann et al., 2020; Weigelt et al., 2021), and relationships between traits and environments (Laughlin et al., 2020a). Phylogenetic structure of trait relationships always yielded insight into trait evolution and methods for incorporating phylogenies into models of trait distributions across environmental gradients are becoming increasingly sophisticated (Li and Ives, 2017; Braga et al., 2018; Beck et al., 2020).

Not all ecologists think that phylogenetics is needed to improve our understanding of plant strategies. At least one ecologist has consistently registered his discontent—like a voice crying out in the wilderness where no one seems to be listening. It is instructive to review the debate between Mark Westoby (Westoby et al., 1995a; Westoby et al., 1995b; Westoby and Leishman, 1997; Westoby et al., 1997; Westoby, 2007) and virtually everyone else over the relevance of phylogenetic information in models of trait–trait and trait–environment relationships (Harvey and Pagel, 1991; Ackerly and Donoghue, 1995; Harvey et al., 1995; Silvertown and Dodd, 1997; Ackerly, 2009; Revell, 2012; Cadotte and Davies, 2016; Orme et al., 2018; Swenson, 2019). Westoby baulks at interpretations that assume ancient trait divergences have no functional value today. Gymnosperms and graminoids occupy different climatic environments precisely because they exhibit different functional traits. In agreement, Hansen (2014, p. 359) wrote "if related species tend to occur in similar environments (i.e., having similar values of their predictor variables), then we still expect a phylogenetic signal in the response variable. Correcting for phylogeny in this situation is throwing the baby out with the bathwater." Others have also pointed out problems with phylogenetic analyses when asking questions about adaptation (Uyeda et al., 2018).

Westoby's primary source of aggravation comes with respect to phylogenetic analyses of trait–environment relationships that test for adaptive value, not necessarily the trait–trait relationships that test for phenotypic constraints. He argues that phylogenetic constraints are not a source of confounding or error that requires correction when we invoke natural selection and functionality in our interpretations of present day patterns of trait variation among species (Westoby, 2007). Species can share traits both because they are related and because they are subject to similar forces of natural selection. Species within a clade can continue to share similar traits because of continuing stabilizing forces of selection that continue to matter in the present day, not just because they share ancestors. Natural selection is at least as likely an explanation for trait similarity among related species than phylogenetic constraint. Westoby also urges caution regarding the assumptions of independence: "Selection for ecological functionality has inherently been confounded with phylogeny during the history of evolution, and statistical corrections are not capable of converting that inherently confounded history into the ideal experiment in which phylogeny is orthogonally crossed with present-day function" (Westoby, 2007, p. 695). Westoby concluded that phylogenetic similarity should be leveraged to select species smartly across key clades for robust comparative analysis (e.g., Bazzaz et al., 1987; Garnier, 1992; Ji et al., 2020).

The last few decades in trait-based analyses of phylogenetic trees have heightened our awareness of the fact that closely related species tend to share similar traits. But does the fact that they are closely related disqualify them from sharing a similar ecological strategy? According to Westoby, the answer is clearly no. In Chapter 1 I gave the example of the widespread occurrence of subalpine forests (recall Figure 1.3)—each dominated by different combinations of spruces (*Picea* spp.) and firs (*Abies* spp.). Clearly the spruces and firs represent a strategy that achieves maximum relative fitness in the subalpine zone. But does this mean that phylogenetics should not routinely be incorporated into trait-based studies? Based on their potential to yield increased clarity about evolutionary processes, the answer is evidently no. Phylogenetics is necessary for enriching our understanding of trait evolution (de Bello et al., 2015). For example, phylogenetics improved our understanding of fine root trait evolution, where thinner roots evolved over time among angiosperms so that species could acquire limiting nutrients by themselves rather than associate with mycorrhizal fungi (Ma et al., 2018). Phylogenetics are essential for understanding xylem evolution (Carlquist, 1975; Zanne et al., 2014), leaf evolution (Flores et al., 2014), and seed evolution (Moles et al., 2005a; Moles et al., 2005b; Linkies et al., 2010). Functional ecologists must continue to collaborate with systematists that hold the keys to computing divergence times to learn more about the evolution of continuous traits.

When traits diverge within a clade and species with different traits go on to occupy different environmental niches, this is viewed as strong evidence for an adaptive radiation. Moreover, if we discover another clade that underwent a similar radiation and find that two species from these two separate clades evolved the same traits to optimize fitness in the same environment, then this is strong evidence for convergent evolution. Recall that convergence in traits was the original inspiration for theories of plant strategies, but trait convergence is not a requirement for the existence of a strategy. Phylogenetic clades, such as conifers and angiosperms, retain important morphological differences that continue to optimize their fitness in present day environments despite the fact that the divergence occurred in the ancient past. Yes, strategies may sometimes arise orthogonally across a phylogeny (as in cases of convergent evolution), but also sometimes entire clades retain key traits that allow them to dominate a certain habitat. For example, clonal traits are very strongly represented in monocots, which allows them to dominate wetland habitats (Van Groenendael et al., 1996). We should continue to learn about trait evolution using modern phylogenetic methods, but we should not disregard trait differences among clades as unimportant because the divergence occurred in the past. Natural selection continues, and it is significant that certain clades with certain traits still dominate particular environments.

6.13   Chapter summary

- Modules are the basic architectural building blocks of a plant. There exist numerous definitions of what constitutes a module, but this book defines a module as the phytomere: a unit of construction produced by apical meristems that consist of a node, an internode, a leaf, and an axillary bud.

- Roots anchor plants in place, provide support for growing stems, acquire water and mineral nutrients from the soil, and transport water and minerals to aboveground tissues. Roots perform many complex functions that make life on land possible.

- Clonality is a neglected aspect of plant form and function. Lateral spreading distance from the mother to the young ramet, multiplication rate (also known as offspring number), and persistence of connection are important clonal traits in herbs. Bud bank size and depth can also be measured on continuous scales and, given the central role dormant meristems play in plant survival, bud bank traits should be important predictors of plant fitness.

- Early land plants separated the functions of support and photosynthesis into stems and leaves. But stems serve multiple functions. Not only do they support large, heavy canopies of leaves and branches, but also they transport and store water, minerals, and carbohydrates. It is impossible for stem tissue to optimize each function simultaneously.

- The vasculature of plants can be thought of as an elaborate plumbing system. Species with wide conduits and thin walls exhibit fast rates of water flow through their pipes but lack resistance to embolism. Species with narrow conduits and thick walls exhibit low flow rates through their pipes but greater resistance to embolism.

- Leaves face a fundamental constraint that limits their rate of photosynthesis. Every time leaves open their stomata to allow CO_2 to diffuse into the mesophyll, water vapor is released into the atmosphere from the inside of the leaf. Plants must lose water to gain carbon.

- The leaf economics spectrum describes a fundamental trade-off between rate of carbon fixation and cost of leaf tissue construction.

- Phenological traits such as pollination date, flowering date, bud burst, leaf out date, and peak biomass production differ across species. It is still unclear whether phenology is a distinct leading dimension of trait variation or if it is linked to other leading dimensions such as height or leaf economics.

- Defining globally comparable phenological traits is an important unsolved problem. It is easy to define phenological traits, such as date of first flowering, within a given climate zone and seasonality, but it has proven much more difficult to define phenological traits that are sensible to compare across all climate zones worldwide.

- The innovation of seeds revolutionized plant reproduction because fertilization was no longer limited to wet environments, and it enabled the young plant to travel in a dormant state to find more favorable opportunities for colonization.

- Functional traits are morphological, physiological, or phenological traits that potentially impact fitness in a given environment. Our understanding of the dimensionality of functional traits is rapidly growing, but at least five dimensions are important: an economics spectrum, height, rooting depth, belowground collaboration with fungi, and a seed-or-resprout spectrum.

- In theory, it should be possible to predict the life history strategy of a species based on the combinations of functional traits expressed in the multidimensional phenotype and the environment in which it is growing.

- Species are not independent. Evolution is a branching process and species that are close together on the evolutionary tree are more likely to share similar phenotypic traits because they are more closely related, especially when traits are strongly conserved. However, phylogenetic conservatism is not the only explanation of trait similarity among related species because natural selection can also explain it. Phylogenetic clades, such as conifers and angiosperms, retain important morphological differences that continue to optimize their fitness in present-day environments, despite the fact that the divergence occurred in the ancient past.

6.14   Questions

1. What was the fundamental design constraint facing early land plants that were essentially photosynthetic stems?

2. How is rooting system type related to phylogeny?

3. When should traits be measured on a plant? At the seedling stage or at reproductive maturity?

4. Can you think of a floral trait that can generalize across the high morphological diversity of vascular seed plants?

5. If analyses of global trait databases were able to emphasize physiological rather than morphological traits, would our understanding of plant trait dimensionality be different?

6. Do you agree with the short list of five dimensions to represent the global variation in plant form and function? What traits are still missing?

7. Are life history traits determined by the multidimensional phenotype?

8. When should phylogenetic trees be included in analyses of functional and life history traits?

7

Plant Strategies Along Resource, Disturbance, and Temperature Gradients

Arrangement of species or populations along a single axis of more or less disturbance is unlikely to be adequate in any scheme intended to have world-wide application.

—Peter Grubb (1985, p. 605), "Plant populations and vegetation in relation to habitat, disturbance and competition: problems of generalization," in *The Population Structure of Vegetation*

I am sure I am not alone in feeling a sense of horror in attempting to predict vegetation change of this magnitude.

—William Bond (1997, p. 190),"Functional types for predicting changes in biodiversity: a case study of Cape fynbos," in *Plant Functional Types: Their Relevance to Ecosystem Properties and Global Change*

7.1 The scale of the habitat can cause mismatches between data and hypotheses

Plant strategy models have historically tended to emphasize life history traits to explain adaptations to disturbance and functional traits to explain adaptations to resource limitation. One of the goals of this book is to unify our understanding of how functional and life history traits simultaneously explain fitness differences along gradients of resource availability and disturbance regimes. Other goals, as they relate to the epigraph from Grubb (1985), are to decompose disturbance gradients into types, severity, frequency, and extent, to decompose resource gradients into limitation of water, nutrients, and light, and to recognize temperature as a distinct regulator of biological processes. Therefore, the objective of this chapter is to review and generate hypotheses about how functional and life history traits determine the shapes of fitness landscapes when resources are abundant, when resources are limited, when disturbances are frequent or large, when disturbances differ by type, and in cold and hot climates. The chapter emphasizes the

Plant Strategies. Daniel C. Laughlin, Oxford University Press. © Daniel C. Laughlin (2023). DOI: 10.1093/oso/9780192867940.003.0007

five dominant functional trait dimensions and the three dominant life history trait dimensions, but it also discusses other important traits (e.g., bark thickness) that are not yet included in these major dimensions.

The scale of the habitat is one of the greatest challenges when building a conceptual theory for how phenotypic traits govern fitness differences across environmental gradients. Grime's framework was built on variation in habitat duration and habitat productivity, but the scales of these habitats were not precisely defined. The concepts were built from Southwood (1977) and Greenslade (1983) "who classified habitats by arranging them on a plane defined by variation in productivity and rate of biomass destruction" (Grime, 2001, p. 8). These scales are implicitly set at the level of the community or landscape. Indeed, Grime often compared traits of plants growing in "stressful" habitats (i.e., Arctic and alpine, arid, shaded, and nutrient-deficient habitats) to productive habitats. These habitat qualifiers (e.g., "nutrient-deficient") are clearly characteristics of a community or a landscape. However, Grime later clarified his concept of the habitat by stating "rates of plant biomass construction and destruction show small-scale variation and operate locally upon individuals and not at scales and mean habitat values convenient for the landscape ecologist" (Grime, 2001, p. 9). This shift in emphasis implies that the habitat is at the scale of the plant, rather than a landscape.

There is no doubt that individual plants only experience the conditions present within the scale of their own canopy and root system. This can clearly be seen by examining the plurality of strategies that have evolved in response to drought (Volaire, 2018). Within the same habitat, one can find drought escapers (annual plants), drought avoiders (phreatophytes that tap into deep ground water), and drought tolerators (plants that continue to function at low water potentials). As described in Chapter 2, escaper, avoider, and tolerator strategies exist for other resources as well, not just water. Attempts to explain evolutionary convergence in vegetation structure at the biome scale must also acknowledge the diversity of vegetation structures that exist within a biome. It has been recognized for quite some time now that interspecific trait variation within a habitat rivals the variation in traits across habitats (Reich et al., 2003). Westoby (1998, p. 214) proposed that a plant strategy "needs to be thought of over a series of generations," implying that the environment to which the strategy is optimized be measured at a population-level scale, which is more akin to a landscape. However, this needs to be reconciled with the perspective that multiple strategies coexist within a habitat due to disturbance-driven dynamics and within-habitat environmental heterogeneity (Westoby et al., 2002; Reich et al., 2003). Resource availability fluctuate over time in response to disturbances that create gaps in canopies, and these dynamics change optimum trait values over time.

The scale of our data dictates the signals that our models can detect. Plant-scale data are powerful in models of individual growth, survival, and reproduction (Worthy et al., 2020), and are more closely aligned with Grime's recent emphasis on plant-scale environmental conditions. However, such data are extremely rare because they are time consuming and more expensive by orders of magnitude. Recent attempts to link variation in traits to climate exposed some of the challenges with a broader-scale approach. Global-scale analyses of trait-environment relationships are often set at the scale of a vegetation plot, which are akin to the "mean habitat values convenient for the landscape

(a)   (b)

Figure 7.1 *(a) Environmental conditions, such as soil temperature or water content (indicated by the range in blue to gold colors), measured at high resolution (i.e., the plant scale) can vary within a landscape extent. In this example, two plants experience different environmental conditions within this landscape, and because of this the larger plant exhibits better performance in the gold-colored environment. (b) However, most global datasets are by necessity average landscape-scale measurements (indicated by the solid 'average' green color across the whole landscape). In this example, we would fail to understand the fitness differences of these two plants because our environmental data is too coarse to capture the local environmental differences. Matching theories to the appropriate data resolution is an important challenge when testing predictions of plant strategy theory.*

ecologist." In such analyses, plot locations are intersected with global rasters of environmental variables such as climate, which are often at the kilometer scale or larger. Bruelheide and colleagues (2018) reported low explained variation of traits by macroscale climate data and speculated that much of the unexplained variation could be driven by small-scale disturbances, soil conditions, and variation in light availability. The actual values of environmental parameters will be within the range of variation in the landscape, but they could be far from the mean (Figure 7.1). As we develop testable predictions for which phenotypes optimize fitness along environmental gradients, we must keep the scale of habitat in mind and acknowledge that the scale of our data may be mismatched with the scale of the plant's immediate surroundings. Environmental data may need to catch up to the appropriate resolution in many circumstances.

7.2   Rugged fitness landscapes and the coexistence of plant strategies

Sewall Wright (1932) proposed that genotypic selection can be thought of as a landscape where the coordinates are defined by genotypes and peaks and valleys correspond to high and low fitness, respectively. Natural selection would tend to drive populations toward the peaks because individuals with optimal trait combinations would survive and reproduce more often. Comparative ecologists have extended these fitness landscapes to communities of species to explore how fitness differences among species are determined by their multidimensional phenotypes within and across environmental gradients (Marks and Lechowicz, 2006; Laughlin and Messier, 2015; Falster et al., 2017).

The classic fitness function of a trait is a smooth unimodal curve, where there is a single trait value that optimizes fitness in a given environment and traits to either side of the peak confer weaker fitness benefits (Figure 7.2). This can be estimated using a Gaussian function where the mean represents the optimum trait value, and the variance represents the steepness of the selection gradient to either side. These can be extended to multivariate Gaussian fitness landscapes when fitness depends on multiple traits. This is currently the most tractable way to empirically estimate fitness landscapes using multiple traits, but even these models are in their infancy. If it was possible to know everything about the environment at the scale of a plant, then it is theoretically possible that a single combination of trait values optimizes fitness in that environment.

But we know that multiple strategies coexist within a landscape, which would imply a rugged landscape of multiple peaks and valleys. If our data are at the landscape scale (which they typically are), then we should expect there to be multiple combinations of trait values that simultaneously optimize fitness in that landscape. This changes the unimodal fitness function into a multimodal fitness function. Multimodal functions can be estimated as weighted combinations of Gaussian functions with multiple means and multiple variances. The complexity of the functions get out of hand quickly as soon as we generalize to multiple trait dimensions. These rugged fitness landscapes are possibly the most realistic conceptualization of the adaptive value of traits within a real environmental landscape, but the thorny mathematics hinders their rapid estimation (Laughlin, 2018).

Now that the five strategy spectrums that capture the variation and covariation in plant traits have been articulated, it is time to generate testable hypotheses about how these spectrums optimize fitness along gradients of resource limitation and disturbance regimes. While Grime (2001) considered disturbance and resource stress to be orthogonal agents of natural selection driving plant evolution, we will revel in the multidimensionality and inter-relatedness of resource limitation and disturbance regimes (Grubb, 1985, 1998a), about which ecologists know a great deal already, especially with respect to economics and resource limitation; however, hypotheses for other strategy spectrums are less well developed and will be necessarily more speculative. In other words, please consider much of what follows to be hypotheses to be tested using the approaches discussed in later chapters.

7.3   Resource-rich habitats select for acquisitive species

Water, nutrients, and light availability limit plant growth. What phenotypic traits would optimize fitness in an environment where neither water, nutrients, nor light are limiting? Recall that if soil resources are abundant, then light limitation is inevitable because the vegetation will develop a closed canopy (Tilman, 1988). Therefore, abundance of water, nutrients, and light would only exist for a limited period of time immediately after a

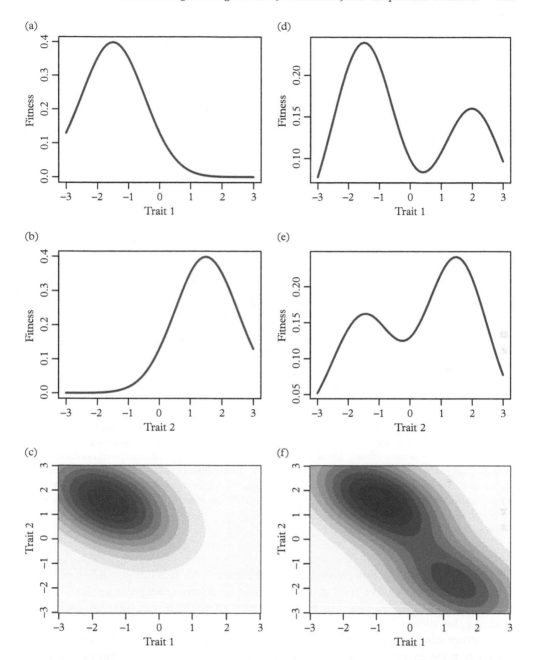

Figure 7.2 *Fitness landscapes for two hypothetical traits. (a–b) Unimodal fitness landscapes for traits 1 and 2. (c) Fitness landscape of both unimodal traits simultaneously represented as a contoured landscape where dark colors represent high fitness and light colors represent low fitness. (d–e) Bimodal fitness landscapes for traits 1 and 2. (f) Fitness landscape of both bimodal traits simultaneously represented as a contoured landscape.*

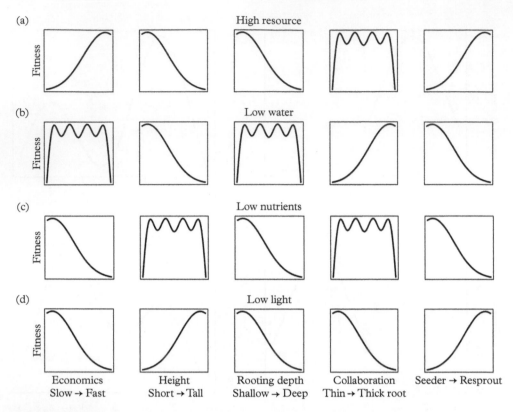

Figure 7.3 *Hypothesized fitness landscapes for each of the five functional trait spectrums in (a) high-resource environments, (b) low-water environments, (c) low-nutrient environments, and (d) low-light environments. Fitness landscapes with four evenly spaced peaks are "flat" landscapes to indicate that multiple trait values are equally fit within the environmental condition that permits coexistence of multiple strategies. These fitness landscapes should be viewed as qualitative, rather than quantitative, hypotheses to be tested with empirical data. For simplicity, these illustrations ignore the potentially important trait-by-trait interactions that likely influence plant fitness in a given environment.*

disturbance in a closed canopy system (recall Figure 3.5). It is not possible to illustrate fitness landscapes for five trait spectrums simultaneously (recall Figure 3.1), so here I illustrate hypothesized peaks in fitness for each of the five trait dimensions in high-resource, low-water, low-nutrients, and low-light conditions (Figure 7.3). It is my hope that further study will advance our understanding of how trait effects on fitness depend on other traits given phenotypic constraints (Pigliucci, 2003; Laughlin and Messier, 2015).

Economic theory would predict that conserving resources would be less important in resource-rich habitats, but rapid acquisition of resources would be critical for maintaining the high growth rates that are required to compete in such an environment. The successful phenotype must be able to exploit resources through rapid uptake in the presence of competing vegetation (Grace, 1990). Reich (2014, p. 277) hypothesized that "being fast at acquiring and processing carbon, water or nutrients in leaves, stems or roots is *advantageous only* when acquiring and processing of all resources is fast for all organ systems, because otherwise plants will possess excess capacity which is costly and wasteful". If this hypothesis is true, then the phenotype with the highest fitness in this circumstance would exhibit fast leaf, stem, and root economics: low LMA, high tissue nutrient concentrations, and low-density stems and roots. My colleagues and I tested this idea on 56 functionally diverse native New Zealand tree species. We grew seedlings in separate pots and were spaced far enough apart to not compete for light. Our analysis of tree seedlings grown in high-resource conditions in the absence of competition (density-independent) supported this hypothesis: the fastest growing species exhibited low LMA, low wood density, and low root tissue density (Figure 7.4). However, these strong correlations vanished if we measured them on wild plants growing in the field under conditions of density and frequency dependence, suggesting that the whole-plant economic spectrum is most easily detectable under conditions of density independence (Laughlin et al., 2017a).

Plants will have relatively shorter maximum heights in productive environments because the respiratory demands of maintaining tall canopies is only advantageous in light-limited environments (Falster and Westoby, 2003). Rooting depth will tend to be shallow if there is little need to develop a deep root system (Figure 7.5). Deep roots tend only to be found in arid climates where deep roots can tap into ground water (Fan et al., 2017). Obtaining nutrients through thin root proliferation may be more beneficial than the carbon costs associated with collaborating with mycorrhiza in a resource-rich environment; the fungal partnership would be more parasitic than mutualistic (Johnson et al., 1997; Bergmann et al., 2020). However, based on a large-scale empirical analysis, it appears that species with any root diameter can occur where water is not limiting (Laughlin et al., 2021), so the fitness landscape associated with root collaboration could be bimodal. Finally, the resprouter strategy appears to gain an advantage in more productive environments with denser vegetation, whereas the seeder strategy is favored in less-productive, more open environments (recall Figure 2.5) (Clarke et al., 2005; Pausas and Bradstock, 2007; Pausas and Keeley, 2014; Clarke et al., 2015).

7.4 Water limitation imposes extreme negative pressure in the xylem

Water limits plant growth over much of the terrestrial land surface, and is regarded as one of the most important filters on species pools globally (Keddy and Laughlin, 2022). Drylands are characterized by co-limitation of water and nutrients (Hooper and Johnson, 1999) and high light availability. But such unlimited access to light is wasted if drought drives xylem pressure potentials to extreme tensions leading to cavitation of the water column. Without continuous access to water, leaf stomata are forced to close, thereby ceasing uptake of CO_2, and causing photosynthesis to grind to a halt.

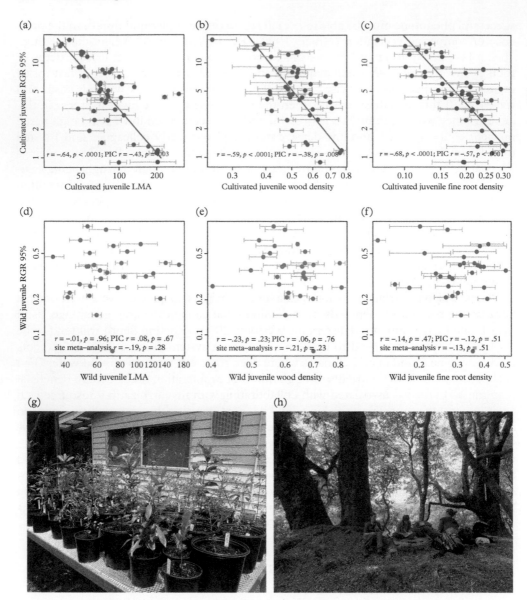

Figure 7.4 *Maximum (95th percentile) relative growth rates (RGR) of indigenous Aotearoa New Zealand seedlings cultivated in a glasshouse with optimal water, nutrients, and light as a function of (a) leaf mass per area (LMA; mg mm^{-2}), (b) wood density (mg mm^{-3}), and (c) fine root tissue density (mg mm^{-3}). However, maximum (95th percentile) RGR of the same species growing in indigenous habitats in competition with neighbors were uncorrelated with (d) LMA, (e) wood density, and (f) fine root tissue density. Phylogenetic independent contrasts (PIC) yielded the same results. (g) Photo of tree seedlings growing free from competition in high light, high water, and high nutrient conditions at the University of Waikato glasshouse facility, Kirikiriroa Hamilton, Aotearoa New Zealand. (h) Field crew measuring plant traits under natural conditions in an old-growth podocarp forest in Whirinaki Te Pua-a-Tāne Conservation Park, Aotearoa.*

(a–f): Reproduced from Laughlin et al. (2017a) with permission. (g–h): Photos taken by the author.

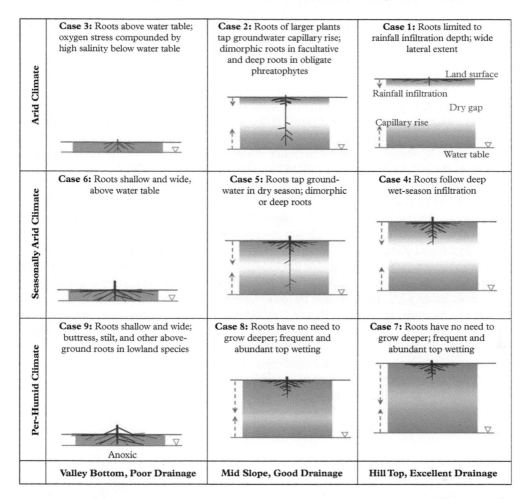

Figure 7.5 *A hydrologic framework of plant rooting depth along a climate gradient (vertical) and land drainage gradient (horizontal).*
Reproduced from Fan et al. (2017) with permission.

Adaptations to water limitation involve multiple solvent strategies, including escaping, avoiding, and tolerating water shortages (Volaire, 2018). If multiple strategies coexist in dry land, then we should observe multiple rugged peaks in the fitness landscapes (Laughlin, 2018). Conservative leaf economics exemplify drought tolerance, but the drought escapers (i.e., annual plants) limit their phenology to brief windows when water is abundant, and the drought avoiders (i.e., phreatophytes) tap into deep water reserves. This means that both conservative and acquisitive leaf economic traits can optimize fitness in drylands, depending on how the leaf traits are combined with root traits. Most phenotypes in drylands are short in stature but depending on their drought strategy vary in seed size and wood density.

Water stress imposes immense negative pressure on vascular conduits, and the collapse of the water column can ultimately lead to plant mortality. Neither light limitation nor nutrient limitation impose such physical stress on a plant. Drought tolerant phenotypes exhibit low vulnerability to embolism in their xylem (Figure 7.6), indicated by more negative P_{50} values in both stems (Choat et al., 2012; Gleason et al., 2016) and leaves (Nardini and Luglio, 2014; Ocheltree et al., 2016). Species that exhibit low P_{50} can maintain hydraulic conductance under lower negative pressure imposed by drought (Choat et al., 2012; Gleason et al., 2016) and are therefore more likely to occur in drier regions (Laughlin et al., 2020a) and in high and well-drained upland plateaus in tropical forests (Oliveira et al., 2019). Many anatomical traits are associated with a plant's vulnerability to embolism. Thicker cell walls in the xylem combined with narrow tracheid or vessel widths are associated with lower vulnerability to embolism (Blackman et al., 2010; Tyree and Zimmerman, 2013). Light-demanding trees tend to have dense root tissue to tolerate the wide range of water availability that occurs in the absence of shade (Zadworny et al., 2018). Species that can maintain leaf turgor and continue to photosynthesize under drought are indicated by lower leaf turgor loss points, which are also lower in dryland habitats worldwide (Figure 7.6) (Bartlett et al., 2012b). Low-water species also exhibit high nitrogen concentrations per unit of leaf area which generates higher photosynthetic rates per area and lower internal CO_2 concentrations, which ultimately increases photosynthetic water-use efficiency (Wright et al., 2001; Reich et al., 2003).

Rooting depth is bimodal across phenotypes in drylands: drought escapers have shallow roots to utilize water at the soil surface after it has rained, drought tolerators also

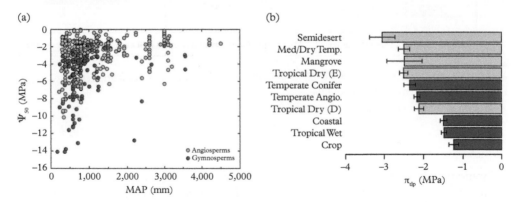

Figure 7.6 *Variation in* P$_{50}$ *(xylem vulnerability to embolism) in angiosperms and gymnosperms along a gradient in mean annual temperature (MAP). (b) Average differences in leaf turgor loss point (π_{tlp}) across different vegetation types. Light blue bars are dry biomes and dark blue bars are wet biomes. Tropical dry forests were separated into evergreen (E) and deciduous (D) types.*

(a): Reproduced from Choat et al. (2012) with permission. (b): Reproduced from Bartlett et al. (2012b) with permission.

tend to have shallow roots, but drought avoiders can send down extremely deep roots to tap into deep water tables (recall Figure 3.6). It was once thought that mycorrhiza had minimal effect on plant water relations, but phenotypes with thicker roots are more likely to occur in dry environments because arbuscular mycorrhizal fungi that inhabit thicker roots can confer drought tolerance to plants through enhanced stomatal conductance and water-use efficiency (Augé et al., 2015; Laughlin et al., 2021). The seeder strategy optimizes fitness in drylands because the open vegetation is easier to colonize (Clarke et al., 2005; Pausas and Bradstock, 2007; Pausas and Keeley, 2014; Clarke et al., 2015). Succulent columnar cactus and euphorbia species are emblematic of drylands worldwide. Many succulents have reduced leaves, photosynthetic stems with high LMA, are short in stature, and produce shallow short-lived roots. Most of these species use CAM photosynthesis, which separates the light and light-independent reactions of photosynthesis temporally by only acquiring CO_2 through stomata at night. This very effective strategy minimizes water loss and increases water-use efficiency (Edwards and Ogburn, 2012).

7.5   Nutrient limitation selects for conservative leaf economics and variable root diameters

Soil infertility is caused by a variety of factors, including soil particle size distribution, pH, low organic matter, cold temperatures, and high rates of leaching caused by high precipitation, but plants of infertile habitats have many traits in common. Nitrogen and phosphorus are two of the most abundant mineral nutrients in plant tissue and they co-limit plant growth in terrestrial ecosystems globally (Elser et al., 2007). Phenotypes that tolerate nutrient limitation share many similarities with phenotypes that tolerate water limitation. This is in part because water-limited systems are also nutrient limited, likely because mass flow and diffusion rates are lower in dry soil (Hooper and Johnson, 1999) and possibly because of the diminished ability of roots to uptake nutrients at low water potential. The difference between water and nutrients is that water must be lost from the plant to photosynthesize, whereas nutrients can be reused and recycled (Craine, 2009). If water is in sufficient supply to drive evapotranspiration, then nitrogen tends to be more available because plants acquire nitrate through mass flow. Nitrate (NO_3^-) is particularly mobile in the soil solution because it has a negative charge and is thus not attracted to negatively charged soil colloids. Uptake of phosphorus, however, is more limited by rates of diffusion (recall Figure 6.8). Diffusion shells form around roots that have exhausted their phosphorus supply, and so the only way to obtain more phosphorus is to either proliferate roots into resource-rich patches using thin roots with low carbon cost or specialized cluster roots, or outsource nutrient acquisition to mycorrhiza (Laliberté et al., 2015; Bergmann et al., 2020). Species adapted to low-nutrient habitats must not merely tolerate low nutrients—they must actively grow well at low supplies, and therefore the production of high root length density is a key trait for do-it-yourself species that actively acquire nutrients by themselves (Berendse and Elberse, 1990; Goldberg, 1990;

Craine, 2009). Increasing resource uptake can also occur by exhibiting thick roots with more cortical tissue to enhance mycorrhizal colonization (Bergmann et al., 2020).

Fitness in low nutrient habitats can be optimized by (1) increasing resource uptake, (2) decreasing resource loss, and (3) increasing the efficiency of resource use (Berendse and Elberse, 1990; Goldberg, 1990; Berendse et al., 1992). High LMA leaves increase the efficiency of resource use and decrease resource loss by exhibiting higher nutrient-use efficiency, which is defined as the rate of carbon fixation per unit leaf nutrient, by exhibiting higher resorption efficiency, which is the rate of resorption of nutrients from senescing leaves into storage, and by exhibiting longer leaf lifespans (Aerts and Chapin, 1999). Observations of phenotypes with conservative leaf economic strategies are very common in low-nutrient habitats (Grime, 2001; Ordoñez et al., 2009; Jager et al., 2015; Simpson et al., 2016), but fewer studies have linked leaf traits to population dynamics. One long-term demographic study demonstrated that herbaceous perennial species with high LMA (low SLA, and therefore long leaf lifespan) exhibited higher survival rates in nutrient-poor soil and species with low LMA (short leaf lifespan) exhibited higher survival rates in fertile soil (Figure 7.7). In contrast to dry habitats, plants can grow tall in nutrient-limited habitats as long as water is available because nutrients can be

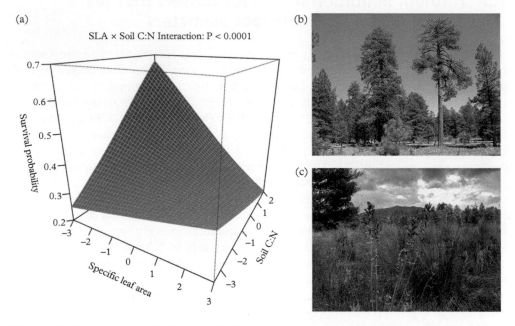

Figure 7.7 *The effect of specific leaf area (SLA) on survival rates was slightly positive in fertile soil (low carbon-to-nitrogen ratio) and strongly negative in infertile soil (high C:N ratio), suggesting that herbaceous perennial species with low SLA exhibit greater survival (and likely greater fitness) in low-fertility soil. (b) Ponderosa pine forest near Flagstaff, Arizona, USA. (c) Herbaceous perennial vegetation including species that were included in the long-term demographic study (e.g.,* Festuca arizonica, Bromus ciliatus, Oxytropis lambertii, Castilleja *sp.).*

(a): Reproduced from Laughlin et al. (2018b) with permission. (b–c): Photos taken by the author.

stored in the tissues (Craine, 2009; Jager et al., 2015). Making deep roots does not assist with nutrient uptake because most nutrients are found in the upper soil layers (Weigelt et al., 2021). Finally, the seeder strategy appears to be more advantageous than the resprouting strategy in less-productive habitats but this may be more related to water limitation than nutrient limitation (Pausas and Keeley, 2014). Clonality is advantageous in infertile soil because resources can be stored, clonality enhances persistence when resources are scarce, and clonal plants can literally move from nutrient-poor patches into resource-rich patches (Van Groenendael et al., 1996).

7.6 Light limitation selects for larger and slower plants

The canopy in productive environments will inevitably close, given sufficient time after a disturbance, which leads to light limitation, which selects for taller species (Falster and Westoby, 2003) with lower whole-plant light compensation points, lower dark respiration rates, and higher LMA. Species with lower light compensation points achieve higher survival rates in the dark but slower growth rates in high light (Lusk and Jorgensen, 2013), and in general shade-tolerant plants are slow-growing species. This was confirmed by Baltzer and Thomas's (2007b) pioneering Bornean rain forest study, which was one of the first to actually measure whole-plant relative growth rates along a light availability gradient. *Dipterocarpus grandiflorus* (Dipterocarpaceae) is an example of a slow-growing shade-tolerant species, and *Macaranga hypoleuca* (Euphorbiaceae) is an example of a fast-growing light-demanding species (Figure 7.8).

Rooting depth and resprouting strategies are likely unchanged in light-limited environments. Large-seeded phenotypes tend to inhabit light-limited environments for a variety of reasons: large-seeded phenotypes tend to germinate at lower red:far-red light ratios, which are common in forest understories (Jankowska-Blaszczuk and Daws, 2007), and large-seeded phenotypes often exhibit higher seedling performance in the shade (Westoby et al., 1997). Relationships with mycorrhiza in light-limited environments may be more parasitic than mutualistic if the plants are starved of carbon (Johnson et al., 2015; Ballhorn et al., 2016), which possibly suggests that the thin-rooted, do-it-yourself strategy would be selected in light-limited conditions.

Some phenotypes that persist in low light have also evolved specialized leaves with a high number of spongy mesophyll cells that scatter light within the leaf to increase absorption of diffuse and far-red wavelengths (DeLucia et al., 1996). Some plants have epidermal cells that act as a lens to focus light onto photoreceptors. Pigments on the lower leaf surface of Indo-Malesian tropical rain forest understory plants, such as *Begonia pavonina*, *Allomorphia malaccensis*, *Forrestia mollis*, and *Piper porphyrophyllum*, have been shown to reflect light back up into the chloroplasts (Lee et al., 1979). Shade-tolerant species also tend to minimize leaf overlap to prevent self-shading (recall the Fibonacci sequence from Chapter 6). The upshot is that shade-tolerant species have lower relative growth rates than light-demanding species in the short term, but actually have higher relative growth rates in the long term in low-light conditions (Sack and Grubb, 2001).

Figure 7.8 *Relationships between whole-plant light compensation point (WPLCP) and various physiological and morphological traits, including leaf-level light compensation point (LLCP), mass-based dark respiration (Rmass), leaf nitrogen concentration (%), mass-based photosynthetic rate (Amass), leaf lifespan, and leaf mass per area (LMA). Closed/open symbols correspond with alluvial/sandstone populations while circles/triangles denote specialists/generalists. (b) Photo of the slow-growing and shade-tolerant* Dipterocarpus grandiflorus *(Dipterocarpaceae), a species with a low WPLCP. (c) Photo of the fast-growing and light demanding* Macaranga hypoleuca *(Euphorbiaceae) clamped in a Licor chamber (a difficult task since this myrmecophytic species is defended by colonies of ants).* Macaranga hypoleuca *exhibits a high WPLCP.*

(a): Reproduced from Baltzer and Thomas (2007b) with permission. (b–c): Photos provided by Jennifer Baltzer with permission.

7.7   On life history traits as predictors of fitness

Disturbances drive cyclic sequences between times when light is available and times when light is consumed by a closed canopy. Plant strategy models that emphasize life history traits have been useful for explaining vegetation dynamics in response to disturbance regimes (recall Chapter 2). Within any given habitat, species that rapidly reach reproductive maturity are the best colonizers of recently disturbed sites, and they are later overtaken by species that reach maturity more slowly (van der Valk and Davis, 1978; Noble and Slatyer, 1980; Rüger et al., 2020). Patch dynamics in forests are driven by the creation of canopy gaps, and gap size can vary from single trees (which often pull down several trees during their fall) to large blow downs during hurricane-force wind events. There can be an ample supply of propagules in the seed bank and seed rain and a large pool of surviving meristems in the belowground bud bank to drive fast recruitment into the gap. High-resource environments immediately following a disturbance would select for a combination of life history traits that align with faster lifestyles (Figure 7.9). Traits associated with the fastest lifestyles include faster times to sexual maturity, greater degree of semelparity, and shorter lifespans. However, as soon as resources become limiting, other life history strategies may become more advantageous. Slower lifestyles could be selected in resource limited conditions, corresponding to longer times to sexual maturity, a greater degree of iteroparity, and longer lifespans. Iteroparity is a form of bet hedging and so has likely been selected in resource-limited habitats, whereas semelparous plants will be most successful if trying to reproduce in a resource-rich environment.

High-resource conditions following a disturbance will select for species with fast pace of life, low degrees of iteroparity, and short lifespans. When resources become limiting over time and vegetation is dense, then species with slower pace of life, high degree of iteroparity, and long lifespans will be selected (Figure 7.9).

The life history theory discussed in Chapter 5 suggests that iteroparous perennials are favored when juvenile survival is low or unpredictable, and that semelparous annuals are favored when juvenile survival is high and when adult survival is low (Stearns, 1992; Friedman, 2020). Grime's model proposed that disturbance frequency selected for herbaceous species with fast growth rates and high rates of reproduction, but this incorrectly ignores adaptations to disturbance exhibited by long-lived species. While it is true that annual and biennial species thrive after recent disturbances, long-lived species can also persist under frequent disturbance regimes by either escaping the effects of the disturbance or, if they are top killed, by resprouting. For example, the giant sequoia (*Sequoiadendron giganteum*) can experience hundreds of fires throughout its lifetime yet live for thousands of years because its thick bark prevents the heat of frequent low-severity fires from damaging its cambium (Harvey et al., 1980). Species in the Brazilian Cerrado, such as *Davilla elliptica*, *Kielmeyera rubriflora*, and *Curatella americana*, protect their meristems beneath thick bark and resprout after being top killed by fire (Corrêa Scalon et al., 2020). Therefore, disturbance frequency appears to select for both ends of the continua across all three life history strategies. In contrast, large disturbances may select for species that achieve reproductive age quickly, have a high degree of semelparity, and short lifespans given the availability of colonization sites. Semelparity is advantageous in large disturbance extents where large allocations to reproduction will pay off if colonization sites are abundant in time and space.

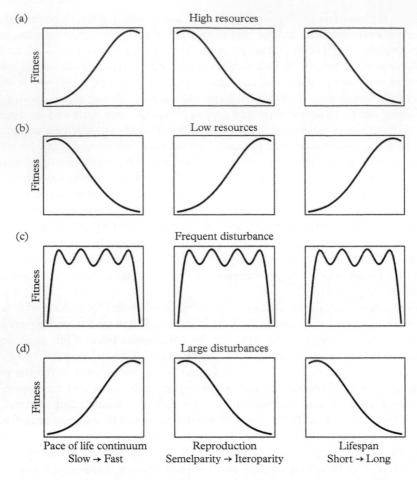

Figure 7.9 *Hypothesized fitness landscapes as functions of life history traits in (a) high-resource environments, (b) low-resource environments, (c) frequently disturbed environments, and (d) large disturbance extents. Fitness landscapes with four evenly spaced peaks are "flat" landscapes to indicate that multiple trait values are equally fit within the environmental condition. These fitness landscapes should be viewed as qualitative, rather than quantitative, hypotheses to be tested with empirical data. For simplicity, these illustrations ignore the potentially important trait-by-trait interactions that likely influence plant fitness in a given environment.*

7.8   The world is green partly because plants are well defended against herbivores

Disturbance types span two primary dimensions (recall Figure 3.7). The first dimension spans a gradient from high-frequency–low-severity disturbances (e.g., herbivory, individual falling trees, animal activities, erosion) to low-frequency–high-severity disturbances (e.g., landslides, hurricanes, sand dunes, rock outcrops, earthquakes, tornados, glaciers, and volcanic eruptions) (Walker and del Moral, 2003). Spatial extent is orthogonal to this first disturbance gradient, indicating that disturbance types span a two-dimensional plane of disturbance attributes. Let us now explore traits associated with particular disturbance regimes, with an emphasis on herbivory, fire, tree-fall gap dynamics, volcanic eruptions, and glacial retreat, with the caveat that a strong synthesis of plant traits and disturbance regimes has yet to be conducted (Prach and Walker, 2020). The potential for plants to adapt to any particular disturbance is greatest when the disturbance occurs frequently but does not incur high mortality (Harper, 1977). We therefore start with herbivory because it meets these criteria: herbivores chronically, yet often non-lethally, ingest live plant material. Herbivory is one of the greatest agents of natural selection in plant evolution.

Why is the world green? This classic question in ecology has two possible answers: the world is green either because predators keep herbivore populations in check or because plants are so well defended that they are poor food for herbivores (Hairston et al., 1960; Polis, 1999; Wilkinson and Sherratt, 2016). There exist several plant defense theories, each starting with the basic idea that plants have responded to this intense selection pressure from herbivores by investing in less-palatable tissues at the expense of reduced rates of growth (Stamp, 2003). One of the difficulties of developing a generalizable theory of plant defense is the diversity of different herbivores that attack plants (McArthur et al., 1991). Moreover, the capacity of a trait to be defensive depends on what the focal plant's neighbors are doing (Grubb, 1992).

Plants would be far more efficient in the absence of herbivores, but the benefits of defending themselves against being eaten are greater than the costs. Indeed, when a species escapes its predators by invading a different continent where the predator is absent, this release from enemies can be one of the most important reasons for its invasion success. Animals, including mammals, birds, and insects, eat plants to obtain energy and protein. The quality of plant material is determined by (1) the amount of labile carbon within plant cell contents (as opposed to within the cell walls), (2) the amount of protein within the cell, and (3) the types and quantities of secondary metabolites (Craine, 2009). Insect herbivory is often specialized to specific plant species, but vertebrate (and specifically mammal) herbivory tends toward generalist feeding on multiple plant species. Plant defenses can be organized hierarchically where the first major split is structural versus chemical defenses (Figure 7.10). Structural defenses are constitutive, that is, they are always built by the genetic code even in the absence of herbivores, in contrast to inducible defenses that are created in response to an attack. These include lignin-rich organic structures such as thorns, spines, and prickles, which greatly slow the rate of herbivory forcing the herbivore to be more careful and selective in their browsing. Note that spines can have other functions as well, including reducing net radiation flux and climbing other vegetation (Grubb, 1992). The divaricate wireplant growth form also slows the rate of browsing and reduces herbivore damage (Lusk et al., 2020). Thicker cell walls also reduce the quality of forage by reducing labile cell contents and increasing the concentration of lignin (Van Soest, 1982). Cell walls can also be made even less palatable by making them more rigid with inorganic substances like silica and other minerals.

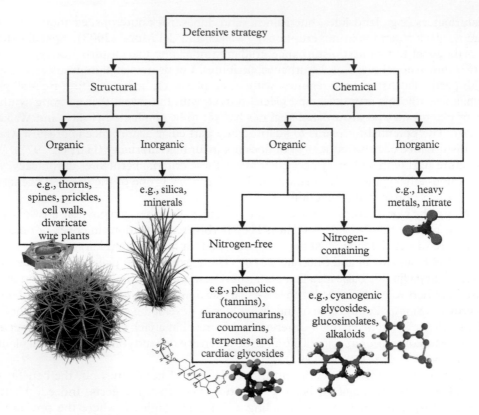

Figure 7.10 *A proposed hierarchy of plant defenses.*
Adapted from Craine (2009) with permission.

Chemical defenses can be either constitutive or inducible. Inorganic metals and nitrate can accumulate in vacuoles, but most defensive chemicals are organic products of secondary metabolism that can either contain or do not contain nitrogen. Nitrogen-free compounds include tannins, furanocoumarins, coumarins, terpenes, and cardiac glycosides (Agrawal and Fishbein, 2006; Agrawal et al., 2012), and nitrogen-containing compounds include cyanogenic glycosides, glucosinolates, and alkaloids. You may not have heard of any of these compounds before, but you will no doubt have heard about a few famous alkaloids: caffeine and nicotine. Yes, humans can become addicted to plants that produce compounds meant to prevent their very ingestion!

Defenses cannot completely prevent all herbivory from happening, but they do slow the rate such that growth can just outpace being eaten. The leaf economics spectrum is often viewed as a gradient in constitutive investment in plant defense. Slow-growing species adapted to low-resource habitats construct less-palatable, long-lived leaves because they cannot afford to lose leaves to herbivores, whereas fast-growing species with nutrient-rich leaves can afford to lose some leaves to herbivores (Coley, 1983; Coley et al., 1985). However, all plants are defended to some degree. It has been proposed that plants from high-resource habitats tend to use inducible defenses or constitutive nitrogen-containing compounds because they can afford to use nitrogen on defense. Plants from nitrogen-limited environments cannot afford to waste nitrogen on defense and so tend to opt for nitrogen-free compounds (Craine, 2009).

We are now starting to realize the general importance of consumers to the global distribution of vegetation. According to Bond (2005), the world is not just green—it is also brown and black. Large parts of the planet fall within a climate space that yield uncertain vegetation types: the site could be grassland, savanna, or forest. It is "unsettling" for a vegetation ecologist to consider that climate is not the driver of vegetation, but in these uncertain situations, it is likely that two important consumers, herbivores and fire, are the primary factors determining the structure and composition of vegetation (Bond, 2019, 2021). The concepts of "pyromes" (fire regimes), "herbivomes" (large mammal herbivory regimes), and "anti-herbivomes" (herbivore-defense regimes) have been proposed in response to these important discoveries (Archibald et al., 2013; Hempson et al., 2015; Dantas and Pausas, 2022). The world is brown in herbivore-controlled systems where traits like spines and unpalatable leaves have been selected to slow down or prevent consumption by beast (Charles-Dominique et al., 2015b; Charles-Dominique et al., 2018). The world is black in fire-controlled systems where protection of buds can prevent damage or allow species to resprout, or else flowers and seeds are promoted by the fire to encourage a new generation of seedlings. Section 7.9 specifically discusses trait selection by fire.

7.9   Fire regimes vary from infrequent-high severity to frequent-low severity regimes

Fire is perhaps the second-most widespread and most important disturbance type on Earth. Fires have consumed vegetation since the Silurian origin of land plants 443 million years ago (Pausas and Keeley, 2009) and has been likened to a "global herbivore" (Bond and Keeley, 2005) because experiments and dynamic global vegetation models show that fire reduces the global extent of closed canopy forests by 50% (Bond et al., 2005)! Grime's CSR model predicts that species with fast life histories would be the most dominant phenotype in frequently burned ecosystems, but long-lived species have clearly evolved a variety of traits to persist in fire-prone habitats. Not all fire-adapted traits are included in the global spectrums of plant form and function (Table 7.1).

Fire is a heterogeneous disturbance with patchy effects (Prach and Walker, 2020). Even severe crown fires leave islands of unburned vegetation, which provide a source of propagules for future recruitment. Fire was located toward the frequent end of the range of disturbance types (Walker and del Moral, 2003), but fire itself exhibits tremendous variation along a gradient of frequency and severity. Fire ecologists distinguish ecosystems based on the severity of burn events ranging from high-severity crown fire regimes to low-severity surface fire regimes (Agee, 1996; Pausas and Keeley, 2014), and patterns of fuel consumption, spread, and seasonality differ among these regimes (Gill, 1975; Bond and Keeley, 2005). Keeley and colleagues 2011, p. 406) have argued that "no species is fire adapted but rather is adapted to a particular fire regime."

Table 7.1 *Woody plant strategies and adaptive traits within contrasting fire regimes.*

Fire regime	Strategies	Traits
Crown fire regimes	Obligate resprouters	Bud bank belowground or aboveground No seed bank Iteroparous reproduction Recruitment between fires Overlapping generations
	Facultative seeders	Bud bank Fire-resistant seed bank Iteroparous reproduction Recruitment after fire Overlapping generations
	Obligate seeders	No bud bank Fire-resistant seed bank Semelparous reproduction Recruitment after fire Non-overlapping generations
Surface fire regimes	Fire escapers	Thick bark Self-pruning lower branches Tall heights Iteroparous reproduction Overlapping generations

Adapted from Pausas and Keeley (2014) with permission.

There are two general mechanisms of persistence in crown fire regimes: resprouting from dormant meristems or seedling recruitment (Figure 7.11) (Whelan, 1995; Bond and van Wilgen, 1996; Pausas et al., 2004; Scott et al., 2014). There is an expected trade-off between resprouting capacity and the capacity for seeding as a mechanism of regeneration following disturbance (Bellingham and Sparrow, 2000). Pausas and Keeley (2014) proposed an evolutionary framework that emphasized the relative survivorship of adults to juveniles (Charnov and Schaffer, 1973) and suggested three main strategies: obligate resprouters, facultative seeders, and obligate seeders (Table 7.1). Resprouting is an ancient trait in woody plants and is advantageous in a broad array of circumstances, not just after crown fire. However, a phylogenetic analyses suggested that thick corky bark and root sprouting are adaptations to fire regimes and not just exaptations that happen to be useful in a fire-prone world (Simon and Pennington, 2012). Resprouting was likely selected for in fire-prone habitats that were mesic and fertile. Indeed, Dalgleish and Hartnett (2006) showed a positive relationship between bud bank size and water availability among grasslands.

Resprouting is a key functional trait and can arise from belowground, basal, or aerial meristems. Within each of these three types, the origin of the bud and the source of

Southwestern Australian ecosystems   Cerrado of South America

Figure 7.11 *Plant adaptations in fire-prone ecosystems such as (a–f) southeastern Australia and (g–j) the Cerrado of South America. Australia: (a) Basal sprouting in* Eucalyptus *sp., (b) epicormic resprouting in* Angophora *sp., (c) aboveground resprouting in* Xanthorrhoea *sp., (d) seed germination in* Eucalyptus *sp., (e) serotinous cone of* Banksia *sp., (f) post-fire flowering in* Xanthorrhoea *sp. Cerrado: (g) Thick and corky bark of* Machaerium opacum, *(h) seedling recruitment of* Hippeastrum goianum, *(i) xylopodium of* Anemopaegma *sp. enables resprouting after fire, (j) thick terminal branch of* Kielmeyera coriacea *is fire resistant.*

(a–f): Photos taken by Rachael Nolan and reproduced from Nolan et al. (2021) with permission. (g–j): Reproduced from Simon and Pennington (2012) with permission.

energy to fund the regrowth varies (Clarke et al., 2013). Species that resprout from belowground rhizomes and roots exhibit the highest probabilities of survival following fire (Figure 7.12). Most resprouters live longer, reach age at maturity slower, and allocate more resources to buds and storage organs than obligate seeders (Pausas et al., 2004). Resprouting potential can be measured as a continuous trait using the density of the bud bank on the plant (Dalgleish and Hartnett, 2006; Klimešová and Herben, 2015), and the size of the bud bank and capacity for clonal growth are highly correlated across species (Klimešová and Herben, 2015). Resprouting from belowground buds is the dominant fire adaptation in the Brazilian Cerrado (Corrêa Scalon et al., 2020; da Silva et al., 2021; Zupo et al., 2021). Some species evolved the ability to resprout from their burned stems aboveground, known as epicormic resprouting. Examples of epicormic resprouters include several Australian gum trees (*Eucalyptus* spp., *Corymbia* spp., and *Angophora* spp.), cork oak (*Quercus suber*), and Canary Island pine (*Pinus canariensis*) (Burrows, 2013; Pausas and Keeley, 2017).

Some plants lost the ability to resprout when they evolved a post-fire seeding strategy. Obligate seeders have been found to be more drought tolerant and more abundant in drier, less fertile habitats. Pausas and Keeley (2014) hypothesized that facultative

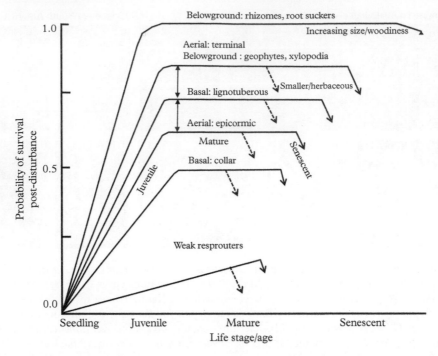

Figure 7.12 *Conceptual model of survival after severe disturbance among different resprouter types. Broken lines represent smaller growth forms that mature more rapidly and decline sooner than longer-lived woody plants. Double-headed arrows indicate that many woody resprouters have both basal and epicormic resprouting.*
Reproduced from Clarke et al. (2013) with permission.

seeders evolved from resprouters by developing a fire-resistant seed bank, and that obligate seeders evolved from facultative seeders by losing the ability to resprout. Seeders restrict recruitment to years immediately after fires, and seedlings germinate either from the soil seed bank or from seeds held captive in aboveground serotinous fruits (Keeley et al., 2011). Serotinous fruits (or cones) only open after exposure to heat. Serotiny is a remarkable adaptive trait for light-demanding species because it prevents seeds from germinating in shaded understories where they cannot grow. High-latitude conifer forests in the northern hemisphere dominated by trees such as lodgepole pine (*Pinus contorta*) and jackpine (*Pinus banksiana*) are shorter and cannot escape the ground fires, and these trees exhibit thin bark, retain their lower branches, and produce serotinous cones in which the seeds remain locked until the heat of a fire forces the cone to open (Keeley et al., 2011). Serotiny is common in obligate seeders in Mediterranean ecosystems as well, and is most prevalent in regions of low soil fertility—possibly because aerial storage in serotinous fruits protects high-nutrient seeds from herbivory in the soil seed bank (Keeley et al., 2011). Remarkably, some seeds will only germinate after heat-shock from fire or being exposed to smoke (Dixon et al., 1995). In fact, you can trick some seeds into germinating by exposing them to liquid smoke that you buy in the grocery store to flavor your meat and veggies on the barbeque.

Given the importance of fire to the life cycle of seeders, it has been argued that plant flammability may be an adaptation toward increased frequency and intensity of fires that reduce competition from fire-sensitive neighbors (Mutch, 1970). For example, in longleaf pine forests dominated by *Pinus palustris*, pines produce flammable litter that ignites and often kills fire-sensitive broadleaf species, such as turkey oak (*Quercus laevis*) and sand live oak (*Quercus geminata*), that invade pine forests (Williamson and Black, 1981). It is also possible that if a species restricts reproduction to post-fire years then being flammable could be adaptive (Keeley et al., 2011). Development of small leaves, retaining dead leaves and branches, and producing volatile compounds enhance flammability. Flammability traits have been positively associated with serotiny in pines, suggesting that these traits in combination are indicative of a post-fire seeding strategy (Schwilk and Ackerly, 2001).

In tropical ecosystems, fire and herbivory interact to influence the distribution of grasses and trees in savannas. A biogeographical comparison of traits between Neotropical and Afrotropical savannas and forests yielded some interesting differences driven by the fact that fire intensities are higher in Neotropical savannas, whereas herbivory by megafauna controls fuel loads and reduces fire intensities in Afrotropical savannas. They contrast "corky" and "lanky" strategies. The corky strategy (thick outer bark and shorter maximum height) is more adaptive in the lightly browsed Neotropical savannas that experience more intense fire, and the lanky strategy (thin bark and taller maximum height) is more adaptive in the Afrotropical savannas that are more heavily browsed (Dantas and Pausas, 2013; Maracahipes et al., 2018). Fire kills trees by destroying the vascular cambium but thick bark protects the vascular cambium from the heat of the fire (Uhl and Kauffman, 1990). These comparisons also hold in Australia, where height growth was greater in rain forest species but bark thickness was greater in savanna species (Ondei et al., 2016). A larger comparison of traits demonstrated that fire in Neotropical savannas selected for increased bark thickness for enhanced insulation of shoots and buds, and belowground storage organs typified by the geoxyles (Figure 7.13) (Burrows, 2002; Tomlinson et al., 2012; Dantas and Pausas, 2013; Charles-Dominique et al., 2015a; Dantas and Pausas, 2020). However, fire selected against high wood density, fast height growth rate, tall maximum height, high SLA, and divarication (wire plant caginess) (Archibald and Bond, 2003; Dantas and Pausas, 2013, 2020). In contrast, megafaunal herbivory selected for higher wood density, height growth rate, spines, and divarication (Archibald and Bond, 2003; Dantas and Pausas, 2013, 2020), but selected against maximum height and SLA (Dantas and Pausas, 2020). Competition in forests is associated with greater SLA and height (Dantas and Pausas, 2020) and large fruits are more common in forests where megafauna occur (Mack, 1993; Guimarães et al., 2008). One trait that has received far less attention is post-fire flowering rates, but they appear to be very important in the Cerrado (Fidelis and Zirondi, 2021).

The "corky" and "lanky" strategies appear to merge among trees in montane forests that use the fire "escape" strategy (Table 7.1). One of my favorite forest types is the ponderosa pine (*Pinus ponderosa*)–bunchgrass forest ecosystem of the southwestern USA (Figure 7.7), which has persisted for centuries under a frequent-low severity surface fire regime where fires historically occurred every 2–10 years (Fulé et al., 1997; Moore et al., 1999). Flashy fires were driven by fine fuels in the grassy understory and the scattered clumps of old-growth pines were mostly untouched. Ponderosa pine and Douglas fir

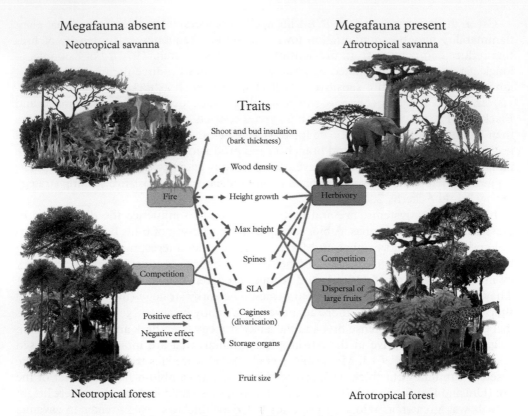

Figure 7.13 *Summary of the effects of megafaunal history and the effects of fire and herbivory on plant traits of tropical woody species in Neotropical and Afrotropical forests and savannas.* Adapted from Dantas and Pausas (2020) with permission.

(*Pseudotsuga menziesii*) evolved in an evolutionary environment of more frequent surface fires. These trees evolved thick outer bark, and the oldest among them are often protected by a bark layer > 10 cm thick! Bark thickness is greatest in forests with high fire frequency and low fire intensity (Laughlin et al., 2011a; Pausas 2015). Curiously, trees with thick outer bark also appear to be more abundant in infertile soil, an observation that is common enough to require more investigation (Jager et al., 2015; Richardson et al., 2015; Vleminckx et al., 2021). Many fire-escaping conifers in these ecosystems also self-prune their lower branches which prevents surface fires from climbing into their canopies. Growing tall allows their canopies to escape the heat of the surface fires. But the bunchgrasses coexist in the same frequent-fire disturbance regime. What strategies are used by the grasses to maintain populations in the face of frequent fire?

Archibald and colleagues (2019) proposed a fresh perspective to consider how plant strategies have been molded by fire and herbivory along three key dimensions: avoidance, resistance, and tolerance (Figure 7.14; Table 7.2). First, the "avoidance–attraction

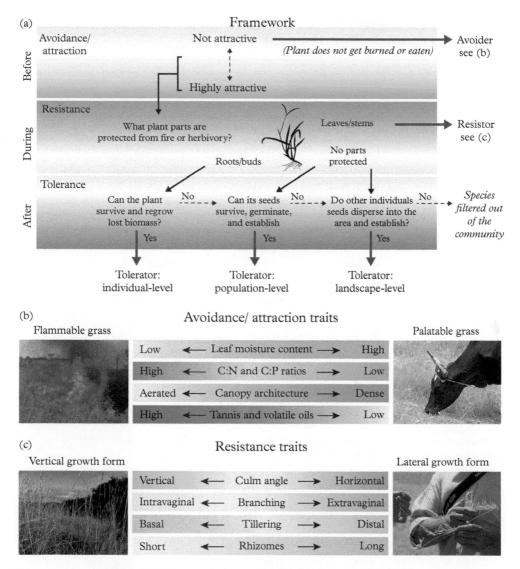

Figure 7.14 *(a) An integrated framework that considers how grasses avoid being consumed by herbivores or fire before the event, how they resist being consumed during the event, and how they tolerate the disturbances and persist after the event. (b) Key to this model is the fact that traits associated with flammability are opposite to those associated with palatability. (c) Fire-adapted grasses tend to grow vertically to resist being consumed by fire, whereas grazer-adapted grasses tend to grow laterally to resist being consumed by grazers.*

Reproduced from Archibald et al. (2019) with permission.

Table 7.2 *Grass traits associated with avoidance, resistance, and tolerance of herbivory versus fire.*

	Grazing	Fire
Attraction traits (palatable/flammable)	Low C:N ratio	High C:N ratio
	High bulk density	Low bulk density
	High leaf moisture content	Low leaf moisture content
	Low tannin content	High tannin content
	High P content	Low P content
	Large leaves	High biomass
	High salt content	
	Low silica content	
Resistance traits	Meristems at/below the soil surface	Meristems at/below the soil surface
	Lateral growth (extravaginal branching, prostrate culms, stoloniferous/rhizomatous)	Vertical growth (intravaginal branching, erect culms, short rhizomes)
	Strong root system that prevents uprooting	Distal tillering to move flames away from the basal meristems
	Leaves and culms with low tensile strength (alternatively) Spikey hard culms/spines that protect aerial leaf material	Retain leaf sheaths to protect buds
Tolerance traits (individual-level persistence)	Rapid resprouting/large bud bank. Substantial stored reserves	Rapid resprouting/large bud bank
Tolerance traits (population-level persistence)	Geniculate growth form (flowers not eaten)	Early seed set and release (before fire season)
	Clonal growth (rooting at nodes)	Smoke-stimulated germination
		Seed dormancy
Tolerance traits (landscape-level persistence)	Good dispersal ability (especially ecto- and endozoochory)	Good dispersal ability (especially wind dispersal)
	Rapid germination and establishment	Rapid germination and establishment
	Short generation times	Short generation times

Reproduced from Archibald et al., (2019) with permission.

continuum" acts before plants are defoliated and determines the likelihood that a plant will be consumed. The fundamental trade-off is that traits associated with flammability (high C:N ratio, low moisture content, high tannin concentration) are opposite to those associated with palatability (low C:N ratio, high moisture content, low tannin concentration). Second, the "resistance" continuum acts during the defoliation event and determines how much biomass is consumed. Fire-adapted grasses tend to grow vertically to resist being consumed by fire, whereas grazer-adapted grasses tend to grow laterally to resist being consumed by grazers. Buried meristems allow plants to resist both grazing and fire. Third, the "tolerance/persistence" continuum acts over the lifespan of the plant and determines whether a population can persist in the presence of the consumer. Large bud banks allow plants to tolerate grazing and fire at the individual level. At the population level, clonal growth permits tolerance of grazing whereas seed bank germination permits tolerance of fire. Good dispersal and germination and short generation times permit tolerance of grazing and fire at the landscape level.

Archibald and colleagues (2019) suggest that four possible strategies emerge when fire and grazing traits are compared simultaneously. First, fire resistors/grazer avoiders are likely to be flammable. Second, grazer resistors/fire avoiders are likely to be palatable. Third, generalist tolerators maintain populations under moderate levels of grazing and fire through resprouting. Fourth, generalist avoiders cannot tolerate grazing or fire and are unlikely to be competitive. The authors then identify eight growth forms that exhibit the traits of each of these strategies. This framework can be applied to predict how different grass phenotypes will respond to changing fire and grazing regimes and represents a significant advancement over the classic increaser/decreaser framework in rangeland ecology (recall Figure 2.2). For example, a species could increase under heavy grazing by being unpalatable to avoid grazing, or by being palatable and resistant to grazing. This promising framework should be straightforward to extend to all herbaceous plants and now requires testing in ecosystems around the world.

7.10 The most severe disturbances are the least frequent

We have so far focused on less severe disturbances that often retain biological legacies that jump start vegetation recovery following the disturbance. Some disturbances are so severe that all biological legacies are eliminated, necessitating colonization processes to start over from scratch. Such is the case on volcanic substrates such as a'a and pahoehoe lava—these substrates are infertile (Walker and del Moral, 2003) and select for species with either high nutrient-use efficiency (Chiba and Hirose, 1993) or species that fix nitrogen through bacterial symbiosis (Clarkson and Clarkson, 1995). Primary succession typically requires seed dispersal into the site because the seed bank was destroyed. Early primary succession is like a lottery in which arrival is the most important determinant of early plant establishment (Walker and del Moral, 2003). If the disturbance was large in extent, then species with wind-dispersed seeds tend to arrive first

(Grishin et al., 1996), but the species that spread the fastest exhibit clonal growth (Fuller and del Moral, 2003). Colonists that combine long-distance seed dispersal with vegetation propagation dominate early stages of succession (Bazzaz, 1979; Prach and Pyšek, 1999). Mudflows from volcanos (lahar) produce substrates that are more fertile and are likely closer to remnant patches of undisturbed vegetation that can provide propagules for plant establishment (Walker and del Moral, 2003).

Glaciers are "slow-moving rivers of ice" (Prach and Walker, 2020, p. 77) that scour the landscape and leave behind substrates bereft of any biological legacy. While not completely sterile, vegetation development is still considered to be primary following glacial retreat. Recently deglaciated surfaces can form substrates that span a range of particle size and fertility. Phosphorus is most abundant immediately following the retreat because phosphates are derived from the weathering of rock, but it declines through time because it cannot be recycled and retained (Walker and Syers, 1976). Nitrogen, on the other hand, is lowest following retreat and gradually increases over time as nitrogen accumulates through fixation and recycling. Vascular plants that associate with nitrogen-fixing bacteria are also common on recently deglaciated soil (Walker and del Moral, 2003). Grubb suggested that long-lived species would invade sites with larger particle sizes: grasses would invade sandy and silty soil, herbs would invade gravels, trees would invade rock crevices, and bryophytes would establish on rock surfaces (Grubb, 1987). Annual and biennial species are uncommon, however, because glaciers occur in cold climates where these life forms are much less common (Prach and Walker, 2020). It is difficult to generalize about traits that confer fitness during primary succession given the variation in environmental conditions that exists across sites. Some stressful sites will be low in nutrients and water following a disturbance. These conditions will prevent the establishment of short-lived and fast-growing colonists, and phenotypes adapted to low nutrient and low water will establish first even though they are slow-growing and longer-lived (Grime, 1979).

7.11   Temperature is the ultimate regulator, but few traits directly relate to it

Temperature is one of the most powerful controls on the distribution of vegetation globally (Collinson, 1988; Akin, 1991; Breckle, 2002), and temperature drives both resource availability and disturbance regimes (recall Figure 3.9). However, we still do not understand which trait or set of traits are the master controls on plant responses to temperature. Mark Westoby told me that, "In the era of climate change, this strikes me as the biggest gaping hole in the whole of our knowledge of ecology."

We do know some things. Plants adapted to the coldest inhabited climates on earth are short and limit their growth to narrow windows during the short growing season (Körner, 2016), and as described in Chapter 6, large leaves are adaptive in hot and humid climates but not in hot and dry climates (Wright et al., 2017). Every gardener knows that some species just can't be grown in certain latitudes. Plant hardiness zones have been created to help homeowners and vegetable growers select which species they are going to plant (Keddy and Laughlin, 2022), but those are species-specific classifications and have been not generalized to phenotypic traits. Raunkiaer (1934) life forms as defined by meristem location also relates plant life forms to temperature gradients. Life forms with aboveground buds are dominant in hot climates whereas plants that bury their buds belowground are more common in cold climates (recall Figure 2.1). Grime linked nuclear DNA contents to shoot phenology timing along temperature gradients, but this relationship has not been established across larger species sets globally (Grime, 1983; Grime et al., 1985). Perhaps species can be ranked by the abundance or expression of heat-shock proteins (Augustine, 2016) given that crops such as maize and rice show potential for enhanced drought tolerance when heat shock proteins are expressed (Xiang et al., 2018).

Plant ecologists have not been able to identify the master continuous trait that has been under strong selection by temperature, although many traits are linked to other underlying gradients that are driven by temperature. Many traits appear to be weakly correlated with temperature (e.g., LMA, wood density) because temperature is correlated with so many other variables (e.g., water availability, disturbances). For example, if we assume that plant height is a proxy for primary production, we can see that plant height generally increases with mean annual temperature (recall Figure 3.10) (Del Grosso et al., 2008), but this relationship is contingent on sufficient water supply and disturbance regimes.

We can make some general predictions about how the fitness landscapes for the five primary plant trait dimensions are shaped in cold climates (Figure 7.15). The fitness landscape for leaf economics is likely bimodal in cold climates. Long-lived conifers have extremely slow leaf economics, but herbaceous plants must finish their phenological cycles within a short growing season and so must exhibit fast leaf economics (Carter et al., 2022). Extremely cold climates are dominated by short vegetation that persists in shallow soil, so we can expect short heights and shallow rooting depth to confer fitness in these cold environments. Thin roots are also advantageous in cold climates because as a plant you either mine the soil for limiting nutrients by yourself or you collaborate with mycorrhizal types that tend to prefer thinner roots, such as ectomycorrhiza and ericoid mycorrhiza (Read, 1991; Laughlin et al., 2021). The resprouter strategy is likely advantaged over the seeder strategy where harsh weather damages plants and limits opportunities for successful establishment from seed. In contrast to cold climates, the fitness landscapes in warm climates are likely to be bimodal or flat—a greater diversity of strategies coexist in warm climates.

Tree species distributions have lower temperature limits (Figure 7.16) based on whether the leaf is broad or needle-shaped and evergreen or deciduous (Woodward, 1987; Woodward et al., 2004). Mean annual and maximum temperature do not have causal mechanistic relationships to plant distributions, but minimum temperatures do. Needleleaf trees are only found in climates with a minimum temperature below 15°C. The deciduous needleleaf class (e.g., *Larix* spp.) can only be found in climates with a minimum temperature below −30°C. Broadleaf deciduous trees can be distinguished based on whether they drop their leaves in the cold season (cold-deciduous) or in the dry season (drought-deciduous). Cold-deciduous trees occur at

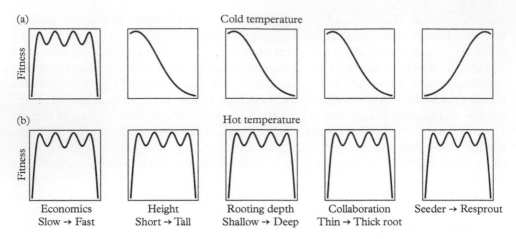

Figure 7.15 *Hypothesized fitness landscapes for each of the five functional trait spectrums in (a) cold climates and (b) hot climates. Fitness landscapes with four evenly spaced peaks are "flat" landscapes to indicate that multiple trait values are equally fit within the environmental condition. These fitness landscapes should be viewed as qualitative, rather than quantitative, hypotheses to be tested with empirical data. For simplicity, these illustrations ignore the potentially important trait-by-trait interactions that likely influence plant fitness in a given environment.*

minimum temperatures of −30 and 5°C, but the drought-deciduous trees occur above a minimum temperature of 7°C.

Herbaceous plants that differ in their photosynthetic pathways also exhibit differences in distribution along minimum temperature gradients (Woodward et al., 2004). C_3 grasses occur in climates where the minimum temperatures range from −50 to 15°C. C_4 grasses only occur in climates where the minimum temperatures range from −20 to 30°C. The underlying mechanism that determines these climatic distributions are different temperature optimums for the enzymes involved in the metabolic pathways (recall Figure 3.8). These boundary conditions could be a starting point for plant strategy models that include minimum temperature (Table 7.3). For example, if the minimum temperature of the site was 20°C, then needleleaf trees, cold-deciduous trees, and C_3 grasses are immediately excluded. Photosynthetic pathway is a key categorical trait that usefully discriminates plant physiological performance along temperature gradients. However, it is not the most generalizable trait because the majority of plant species use C_3 photosynthesis.

This chapter reviewed empirical evidence and theoretical expectations for how the shapes of fitness landscapes in particular environments are governed by functional and life history traits. I wish to emphasize that the illustrated fitness landscapes in this chapter are hypotheses to be tested and should not be viewed as empirical facts. Indeed, the shape of these landscapes may depend on the scale of the habitat (Section 7.1). Chapter 8 describes how these hypotheses can be tested using a variety of empirical approaches to galvanize rapid progress.

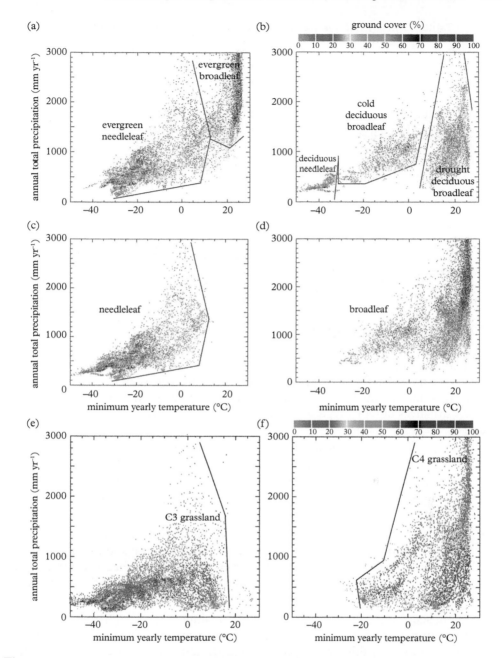

Figure 7.16 *(a–d) Tree cover (percentage of ground cover) along gradients of mean minimum annual temperature and annual total precipitation for (a) evergreen, (b) deciduous, (c) needleleaf, and (d) broadleaf trees. Boundary lines were drawn by eye. (e–f) Grass cover (percentage) along the same gradients for (e) C₃ grasses and (f) C₄ grasses.*

Reproduced from Woodward et al. (2004) with permission.

Table 7.3 *Woody and non-woody plant traits that relate to minimum temperature.*

	Lower Minimum Temperature (°C)	Upper Minimum Temperature (°C)
Woody plants		
Needleleaf	−50	15
Deciduous needleleaf	−50	−30
Cold-deciduous	−30	5
Drought-deciduous	7	30
Non-woody plants		
C3	−50	15
C4	−20	30

Summary data from Woodward et al., (2004).

7.12   Chapter summary

- This chapter reviews and generates hypotheses about how functional and life history traits determine the shapes of fitness landscapes when resources are abundant, when resources are limited, when disturbances are frequent or large, when disturbances differ by type, and in cold and hot climates.

- The scale of the habitat is one of the greatest challenges when building a conceptual theory for how phenotypic traits govern fitness differences across environmental gradients. There is no doubt that individual plants only experience the conditions present within the scale of their own canopy and root system, but most environmental data used to test the effects of traits are at the landscape-scale.

- The classic fitness function of a trait is a unimodal curve, where there is a single trait value that optimizes fitness in a given environment and traits to either side of the peak confer weaker fitness benefits. However, multiple combinations of trait values can simultaneously optimize fitness in a given environment, and this changes the unimodal fitness function into a multimodal fitness function.

- Rapid acquisition of resources is critical for maintaining the high growth rates that are required to compete in a resource-rich environment. The phenotype with the highest fitness in this circumstance would exhibit fast leaf, stem, and root economics: low LMA, high tissue nutrient concentrations, and low-density stems and roots.

- Water stress imposes immense negative pressure on vascular conduits, and the collapse of the water column can ultimately lead to plant mortality. Drought-tolerant phenotypes exhibit strong resistance to embolism in their xylem and low leaf turgor loss points. Rooting depth is bimodal across phenotypes in drylands: escapers have shallow roots to utilize water at the soil surface after it has rained, tolerators also tend to have shallow roots, but avoiders can send down extremely deep roots to tap into deep water tables.

- Phenotypes that tolerate nutrient limitation share many similarities with phenotypes that tolerate water limitation. This is in part because water-limited systems are also nutrient-limited because mass flow and diffusion rates are lower in dry soil. The difference between water and nutrients is that water must be lost from the plant to photosynthesize, whereas nutrients can be reused and recycled. Fitness in low-nutrient habitats can be optimized by increasing resource uptake, decreasing resource loss, and increasing the efficiency of resource use.

- The vegetation canopy in productive environments will close given sufficient time after a disturbance, and canopy closure leads to light limitation. Light limitation will select for taller species with lower whole-plant light compensation points, lower dark respiration rates, and higher LMA.

- High-resource conditions following a disturbance will select for species with fast pace of life, low degrees of iteroparity, and short lifespans. When resources become limiting over time and vegetation is dense, then species with a slower pace of life, high degree of iteroparity, and long lifespans will be selected.

- Plants would be far more efficient in the absence of herbivores, but the benefits of defending themselves against being eaten are greater than the costs. Most animals, including mammals, birds, and insects, eat plants to obtain energy and protein. The quality of plant material is determined by the amount of labile carbon within plant cell contents (as opposed to within the cell walls), the amount of protein within the cell, and the types and quantities of secondary compounds.

- There are two general mechanisms of persistence in crown fire regimes: resprouting from dormant meristems or seedling recruitment. In surface fire regimes, fire escapers grow tall and protect their lateral meristems with thick bark. Herbaceous plant strategies involve traits that align on axes of avoidance, resistance, and tolerance.

- We still do not understand which trait or set of traits are the master controls on plant responses to temperature, but photosynthetic pathways and woody plant leaf habit are strong categorical determinants of distributions along minimum temperature gradients.

7.13 Questions

1. Are there any dimensions of resource availability or disturbance that were not discussed in this chapter that deserve greater recognition for a general theory of plant strategies?

2. Climate is often assumed to be the dominant driver of global vegetation patterns, but fire and herbivores are globally significant consumers of vegetation. How are fire and herbivore regimes determined by climate?

3. Grime's model assumes that habitat duration (disturbance) and habitat productivity (resources) are orthogonal, but how do disturbance gradients influence resource availability? Do traits that confer fitness in disturbance prone habitats also confer fitness in resource-rich habitats?

4. Does bark thickness align with one of the five major trait dimensions, or do you think it will be independent of other traits?

Part IV

The Net Effect of Traits on Fitness

Part IV integrates all that has been discussed so far into empirical and theoretical approaches for explaining and predicting fitness from traits and clarifies how plant strategies are central to conservation and restoration.

Chapter 8 describes a variety of observational and experimental approaches for measuring fitness of multidimensional phenotypes and describes analytical approaches for modeling fitness as interactions between traits and environments. It describes ways to account for density dependence to estimate the propensity of a population to recover from low population sizes to quantify the fundamental niches of phenotypes.

Chapter 9 introduces evolutionary game theory to predict fitness from traits when the fitness of a focal plant depends on what others are doing in the community. It describes how theoreticians explore the effects of density-dependent and frequency-dependent selection of the evolution of plant strategies to quantify the realized niches of phenotypes. Incorporating game theoretical insights into empirical estimates of fitness remains an important challenge in ecology.

Chapter 10 demonstrates how plant strategy theory is integral to effective conservation of rare species, the restoration of degraded habitats, our understanding of invasive species, and the assisted migration of species in a changing world.

8

Empirical Approaches to Infer Fitness from Traits

Vegetation science is unlikely to be able to make useful predictions unless the fundamental niche is known.

—Michael Austin (1990, p. 227), "Community theory and competition in vegetation," in *Perspectives on Plant Competition*

8.1 The fitness of a species varies along environmental gradients

Both ecology and evolution are grounded in the same paradigm that species grow in the habitats to which they are adapted. That is not to say that all traits of an organism are Panglossian perfections (Gould and Lewontin, 1979). Some traits are adaptive and functional, some traits are exaptive, others are just along for the ride, and to assume that all phenotypes have achieved peak fitness in a given environment approaches fantasy (Niklas, 1997). But few botanists would argue with me if I predicted that an organ pipe cactus (*Stenocereus thurberi*) would die if planted in the Tierra del Fuego archipelago. Reciprocally, Magellan's beech (*Nothofagus betuloides*) would perish if planted in the Sonoran Desert. These predictions reflect our expectation that these species are adapted to the environmental conditions within their respective home habitats. But just how far away from these home habitats do we need to go to witness the precipitous fitness decline?

An important premise of this book is that the fitness of a species varies along environmental gradients (Levins, 1968). In the words of John Harper (1977, p. 6), "The intrinsic rate of natural increase is of course a function of a specified environment." We expect the population of organ pipe cactus to be stable in the desert yet decline to extinction in the Magellanic subpolar forest. In Chapter 3 we saw an example of this phenomenon with the Colorado butterfly plant (*Oenothera coloradensis*), where one population was doing well but the other was in decline due to changing environmental conditions (recall Figure 3.2). Virtually all demographic studies of plant populations verify the important concept that population growth rate varies among populations of the

Plant Strategies. Daniel C. Laughlin, Oxford University Press. © Daniel C. Laughlin (2023). DOI: 10.1093/oso/9780192867940.003.0008

same species, but the central challenge is to determine what environmental conditions cause this variation. Expressed as a function,

$$\log(\lambda) = F(e), \tag{8.1}$$

where e represents the environmental conditions. The classic fitness function of a single species is unimodal with an optimal fitness at some point along an environmental gradient (Figure 8.1a).

One classic study showed that population growth rates (λ) of the monocarpic perennial teasel (*Dipsacus sylvestris*) varied from a low of 0.275 to a high of 2.605 (Werner and Caswell, 1977), suggesting that the species was either declining at a rate of 72.5% or increasing at a rate of 160.5%. Patricia Werner and Hal Caswell surmised that the variation was driven by the biotic environments in each field and determined that germination rates and seedling survival rates were depressed in fields with high grass and forb cover. Population growth rates were highest in sites of low total community productivity, indicating that teasel was best suited to sites where more frequent disturbances kept the vegetation open. Only eight populations were studied so it was difficult to draw strong conclusions, but it is possible to imagine building a larger model of the environmental factors that influence teasel population growth to predict population dynamics across fields, wetlands, forests, roadside verges, and parking lots. Understanding the environmental drivers of single-species population dynamics has been a major research agenda in plant demography for the last half century, and it is the first step toward achieving the goal in this book.

Demographic modeling has come a long way since Werner and Caswell's (1977) pioneering work: ecologists are now projecting population dynamics of species across continents. By combining field and greenhouse experiments, Merow and colleagues

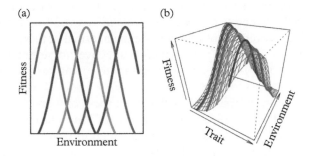

Figure 8.1 *(a) The fitness of each species is classically described as a nonlinear (typically Gaussian) function of the environment. (b) The fitness of each species along the environmental gradient depends on their traits (Laughlin and Messier, 2015). Given the negative interaction between traits and the environment, low trait values only optimize fitness at the high end of the environmental gradient, which is why the red species is found there. Moreover, high trait values only optimize fitness at the low end of the environmental gradient, which is why the orange species is found there.*

Reproduced from Laughlin et al. (2020b) with permission.

(2017) were able to determine the demographic mechanisms that lead to expansion or contraction of invasive species ranges under climate change. The researchers set up field experiments to quantify growth, survival, and reproduction rates of multiple species across a gradient of temperature, precipitation, and light. They demonstrated that high population growth rates across New England could be achieved by the invasive *Alliaria petiolata* (garlic mustard) through demographic compensation (Doak and Morris, 2010): in open habitats with high light, low growth and survival could be compensated by high fecundity, whereas in closed habitats with low light, low fecundity could be offset by high growth and survival (Figure 8.2). Remarkably, the estimated probabilities that a population of garlic mustard could establish (i.e., $\lambda > 1$) corresponded strongly with the southern limit of known populations (Figure 8.2), demonstrating the power of a demographic approach to forecasting population dynamics under climate change (Merow et al., 2017).

8.2 Fitness is determined by traits in a given environment

Plant strategy theory makes a bold proposition that predictions of population growth for any species can be made from its traits. Here, we are concerned with all species, not just one or two. The goal of plant strategy theory is to predict population growth rates of any species in any environment using its trait values as predictors. This goal is grounded in the idea that there is recognizable order to the diversity of plant form and function—that species who share sets of traits respond similarly to the environment and to the strategies of others. Note that the question is not whether the phenotype matters—for what else would fitness depend on? Rather, the question is how much predictive power is contained in the traits that we currently measure and in traits that we aren't measuring?

Our general model of fitness needs to be expanded to include multiple species. Equation 8.1 is too simple. Let us now consider the population growth rates of multiple species simultaneously across an environmental gradient. The environment does not affect each species the same way. For example, aridity will increase the fitness of organ pipe cactus but will decrease the fitness of Magellan's beech. To generalize across species, we need traits in our model to account for this dependency because every species response to the environment depends on their traits. Our model then grows to

$$\log(\lambda) = F(e, v), \tag{8.2}$$

where we model population growth rates of each species as an interaction between trait strategies of the focal species v and the environment e. Note that in Chapter 9 we expand this further to include the strategies and densities of other species within game theoretic models of adaptive dynamics. The classic unimodal fitness function of

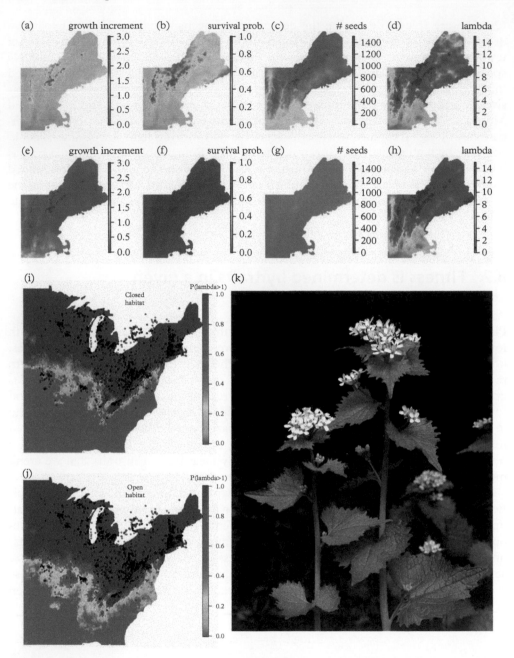

Figure 8.2 *Demographic rates (growth, survival, and fecundity) and population growth rates (λ) for* Alliaria petiolata *in (a–d) open and (e–h) closed habitats. Occurrence records (black dots) of* Alliaria petiolata *overlaid on top of estimated probabilities that a population could establish (λ > 1) in (i) closed and (j) open habitats. (k) Photograph of* Alliaria petiolata *in Grant County, Indiana*

(a–j): Reproduced from Merow et al. (2017) with permission. (k): Photo by Paul Rothrock, reproduced here with permission under a CC BY-SA license (https://swbiodiversity.org/seinet/).

a single species can be expanded to include multiple species fitness functions along the environmental gradient, where the optimum fitness is found along a ridge that varies as a function of species traits (Figure 8.1b). Gaussian functions are often used for mathematical convenience, but skewed functions have been proposed as better descriptors of fundamental niches along environmental gradients (Austin and Smith, 1989; Austin, 1999).

In Chapter 3, I discussed two hypotheses that extended Arnold's (1983) model of organism fitness to plant strategy theory (Figure 3.3). In one hypothesis, life history traits are emergent properties of the effects of functional traits on demographic rates, but they have no direct effect on fitness. The logic to this argument was that since fitness is a direct function of demographic rates, and since life history traits are also a direct function of demographic rates, that it would be double counting to use life history traits as predictors of fitness. The second hypothesis relaxes this assumption and states that both life history traits and functional traits drive fitness differences. The modeling framework in this section of the book conveniently ignores the direct effects of traits on demographic rates, and instead, focuses on modeling the net effect of traits on fitness (Laughlin et al., 2020b). Therefore, we can use both functional traits and the emergent life history traits as predictors of fitness in a given environment. The final chapters of the book explore the implications of a more mechanistic approach to modeling fitness through each demographic rate, in which case we will have no choice but to leave out life history traits, given their derivation from the demographic rates.

In this chapter, I emphasize how traits optimize fitness in the absence of competition (density independence) and discuss how to account for density dependence empirically when demographic rates depend on local neighborhood densities. Optimization by natural selection is a powerful concept for understanding traits in relation to their environment (Mäkelä et al., 2002; Kokko, 2007; Prentice et al., 2014). Chapter 9 explores other modeling approaches that account for density and frequency dependence when fitness depends on what heterospecifics are doing in the community. Despite decades of acknowledging the difference between the fundamental niche (i.e., fitness landscapes in the absence of competition) and the realized niche (i.e., fitness landscapes in the presence of competition), plant ecologists will continue to struggle to make accurate predictions until we understand the fundamental niches of phenotypes (Austin, 1990). The fundamental niche of a phenotype is the range of environmental conditions where growth rates are positive for a trait or combination of traits. It extends the concept of the species niche to phenotypic traits. The realized niche is (most often) a subset of the fundamental niche, and in many cases, it is likely that realized niche boundaries are set by biotic interactions at the more favorable edge and by ecophysiological limits at the unfavorable edge (Westoby, 2022). It is difficult in practice to separate the effects of the environment from the effects of neighbors, so care must be taken to interpret modeling results accordingly (Kraft et al., 2015a). Before we get to the realized niche, let us first examine the fundamental niche and explore a variety of different empirical approaches to advance this research agenda.

8.3   Observational approaches to model the net effect of traits on fitness

We are living in the Big Data revolution. We are now able to merge large datasets of species occurrences spanning global environmental gradients with large datasets of species traits (Bruelheide et al., 2018). Such analyses have yielded valuable insights into the importance of traits as drivers of species distributions. New model-based frameworks can provide strong tests to determine if trait–environment interactions explain species occurrences beyond what the environment explains by itself (Pollock et al., 2012; Jamil et al., 2013; Miller et al., 2018). For example, after accounting for non-linear species distributions across broad temperature gradients, root tissue density and specific root length explain additional variation in probability of occurrence (Figure 8.3) (Laughlin et al., 2021).

Such analyses will continue to yield insights, but it is important to be clear about what they are testing and what inferences can be drawn from them. Such analyses are limited to modeling realized niches, not fundamental niches, so the influence of conspecific and heterospecific densities and frequencies are confounded with the processes of abiotic environmental filtering (Keddy and Laughlin, 2022). Plant strategy models have historically been built on observing species that share similar sets of traits in similar environments (Grime, 2001), but occurrence is not fitness. Knowing just the occurrence or even the abundance of a species tells us nothing about the trajectory of that population, and just because we observe a species somewhere does not mean its traits are well suited to that environment. Populations that are present at a site could be experiencing negative population growth rates due to a lack of fit between their traits and the environment, and absences of a population from a site could be driven more by dispersal limitation or competition rather than a mismatch to the environment (Kraft et al., 2015a). In short, merely observing traits in a given environment is not a strong test of plant strategy theory.

I also emphasize that our ability to draw strong inferences about plant strategies depends on whether demographic failures (i.e., attempts to establish a population that ultimately lead to local extinction) are included in our observations. We need to observe failed introductions in *unsuitable* habitats to learn about which traits promote successful introductions into *suitable* habitats. Observational demographic datasets often lack measurements of population declines because you cannot observe the behavior of a species that has already been filtered out of the habitat. In many cases, experimental introductions into unsuitable habitat will be required. With that word of caution, I now describe several empirical approaches that can galvanize progress toward understanding the net effect of traits on fitness (Table 8.1). These approaches vary in their degree of difficulty and start with the easiest approach first. Common to all these approaches is the requirement to simultaneously analyze fitness of multiple phenotypes across environmental gradients.

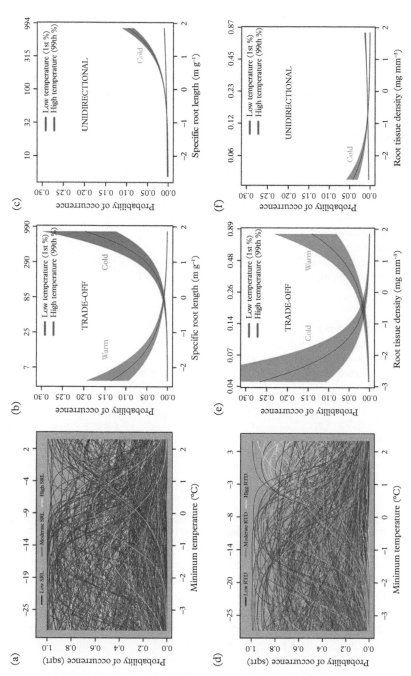

Figure 8.3 *To test for the effects of traits on species occurrence, one must first account for the variation in occurrence that is explained by the environment. Traits can then be added simultaneously to these models to test whether trait effects on occurrence vary along the environmental gradient. (a) Modeled species distributions for 875 species in the model along the temperature gradient (in all figures, the x-axes are scaled to unit variance on the lower edge and displayed in log scale on the upper edge). (b) The effect of specific root length (SRL) on probability of occurrence strongly depends on temperature in forests. (c) In contrast, the effect of SRL exhibits unidirectional benefits in grasslands. (d–f) The same analyses using root tissue density (RTD) on 625 species.*

Reproduced from Laughlin et al. (2021) with permission.

Table 8.1 *Summary of empirical approaches to test plant strategy theory discussed in Chapter 8.*

Approach	Strengths	Limitations	Difficulty
1. Occurrence or abundance models (Figure 8.3)	Data is available at global scales across large gradients. Hierarchical models can isolate the effects of traits on probability of occurrence.	Does not account for density or frequency dependence.	Easiest approach, but occurrence and abundance are not proxies for fitness.
2. Population-level monitoring data (Figure 8.4)	No need to integrate across all demographic rates of individuals. Intrinsic growth rates can be estimated through statistical control of density.	Competition is controlled statistically, not experimentally. Statistical artefacts and error-prone proxies can affect estimates. Does not account for age or size structure.	Easiest approach that estimates fitness.
3. Density-dependent population models (Figure 8.5)	Density dependence can be modeled using regression models of demographic rates.	Competition is controlled statistically, not experimentally.	Moderately difficult and rigorous.
4. Experimental common gardens beyond range boundaries (Figure 8.6)	Competition and environmental gradients are experimentally controlled.	Can be expensive and time consuming. Can be difficult for long-lived species.	Most difficult but most empirically rigorous.

The first approach to modeling fitness leverages widely available population-level monitoring data (Figure 8.4). It does not require population models built from observing individual plants over time. All that is required are annual measurements of population abundance (numbers of individuals, total cover, total biomass, etc.). Consider a dataset where cover estimates of every species in the community are recorded in plots in consecutive years. These abundances can be used to compute the annual growth rate of a population by dividing population size next year by size in the current year (*i.e.*, $\lambda_t = N_{t+1}/N_t$). When studying species where individual genets are rarely counted then cover or biomass of the population could be substituted for population counts (i.e., $\lambda_t = Cover_{t+1}/Cover_t$) (Tredennick et al., 2017). Perhaps the data come from an herbaceous community that exhibits rapid turnover that can be described by the carousel model where species come and go over time (van der Maarel and Sykes, 1993),

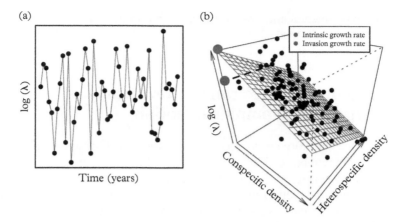

Figure 8.4 *(a) Time series of population sizes can be used to compute annual population growth rates (λ). (b) These population growth rates can be modeled as functions of conspecific (same species) and heterospecific (other species) densities to compute empirical intrinsic and invasion growth rates. In this example, both effects of neighborhood density are negative. Setting C = 0 and H = 0 and solving for log(λ) is an estimate of the intrinsic growth rate (blue dot) of a species in the absence of neighbors. Setting C = 0 and H = mean(H) and solving for log(λ) is an empirical estimate of the invasion growth rate (orange dot) of a species when it is rare but other species are at their average abundance. See text for how this estimation of invasion growth rates differs from modern coexistence theory.*

and this turnover is driven by annual variation in climate. To ask whether traits explain this temporal variation in annual growth rates, you can model λ_{ij} for each species i in each year j as a function of inter-annual climate and species trait values. Traits that enhance population growth in wet years but reduce populations in dry years will be detected by significant trait-by-climate interactions. This approach can either be used to analyze annual growth rates directly as functions of trait-by-environment interactions at the plot-level, or the data can be further processed to compute low-density population growth rates to account for density dependence at the site-level.

This first approach is advantageous because the method implicitly integrates across the individual processes of growth, survival, and reproduction by emphasizing changes in total population abundance. This approach can also be applied to dozens of species that co-occur in widely available monitoring datasets. However, caution is required when using this method for three reasons. First, this approach does not account for age or size structure of the populations, which are important drivers of population dynamics. Second, comparisons of population growth rates across species using changes in total cover may be affected by the fact that species vary in maximum size. Third, statistical artefacts can sometimes affect the estimates of density dependence given that $Cover_t$ is in the denominator of λ (which could mathematically guarantee that the relationship between $\log(\lambda)$ and $Cover_t$ is negative), and so accounting for census error can be a way to account for this and improve estimates (Freckleton et al., 2006; Detto et al., 2019). There are likely thousands of such datasets tucked away in drawers or on computers.

Time series of experimental and observational communities are becoming increasingly available for this sort of approach (Borer et al., 2014; Dornelas et al., 2014; Komatsu et al., 2019; Sperandii et al., 2022).

8.4 Density dependence can be statistically accounted for in empirical models

Contrary to one of the more controversial propositions in Grime's model, most plant ecologists view competition for limited resources to be a ubiquitous process in all vegetation types, from sparse deserts to dense tropical forests. One can almost watch competition between individuals in tropical forests, but it is perhaps more difficult to imagine it occurring in sparse vegetation. Nevertheless, the effects of density on individual growth, survival, and reproduction are clear in practically every competition experiment ever conducted (Tilman, 1988).

Recent meta-analyses indicate that competition for resources tends to be strongest between individuals of the same species (Adler et al., 2018b). These results were a welcome confirmation of the theoretical requirement for coexistence—stronger competition between individuals of the same species permits individuals of different species to coexist. I often ask students at the beginning of my ecology class which type of competition they think would be most fierce: within-species (intraspecific) or between-species (interspecific). The answers at first are a toss-up—50/50—most likely because they are guessing. Then I ask them to consider the jobs that they will be applying for when they graduate. Will they be competing for those jobs against other students or against dolphins, octopi, or orangutans? After considering this amusing thought, they rarely get the question wrong when they see it again on the test.

Demographic rates are often affected by the density of conspecifics, that is, the density of individuals of the same species. These density-dependent effects can be positive or negative. Pollination rates could exhibit positive density dependence because the presence of more individuals of the same species nearby will increase the chance of fertilization, especially in obligate out-crossing species. These positive effects at low populations sizes are called Allee effects (Silvertown and Charlesworth, 2001). But growth, survival, and recruitment rates often exhibit negative density dependence because individuals are competing for the same set of resources. To account for density dependence at the site-level one can fit the following regression model:

$$log\,(\lambda) = \alpha + \beta_1 C + \beta_2 H \tag{8.3}$$

where C is conspecific (i.e., same species) density and H is heterospecific (i.e., all other species) density. Increasing conspecific density almost invariably decreases population growth rates due to density dependence (Morris and Doak, 2002; Caughlin et al., 2015). Increasing heterospecific density can either depress population growth rates in cases of interspecific competition or increase population growth rates in cases of facilitation. Either way, we can use the fitted model to estimate the intrinsic growth rate in the absence of neighborhood competition by setting C and H to zero. The y-intercept of this model represents the intrinsic growth rate of the population because densities of all neighbors are set to zero (Figure 8.4). In this way, intrinsic growth rates measured across multiple phenotypes and environmental gradients can be used to model the fundamental niche of phenotypes—the primary goal of this chapter.

A second quantity can be derived from Eq. 8.3 that is more closely related to concepts in modern coexistence theory, though not exactly. Invasion growth rates are defined as the low-density growth rate (or growth rate when rare) when heterospecific competitors are at their equilibrium abundances (Chesson, 2000; Angert et al., 2009; Barabás et al., 2018; Hallett et al., 2019). Equilibrium abundances are the abundances of the resident species when their growth rates are stable ($\log(\lambda) = 0$). These equilibrium abundances are theoretical quantities that can only be found by simulating population dynamics using the model parameters estimated from the models of competition (Shoemaker and Melbourne, 2016). This quantity is meant to tell us whether a species can invade a community that is fully established. Can we estimate something directly from empirical data that accomplishes the same goal, but does not require a competition model? There are two possible ways to approach this. First, we could approximate equilibrium abundances by estimating the population size of each resident species at which its average $\log(\lambda) = 0$, which is the x-intercept of a model where $\log(\lambda)$ is a function of conspecific density. Equilibrium abundances for each resident species would need to be summed to compute the total resident community density. Second, the long-term average abundance of the residents could serve as a crude approximation of the density of a fully established community. The average abundance of the residents is not identical to the equilibrium abundances of the residents, but it accomplishes a similar objective. This new kind of "invasion growth rate" can therefore be estimated by setting conspecific density to zero ($C = 0$) and heterospecific density (H) to the estimated equilibrium densities (option 1) or the mean of H at the plot-level over time (option 2) (Figure 8.4). While these are not technically the same as the invasion growth rates as defined in modern coexistence theory, they are an empirical approximation that may prove useful for comparing large numbers of species using long-term monitoring data.

This "short cut" to the invasion growth rate of modern coexistence theory is probably most relevant to species that are not dominant in the vegetation. Consider a removal experiment. Removing a rare species may not significantly impact the resident abundances, but removing a dominant species will most certainly lead to changes in resident abundances. More work is needed to understand when the average observed abundances of heterospecifics surrounding a focal species can serve as a short cut to equilibrium abundances.

Accounting for the density effects of each species separately can be even more challenging. Recall that population growth rate λ is estimated as N_{t+1}/N_t. This model allows us to project the size of the population of species i at time $t + 1$ if we know λ and the size of the population in time t:

$$N_{i,t+1} = \lambda N_{i,t}. \tag{8.4}$$

One way to account for the non-linear density dependence of each heterospecific is to divide the predictor on the right side by another set of terms that include the abundances of species i and other species j and their competition coefficients (α_{ij}):

$$N_{i,t+1} = \frac{\lambda N_{i,t}}{1 + \sum_j \alpha_{ij} N_{j,t}}. \tag{8.5}$$

This is the Beverton–Holt model and the competition coefficients (α_{ij}) are scaled by the species abundances. In this case, λ represents the population growth rate in the absence of competition and is therefore another way to compute the intrinsic growth rate, that is, the maximum growth rate at low density. We include a 1 in the denominator to prevent it from going to zero. This model structure is used widely in studies of competition in annual plant communities where individuals can be counted (Watkinson, 1980; Levine and HilleRisLambers, 2009; Shoemaker and Melbourne, 2016; Pérez-Ramos et al., 2019) but other model structures can also be used (García-Callejas et al., 2020). Extending this framework to perennial plant communities is possible by treating N as a measure of cover (or biomass) rather than as counts of individuals in the population. Note that the denominator reduces the rate of increase in the population, so it only allows for negative density dependence.

This model can be used to test plant strategy theory by allowing λ to be a function of trait-by-environment interactions. This framework allows us to understand the forces that optimize low-density growth rates while accounting for negative density dependence. It is also possible to model the competition coefficients (α_{ij}) as functions of traits to partition the effects of traits on habitat suitability versus species interactions (Chalmandrier et al., 2021; Chalmandrier et al., 2022). This approach takes us beyond the world of negative density dependence and into the world of frequency dependence, where the fitness of a strategy depends on the other strategies in the competitive milieu. We address this topic in Chapter 9.

Before moving onto the next approach, we need a reality check. I have so far described a phenomenological approach to account for conspecific and heterospecific density dependence. But individuals compete for raw materials, and these regression-based approaches to modeling competition ignore the mechanisms of resource reduction, the proximate factors that limit population growth. Furthermore, these estimates of intrinsic growth rates as described could be confounded if individuals are only growing by themselves in poor microsites or are only growing in a crowd on rich microsites. This confounding can only be strongly controlled using a careful experimental approach, but experimental control can also be difficult. So, while data are abundant to apply these approaches, there are limitations (Table 8.1).

The second approach to computing intrinsic growth rates applies population models parameterized from observational data on individual demographic rates using data that is standard for plant demography studies (Gibson, 2015; Salguero-Gómez and Gamelon, 2021). Many published population models have reported asymptotic population growth rates, but these estimates ignore density dependence when population size can regulate growth, survival, and reproduction (Crone et al., 2011; Ellner et al., 2016). Models of demographic rates that incorporate the effects of population size can be used to calculate intrinsic growth rates (Fowler et al., 2006; Dahlgren et al., 2016; Ellner et al., 2016). Integral projection models (IPMs) can be especially powerful in this context because they harness the strength of regression analysis to build models of demographic rates as functions of organism size in time *t* and any other covariate, including the density of neighbors (Merow et al., 2014a). This method requires that adequate variation in neighborhood density is observed. There are a variety of ways to include density effects in the demographic rate regression models (Adler et al., 2010; Ellner et al., 2016). Conceptually, one simply includes a term in the demographic rate regression that accounts for conspecific density or some other measure of abundance. In practice, the effects of density can take a variety of forms, both linear and non-linear, and this is an active area of research. In its simplest form, intrinsic growth rates can be estimated by setting conspecific and heterospecific neighborhood density in the demographic rate regression models to zero (Figure 8.5). Similarly, a crude approximation of invasion growth rates can be estimated by setting conspecific neighborhood density to zero and setting heterospecific neighborhood density to its average value. This technique statistically controls for the effects of neighborhood competition (Kunstler et al., 2016; Laughlin et al., 2018b), but it makes the important assumption that neighbor density is a good proxy for resource competition (Detto et al., 2019).

These first two approaches are still considered to be "observational" in nature. They rely on the hope that sufficient variation in intrinsic growth rate can be observed across multiple species across an environmental gradient to capture the trait-by-environment interactions. Angert and colleagues (2009) long-term observational study is a good example when this can work well. In this case, the variation in inter-annual rainfall was large enough to observe large variation among species in population recovery from low density, and they linked this variation in differences in water-use efficiency and relative growth rate. Observational approaches are likely to be important for generating new hypotheses that can be empirically tested in common gardens and for identifying the most important traits to use in models of fitness. However, these approaches can easily fail. After all, "the average plant leaves on average just one descendant" (Harper, 1977, p. 60), which is another way of saying that population growth rates tend to be stable and hover around unity. This is to be expected because, after all, we are observing populations that have established and continue to reproduce naturally by themselves. In other words, they have passed through the environmental filter and seem to be well adapted enough to the site (Keddy and Laughlin, 2022).

If we only include observations of stable populations, it is unlikely we will have sufficient information to test plant strategy theory. In fact, using observational data alone could lead us to conclude that growth rates are not even positively correlated with probabilities of occurrence (McGill, 2012; Thuiller et al., 2014), but I argue that if negative population growth rates in unsuitable habitats were included in such analyses then the correlation would be clear. In other words, we need to observe failed introductions into unsuitable habitats for the empirical approaches to succeed. Observations

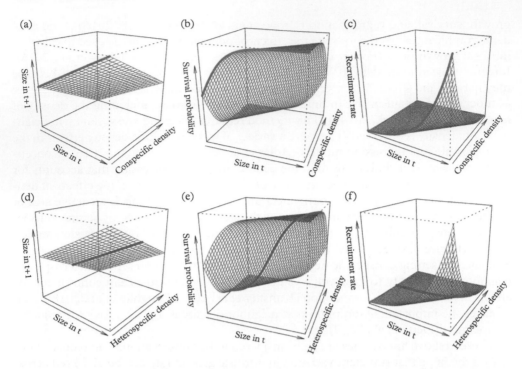

Figure 8.5 *(a–c) Accounting for negative conspecific density dependence in models of (a) growth rate, (b) survival rate, and (c) recruitment rate. The thick blue lines represent the predictions of demographic rates across a range of individual sizes in time* t *when conspecific density is set to zero for computing intrinsic population growth rates in the absence of density dependence. (d–f) Similarly, accounting for heterospecific density dependence in models of (d) growth rate, (e) survival rate, and (f) recruitment rate. The thick orange lines represent the predictions of demographic rates across a range of individual sizes in time* t *when heterospecific density is set to its average for computing empirical invasion growth rates.*

of demographic failures and population declines are just as important as observations of stable and growing populations. Without them, the predictive model surfaces will be flat, and we will have no hope of detecting the trait-by-environment interactions of dynamic fitness landscapes (Laughlin and Messier, 2015). It would be like trying to map the height of the sun in the sky over the course of a year by only sampling on the equinoxes while missing the peak and trough in height that occur on the solstices. Observational demographic datasets may lack measurements of population declines outside their natural range of environmental conditions precisely because the species cannot live in those conditions! You cannot observe the behavior of a species that isn't there. Natural assemblages exhibit many limitations when testing theories in community ecology (Towers et al., 2020) but it is possible that large observational datasets still prove to be useful. Experiments can help to identify the environments in which populations decline by forcing the filtering process to occur.

8.5   Experimental approaches to model the net effect of traits on fitness

Quantifying intrinsic growth rates experimentally is the third and undoubtedly most powerful approach. If observational datasets do not provide the variation in fitness that is needed to test plant strategy theory properly, then species need to be introduced to sites across a broad environmental gradient. These experiments are called common gardens in which known genotypes are grown and monitored. Many common garden studies have replicated gardens of multiple genotypes within a species, but others have involved multi-species experiments that represent a range of plant phenotypes. Research programs in the forest products industry have a long tradition of common garden experiments where multiple provenances of tree species are planted to evaluate genetic and environmental effects on species performance (Wang et al., 2010; Martínez-Berdeja et al., 2019; Dixit and Kolb, 2020). These datasets represent an untapped source of data for evaluating demographic rates of tree species across environmental gradients. The Southwest Experimental Garden Array in northern Arizona is another impressive, ambitious, and priceless experiment that includes multiple species growing at replicated gardens across an elevation gradient (Whitham et al., 2020).

Common gardens are the gold standard because they provide experimental control over environmental conditions and competitive interactions (Angert and Schemske, 2005; Kraft et al., 2015b; Grady et al., 2017; Merow et al., 2017; Pérez-Ramos et al., 2019; Johnson et al., 2022; Schwinning et al., 2022). Intrinsic and invasion growth rates of long-lived perennial species can be precisely estimated because competition can be experimentally controlled. In other words, extra parameters in population models are not needed to account for density dependence in the absence of competitors—it is baked into the data. Population models of perennial plants still require adequate measurements of annual transitions across a range of stages, so planting species from seed will not be the most efficient way forward. Ideally, a range of ages and stages (e.g., seeds, seedlings, vegetative plants, flowering plants) are planted from the beginning of the experiment to start multiple size-classes simultaneously.

Common gardens can be so effective because species can be transplanted beyond their geographical and environmental range limits, which is the only way to accurately quantify the limits of physiological tolerance (Figure 8.6). Transplanting species beyond their range is the most powerful approach because it can determine whether a species can recruit, grow, and survive outside its current range of environmental conditions (Gaston, 2009; Hargreaves et al., 2014; Alexander et al., 2016; Anderson, 2016; Huxman et al., 2022; Schwinning et al., 2022; Sumner et al., 2022). Moreover, the experiment can be done in the presence and absence of interspecific competition, which is an important component of range boundaries (Westoby et al., 2017; Westoby, 2022). Despite the recognition of the importance of such studies for several decades, relatively few studies have been conducted, and even fewer have linked responses and climate adaptation to traits (Paquette and Hargreaves, 2021; Lortie and Hierro, 2022; Westoby, 2022).

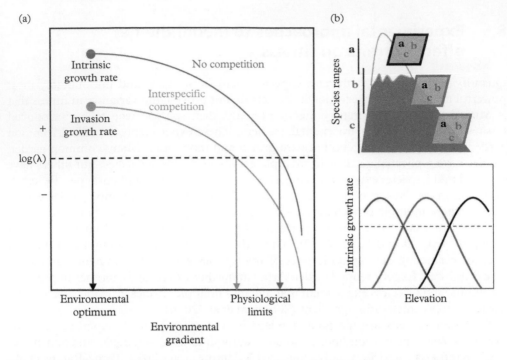

Figure 8.6 *(a) Relationships between population growth rates (log(λ)) in the absence and presence of competition along an environmental gradient. The vertical arrows denote the environmental conditions that represent the environmental optimum for the species and the physiological limits (i.e., where log(λ) = zero) in the presence and absence of competition. Note that the species range is extended in the absence of competition. (b) Common gardens are a powerful method to determine physiological range limits for species because they force demographic failures to occur in sites beyond the physiological limits. In a reciprocal common garden experiment, three species with distinct ranges are grown together in separate gardens across a broad range of environmental conditions.*

Angert and Schemske (2005) evaluated the demographic causes of geographic range limits for two species of monkeyflower in the Sierra Nevada mountains of California. *Mimulus cardinalis* grows at lower elevations and *Mimulus lewisii* grows at higher elevations with just a narrow sliver of elevational overlap between the two (Figure 8.7). They transplanted both species into four common gardens along the full elevation gradient and assessed fitness by monitoring growth, survival, and reproduction. Both species exhibited highest fitness in the center of their ranges and became increasingly maladapted at increasing distances from their range centers. *Mimulus cardinalis* experienced reduced fitness at high elevations beyond its range because of reduced growth and reproduction, whereas *Mimulus lewisii* experienced rapid mortality possibly related to heat stress beyond the lower elevation range boundary. This is a fine example of pushing species into uncharted territory to determine where the physiological limit to the fundamental niche occurs. The authors were unable to determine whether trait differences explained

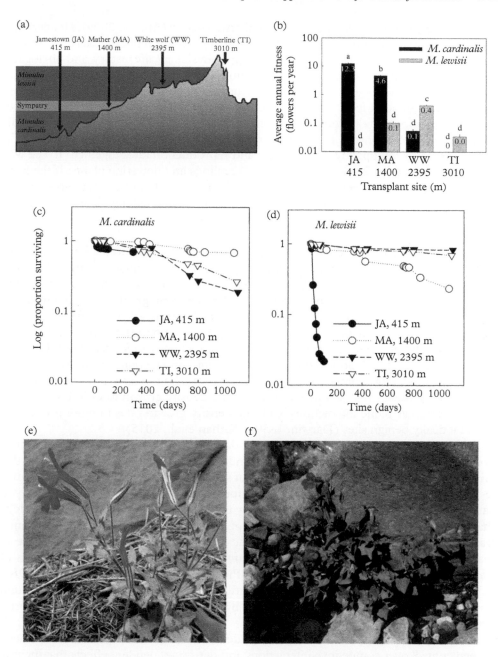

Figure 8.7 *(a) Natural ranges of two monkeyflowers:* Mimulus cardinalis *grows at lower elevations and* Mimulus lewisii *grows at higher elevations, with a narrow band of sympatric overlap between the two. The two species were each grown in the four locations shown along the elevation gradient within common gardens. (b) Number of flowers produced by each species in each common garden. Survival rates of (c)* Mimulus cardinalis *and (d)* Mimulus lewisii *in each of the four common gardens. (e)* Mimulus cardinalis *in the West Fork of Oak Creek, Sedona, Arizona. (f)* Mimulus lewisii *in Mount Rainier National Park, Washington.*

(a–d): Reproduced from Angert and Schemske (2005), with permission. (e): Photo by Max Licher. (f): Photo by Paul Rothrock. Photos reproduced here with permission under a CC BY-SA license (https://swbiodiversity.org/seinet/).

the fitness responses with a sample size of only two species (Angert, 2009). Hence, there is a need for experiments where broader sets of species are replicated across common gardens to represent full variation in morphology, physiology, life history, and phylogeny.

Hargreaves and colleagues (2014) led an important meta-analysis of 111 transplant studies beyond the range. Not all of these studies quantified population growth rates, but most measured at least one fitness component. In the cases that measured multiple fitness components, a combined fitness measure could be estimated by taking the product of the components. Hargreaves and colleagues (2014) computed an overall fitness measure by multiplying survival rates and reproduction rates together. To see why this works, consider a 2×2 matrix model of seedlings and flowering plants. If the transition probability from seedlings to flowering plants was 0.4, and the per capita rate of seedling production by flowering plants was 3, then the estimated population growth rate would be $0.4 \times 3 = 1.2$. In other words, the population would increase 20% each year (Ehrlén, 1999).

Using this approach, Hargreaves and colleagues (2014) came to the important conclusion that integrative measures of fitness were superior to single demographic rates at detecting reductions in performance beyond species ranges (Figure 8.8). This result emphasizes the importance of not relying on individual demographic rates but integrating across them to compute population growth rates (Gaston, 2009). These studies also highlight the importance of distinguishing between range boundaries that are driven by dispersal limitation versus deteriorating fitness components. In a later study, Paquette and Hargreaves (2021) provided evidence that biotic interactions are more prevalent on the warm edge of range boundaries than on cool edges. This study supported the long-standing theory that abiotic stress limits low-density growth rates in less productive sites whereas biotic interactions limit low-density growth rates in more productive and abiotically benign sites (Darwin, 1859; Louthan et al., 2015).

Reductions in fitness components are the demographic outcomes that we expect to observe when species are transplanted beyond their native range, yet we still lack this critical demographic data for large numbers of species because most demographic studies are limited to places where the species is found. The reason for this is simple—you cannot observe what isn't there. The only way to quantify the environmental conditions that cause demographic deterioration for a given phenotype is through experimental transplant studies beyond the range. One study merged the COMPADRE database of matrix population models with the TRY plant trait database to test for the effects of trait-by-environment interactions on fitness (Pistón et al., 2019), but the COMPADRE database is not optimal for detecting these effects because it may not include enough demographic failures. This could be tested by evaluating changes in fitness within a species across the environmental gradient.

We need new experiments that span decades to wield greater statistical and experimental control over competitive interactions and density-dependent effects that include demographic failures. We cannot start this endeavor using the currently available global datasets. We need to start locally and regionally and build up the proper datasets from there to empirically test these important ideas. These experiments can get quite complicated, given all the factors that impose constraints of species ranges, including

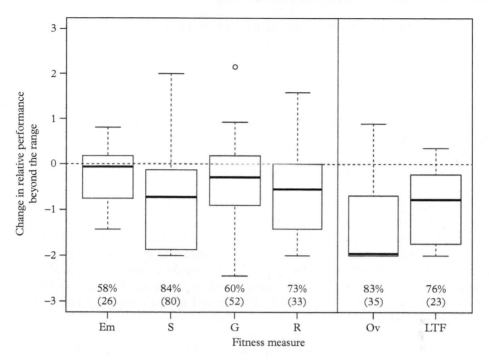

Figure 8.8 *Meta-analysis results (N = 111) for change in relative performance beyond the species range for a variety of demographic measures.*

Em = emergence, S = survival, G = growth, R = reproduction, Ov = overall fitness, LTF = Lifetime fitness. Reproduced from Hargreaves et al. (2014), with permission.

dispersal limitation, competition, resource availability, facilitation, mutualisms, herbi-vores, and pathogens. At a minimum, transplanting seeds and seedlings of dominant species into sites beyond their ranges with and without existing vegetation to test the importance of dispersal limitation, abiotic tolerance, and biotic competition will be most useful for moving this research agenda forward (Westoby, 2022).

8.6 Quantifying the demographic consequences of functional traits in changing environments

Once the hard work of measuring fitness across multiple species and environments is complete, the final step is to model fitness as a function of traits and environments (MacColl, 2011; Laughlin et al., 2018b; Yang et al., 2018). This model is a direct test of the dynamic adaptive landscape model illustrated in Figure 8.1 to determine how the effects of traits on population fitness across multiple species depends on the environment (Laughlin and Messier, 2015). The question is not whether population fitness among species varies along environmental gradients—this has been known for centuries. Rather, it is whether traits explain additional variation in population fitness through an interaction with the environment.

One way to test this explicitly is to compare the empirical support for two models (Laughlin et al., 2021). The first model is an "environment-only" model, where fitness among all species is modeled as a function of an environmental variable. To keep the mathematics simple, consider a linear mixed effects model with random intercepts for each species and a quadratic polynomial to account for non-linear fitness responses to the environment:

$$log\,(\lambda) = \beta_0 + \gamma_{j0} + (\beta_1 + \gamma_{j1})\,E + (\beta_2 + \gamma_{j2})\,E^2, \tag{8.4}$$

where β_0 is the global intercept, γ_{j0} is a random intercept for each of j species drawn from a normal distribution $N(0,\sigma^2_{\gamma j0})$, β_1 is the fixed effect term describing the main effect of the environmental gradient, γ_{j1} is a random slope for each of j species drawn from a normal distribution $N(0,\sigma^2_{\gamma j1})$, β_2 is the fixed effect term describing the quadratic effect of the environmental gradient, and γ_{j2} is a random slope for each of j species drawn from a normal distribution $N(0,\sigma^2_{\gamma j2})$. We typically model log(λ) to normalize the positively skewed distribution of λ. The second model adds traits and a trait-by-environment interaction:

$$log\,(\lambda) = \beta_0 + \gamma_{j0} + (\beta_1 + \gamma_{j1})\,E + (\beta_2 + \gamma_{j2})\,E^2 + (\beta_3)\,T + (\beta_4)\,T \cdot E \tag{8.5}$$

where β_3 is the fixed effect term describing the main effect of traits, and β_4 is the fixed effect term describing the interaction between the trait and the environmental gradient. Model selection statistics such as Akaike information criterion or likelihood ratio tests can be used to compare the empirical support for each model (Laughlin et al., 2021).

The strongest ecological trade-offs will be seen when two conditions are met: the trait-by-environment interaction is both statistically supported and the effect of the environment on fitness changes sign along the range of the trait (Laughlin et al., 2018b). Computing the first partial derivative of fitness with respect to (W.R.T.) traits isolates the fitness response to traits as a function of the environment. The first partial derivative of the fixed effects WRT traits is

$$\frac{\partial log\,(\lambda)}{\partial T} = \beta_3 + \beta_4 E. \tag{8.6}$$

This derivative shows that the slope of the relationship between log(λ) and traits linearly depends on the environment. In order to be considered a strong trade-off, the effect of traits on fitness must switch signs between each end of the environmental gradient (Laughlin et al., 2018b). To illustrate this, plot the first partial derivative of the model WRT the environment ($\partial log\,(\lambda)\,/\partial T$) to demonstrate how the effect of traits on fitness changes along the environmental gradient. A significant positive interaction would be illustrated as a line with positive slope that passes through $\partial log\,(\lambda)\,/\partial T = 0$. Note that the partial derivative can be solved WRT either traits or the environment—it depends on what you wish to emphasize, as they are two sides of the same coin. If you would prefer to view the effect of the environment on fitness across the trait gradient, then you take the derivative WRT the environment. But either way, the partial derivative must switch sign for the interaction to be considered strong and ecologically meaningful.

Linear derivatives may not be the best functional form especially if fitness landscapes are non-linear (Siefert and Laughlin, 2023). The effect of traits on fitness can alternatively be modeled as a Gaussian surface:

$$log\,(\lambda) = exp\,\left(\beta_{dir}{}^{t}T + T^{t}\beta_{nonlin}T\right) \tag{8.7}$$

where t is the transpose, β_{dir} is a vector of directional (linear) performance gradients, and β_{nonlin} is a matrix of non-linear performance gradients. The diagonal elements of β_{nonlin} measure the strength of stabilizing (if β is negative) or disruptive (if β is positive) selection for each trait, and the off-diagonal elements measure the strength of correlational selection between trait pairs (Arnold, 2003). Positive correlational selection means that fitness is maximized by having either high or low values of both traits. Negative correlational selection means that fitness is maximized by having a high value of one trait and low value of the other trait. This function can produce surfaces of various shapes, including a peak, a saddle, a ridge, or a slope. To allow fitness landscapes to vary across an environmental gradient, the performance surface parameters (the elements of β_{dir} and β_{nonlin}) can be modeled as functions of the environment.

Almost 100 years ago, Sewall Wright (1932) introduced fitness landscapes to describe how genotypes and phenotypes determine organism fitness. Multispecies fitness landscapes that quantify how traits affect fitness change shape across environmental gradients because the adaptive value of traits depends on the environmental context (Laughlin and Messier, 2015). Vegetation is composed of individuals of different sizes and ages across many different species, and this population structure of vegetation challenges our ability to model fitness (White, 1985b). Constructing fitness landscapes for entire assemblages of species has been fraught with empirical challenges because the effect of a trait on demographic performance depends on (1) other traits of the organism (i.e., trait-by-trait interactions), (2) the life stage or size of the individual (i.e., trait-by-size interactions), and (3) the local environment (i.e., trait-by-environment interactions). One promising way forward is to model demographic rates (growth, survival, and reproduction) across all species as functions of traits, organism size, and environmental context, and integrate them in a population model to project population dynamics.

Andrew Siefert incorporated all these contexts into a hierarchical Bayesian model to estimate the effects of multidimensional phenotypes and their interaction with size and temperature on survival, growth, and recruitment rates of trees in the eastern USA. Integral projection models were used to integrate demographic performance across the life cycle to quantify the net effects of multidimensional phenotypes on fitness across the temperature gradient (Ellner et al., 2016). We found that the tallest trees with the densest wood survived at the greatest rates (Figure 8.9). Wood density and SLA had an interactive effect on survival, especially of small-diameter trees, where survival peaked for species with either dense wood and high SLA or low wood density and low SLA.

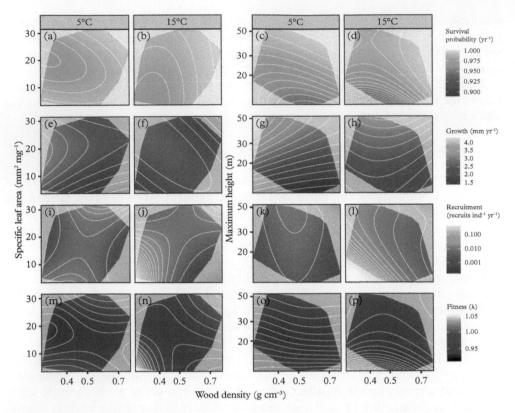

Figure 8.9 *Tree performance and fitness landscapes at low and high mean annual temperatures for trees in the eastern USA. These landscapes were generated using 99 species that spanned the darkened region of the trait space. Landscapes show expected demographic rates (survival, a–d; diameter growth, e–h; recruitment, i–l) and fitness (population growth rate, m–p) for an average tree at different temperatures depending on their trait combinations.*
Reproduced from Siefert and Laughlin (2023) with permission.

Species with the tallest maximum heights and lowest wood density grew the fastest, especially in cold sites. Species with low wood density exhibited the highest per capita recruitment rates (i.e., the rate of new recruits reaching the 2.54-cm diameter threshold per adult tree), particularly in warm sites. These interactions occurred because species with dense wood and tall maximum height had a larger size at onset of reproduction. Importantly, the theoretical expectations for demographic trade-offs emerged naturally from the model (Rüger et al., 2020). We detected a growth–survival trade-off among saplings driven by wood density: species with low wood density had high sapling growth rates but low survival, and species with high wood density had high sapling survival but low growth rates. We also found evidence of a stature–recruitment trade-off mediated by maximum height and wood density: species with low maximum height and low wood density had high recruitment at small sizes but low growth and survival at larger sizes, and species with tall maximum height and dense wood had high growth and survival as large trees but produced few recruits when they were small. These results controlled for tree density as a covariate in each of the demographic rate models, but future applications could build on this general approach to explicitly account for density and frequency dependence (Siefert and Laughlin 2023).

This new approach makes it possible to empirically estimate fitness landscapes for multispecies communities across environmental gradients, leading to improved understanding of the selective forces that drive community assembly and permitting mechanistic predictions of community dynamics in changing environments. Adding species interactions and more complicated disturbance processes are currently ongoing to improve the generality of the approach. This approach can be applied in any ecosystem providing a promising path toward achieving the long-held goal of making ecology more general, mechanistic, and predictive. It fulfils the goal of plant strategy theory by using traits to produce predictions of population growth rate in any environmental context. These statistical models can be pushed to their full potential by testing them on species outside the training dataset, that is, species that were not included in the original model. Indeed, testing these model predictions on species outside the training data would be the most powerful test conceivable for the generality of the adaptive value of traits.

8.7 Categorical strategies versus continuous spectrums in empirical plant strategy models

Rather than being overly prescriptive, this book aims to clarify the boundaries of plant strategy theory, the persistent problems that the theory confronts, and to lay out the most logical ways forward to galvanize progress in the field. This chapter describes a general empirical framework for testing the net effect of traits on fitness, but the model shown in Eq. 8.5 includes only one trait and one environment. It does not take much imagination to realize that the number of trait-by-environment interactions can grow exponentially with the number of traits and environments included in the model. This is not only a logistical challenge, but also directly related to how strategies are defined. Plant strategies have traditionally been defined as sets of correlated traits (Craine, 2009). Working under this assumption, any test of a plant strategy theory should ideally include multiple traits, but this can take at least three forms analytically: (1) multiple traits can be used to classify categorical strategies (Westoby and Leishman, 1997); (2) multiple traits can be used all at once in a single model or used to define reduced trait dimensions or syndromes (Grime et al., 1997b); or (3) a few carefully selected single traits that each represent a key trade-off dimensions can be used (Westoby, 1998).

These three approaches are concerned with how variation in phenotypes are defined, but they have implications for how demographic outcomes for different strategies are tested. Let us look at an example. Imagine that we are interested in determining drivers of fitness along a gradient of soil water potential. To keep it tractable, this toy example will involve four species, two correlated traits, and one environmental gradient. Ecophysiologists have identified two traits that are good indicators of drought tolerance: leaf dry matter content (LDMC) and leaf turgor loss point (TLP). Let species A and B exhibit high LDMC and low TLP, which allows them to maintain leaf turgor and to continue photosynthesizing even in dry conditions. Species A and B could represent the sclerophyll shrubs from the dry Mediterranean biome. They are clearly distinct from species C and D, which have trait values that would make them ill-suited to surviving in a dry climate (Figure 8.10a). Species C and D represent species found in mesic climates like temperate forests.

First, a categorical strategy model would classify species into groups using a classification algorithm (Westoby and Leishman, 1997). Species A and B would be lumped into a group we can name the "low water strategy" and species C and D into a "high water strategy" based on the similarity of their traits (Figure 8.10b). This approach would satisfy ecologists who like to put species into groups, although there is never a guarantee that species will cleanly separate into groups using this approach when more traits and more species are in the dataset.

Second, a continuous strategy model would do one of two things. We could keep track of both sets of trait values and use them as independent predictors of species fitness responses, but given their strong negative correlation, the traits are redundant and there could be strong problems with collinearity in the models. Alternatively, we could simplify the dataset by extracting a single "trait axis" from the correlated trait values using a multivariate data reduction technique, such as principal components analysis (Figure 8.10c). This can be helpful when multiple correlated traits are indicators of the same strategy. Here, high LDMC and low TLP are indicators of drought tolerance.

Third, one could select a single trait to represent the multi-trait trade-off axis. Westoby (1998) proposed the LHS model to increase global comparisons of species traits and so recommended SLA to quantify leaf trait variation because it is relatively easy to measure. In that same spirit, I recommend that we measure LDMC because it is far easier to measure than leaf TLP. So we choose one trait and leave the other one out of the analysis.

Some ecologists would stop there and would be satisfied with just a quantitative description of the phenotypes. But a mere analysis of phenotypes tells us nothing about whether the strategies confer fitness differences across environments. Consider the fitness responses of these four species to soil water potential. Species A and B exhibit their highest fitness at low soil water potential and species C and D exhibit their highest fitness at high soil water potential (Figure 8.10d). This is classic species distribution modeling that cares nothing about explaining why these species are distributed along the gradient in this pattern. Plant strategy theory demands to know why these patterns exist by relating the distributions of the species to traits that evolved as adaptations to dry conditions.

This leaves several possibilities for testing the theory. The first possibility is to test whether the two categorical strategies explain the variation in fitness responses. In other words, does the fitness response to soil water potential differ by strategy? Note how each strategy exhibits a different peak fitness response to soil water potential (Figure 8.10e).

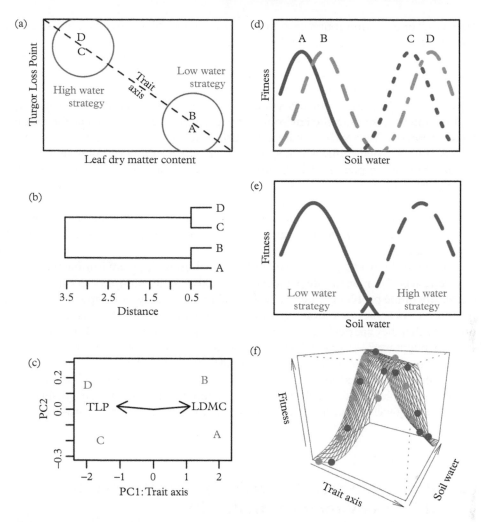

Figure 8.10 *(a) Location of four species (capital letters A–D) in the space of two traits. (b) Hierarchical classification of the four species into two clear groups based on the multivariate dissimilarity values of their traits. (c) Principal components analysis can extract the underlying trait axis. (d) Fitness response curves for each of the four species along an environmental gradient. (e) With a categorical approach to strategies, one can test whether the fitness responses along the environmental gradient depends on the strategy group. (f) With a continuous approach to strategies, one can test whether the fitness response along the environmental gradient depends on the continuous trait axis.*

The remaining possibilities are to test whether fitness can be explained by a continuous surface response model, where the fitness responses to soil water potential depends on continuous variation in both correlated traits, on continuous variation in the reduced "trait axis," or on continuous variation in LDMC alone. Note how the fitness responses to soil water potential depends on the trait axis values (Figure 8.10f).

Which approach is superior? Superiority will depend on several conditions. First, the categorical differences between these two groups are clear in this example, but in reality, species vary continuously in their traits and classification algorithms often struggle to define such clear-cut groups. Second, estimating model coefficients for each group and their interactions with the environment will get more complicated as the number of groups increases. We could compare the statistical support between categorical and continuous models to determine which approach explains the most variation in fitness while penalizing more complex models. Ultimately the strongest test of the theory would evaluate model predictions on species not included in the training dataset. If we measured leaf traits on a fifth species E, then we could use the models to predict its fitness response to soil water availability a priori and test whether our predictions matched reality. This is the promise and potential of a trait-based plant ecology (Lavorel and Garnier, 2002; Shipley et al., 2016).

Testing the ability of traits to predict fitness responses along environmental gradients is the central goal of plant strategy theory. From this simple example you can glimpse the complexity of the problem. The real world will include many more species, more traits, and more environmental gradients. In this example the two leaf traits are so strongly correlated that they are perfect indicators of only one trait dimension. But other traits will be independent from the leaf trait axis, and we would need to include multiple orthogonal trait axes in our models. Likewise, we could also measure other resources to determine whether these species exhibit different fitness responses to shade or soil infertility. The increasing complexity of model structure to account for the true dimensionality of plant strategy theory will demand a large sample size to improve the signal-to-noise ratio in the data.

8.8   Strong directional selection affects the nature of trait–environment trade-offs

Regardless of which of these approaches we use, it is important to understand what these analyses can tell us about ecological trade-offs, which underpin our conceptual understanding of global biodiversity distributions (Futuyma and Moreno, 1988; Fry, 2003; Agrawal et al., 2010; Grime and Pierce, 2012; Grubb, 2016; Agrawal, 2020). Models that use traits to predict the success of a species across an environmental gradient expect to detect trade-offs. Ecological trade-offs arise as a consequence of the adaptive value of a trait, which Grime and Pierce (2012) eloquently describe as "an evolutionary dilemma, whereby genetic change conferring increased fitness in one circumstance inescapably involves sacrifice of fitness in another." As the term "trade-off" implies, the trait effects are "bidirectional," where, for example, low trait values of a species confer adaptive advantage at one end of an environmental gradient whereas high trait values confer benefits at the opposite end of the gradient (Figure 8.3).

Classical ecological theory has long emphasized this bidirectional perspective on trait–environment relationships at the species level (Kneitel and Chase, 2004). Tilman's (1988) resource ratio theory is built on one central trait-environment trade-off: at the high end of the soil fertility gradient, plant species that allocate relatively more carbon aboveground than belowground are predicted to be better competitors for light, whereas at the low end of the soil fertility gradient, plant species that allocate relatively more carbon belowground than aboveground are predicted to be better competitors for soil nutrients. The prediction is clear.

Empirical evidence for such a trade-off at the species level would be provided by showing that the effect of a trait on fitness switches sign (i.e., changes direction) along an environmental gradient (Laughlin et al., 2018b). This directional switch in sign is fundamental, but detecting the switch empirically is nontrivial because it cannot be observed through a simple trait-environment correlation. The absence of a switch in sign would indicate that a trait only confers an adaptive advantage at one end of this gradient, thereby exhibiting a mere "unidirectional benefit." Recent evidence suggests that unidirectional benefits are more prevalent than classic trade-offs.

One international collaboration synthesized massive datasets of root traits and relevés to test the effects of root traits on species occurrence (Laughlin et al., 2021). The authors analyzed the effects of four root traits (specific root length, root tissue density, root nitrogen, and root diameter) on species occurrences across global gradients in climate (minimum temperature and an index of aridity). They did detect a few classic trade-offs, for example, the relationship between SRL and temperature in forests was a clear trade-off. Low-SRL species associated with AM fungi, such as Chinese fir (*Cunninghami lanceolata*), were more likely to occur in warmer climates, whereas high-SRL species associated with ericoid mycorrhizal fungi, such as lingonberry (*Vaccinium vitis-idaea*), were more likely to occur in colder climates (Figure 8.3). In grasslands, however, not every trait exhibited a classic bidirectional trade-off. Low root tissue density and high SRL increased probabilities of occurrence in cold grasslands, but these traits had no effect on occurrence probability in warm grasslands. This empirical result indicated that traits can sometimes only exhibit unidirectional benefits rather than the classic bidirectional trade-off. In fact, the majority of trait-environment interactions that were tested were not classic trade-offs, but rather were unidirectional benefits. These results agree with Grubb's (1985, p. 600) insight that "the concept of a trade-off, which implies that being suited to one condition necessarily involves not being suited to the opposite, is widely diffused in the current literature but is not universally applicable."

In this study, unidirectional benefits were consistently associated with the more extreme cold and dry climates that are more resource-limited than warm and wet climates. This supports the idea that environmental filtering increases in intensity where resources are more limited (Butterfield, 2015). Single optimum traits were observed in cold and dry climates, while single trait optima were not observed in warmer and wetter climates. In other words, warm and wet climates exerted no clear directional selection on root traits. This is similar to Grime's idea that at very stressful sites only stress-tolerators are dominant, but in productive sites both competitors and stress-tolerators occur because competitors render life stressful for others. There are also examples of unidirectional benefits in aboveground traits. Plant communities in New Zealand exhibit trait convergence toward low leaf nitrogen concentration in phosphorus-poor soil, whereas in phosphorus-rich soil communities display wide divergence of leaf nitrogen concentration (Mason et al., 2012). This suggests that low leaf nitrogen is adaptive in phosphorus-poor soil to maintain a balanced leaf nutrient stoichiometry, whereas high leaf nitrogen is not adaptive in phosphorus-rich soil.

Predictive models that use sets of continuous traits as predictors of fitness need to know whether a trait exhibits trade-offs or unidirectional benefits along an environmental gradient. Plant strategies could be generated by a combination of trade-offs for some traits, for example, light compensation point along light gradients (Lusk and Jorgensen, 2013) and unidirectional benefits for others, for example, leaf nutrient concentrations along soil fertility gradients (Mason et al., 2012; Laughlin et al., 2015a), which inevitably makes the task of predicting fitness using continuous traits more difficult than previously anticipated. An expanded theory of trait–environment interactions that incorporates unidirectional benefits will advance our understanding of the adaptive value of traits in community assembly and may improve predicted responses to climate change.

Pursuing the truth about plant strategies will require hard work and creative solutions. Is it worth it? Physicists in search of the Higgs boson did not hang their heads and admit that it was too hard. Rather, they developed strong international collaborations, convinced their government funding agencies to support their bold ideas, spent billions of dollars, and did the hard experimental work and found evidence for a particle that is "consistent" with the Higgs boson. As our belts are continually tightened, society will need to decide what science to fund. We can chart the path of stars thousands of light years away, yet we struggle to chart the path of populations on our own imperiled planet. We can do better.

8.9   Chapter summary

- This chapter focuses on methods for estimating the effects of traits on fitness across environmental conditions to approximate the fundamental niche of a phenotype. Chapter 9 expands this focus to include the effects of neighbors to approximate the realized niche of a phenotype. It is difficult in practice to separate the effects of the environment from the effects of neighbors.

- The fitness of a single species varies along environmental gradients. The fitness of multiple species along environmental gradients can be modeled as a non-linear interaction between their traits and the environment.

- Changes in demographic rates will determine whether species ranges shift in response to climate change.

- Observational approaches to modeling fitness can leverage long-term monitoring datasets and population models. Intrinsic and invasion growth rates can be computed after accounting for conspecific and heterospecific neighborhood densities, although the methods described here differ from those in modern coexistence theory.

- Observations of failed introductions into unsuitable habitats are just as important as observations of stable and growing populations. Observational demographic datasets often lack measurements of population declines outside their natural range of environmental conditions precisely because the species cannot live in those conditions.

- Experiments are useful to identify the environments in which populations decline by forcing the filtering process to occur, although large observational datasets may have enough variation in them to still be useful. Common gardens beyond species ranges are the gold standard because they provide experimental control over environmental conditions and competitive interactions.

- Integrative measures of fitness are superior to single demographic rates at detecting reductions in performance beyond species ranges.

- Categorical and continuous strategy models require different analytical approaches to test the theory.

- Traits may sometimes not exhibit classic trade-offs, but rather may exhibit unidirectional benefits. In other words, trait values may influence fitness at one end of the environmental gradient but have neutral effects at the other end of the gradient.

- Setting up new experiments to develop robust tests of plant strategy theory will be expensive and difficult, but gains in understanding will be well worth the effort.

8.10 Questions

1. Why are most published population models not always useful for testing plant strategy theory?

2. Can you find other examples of common garden experiments in the literature that have forced population declines by planting species beyond the range?

3. How does the invasion growth rate as specified in this chapter differ from the invasion growth rate as defined in modern coexistence theory?

4. Is it empirically possible to quantify the realized and fundamental niche of a phenotype, or is it just more of a thought experiment?

5. Compare the workflow for a categorical versus a continuous approach to testing plant strategy theory. What factors would make one approach superior over the other?

9

Game Theoretical Approaches to Infer Fitness from Traits

The ability of some plant species to form fragmented phenotypes of a single genotype is just one of the variety of successful ways of playing the game of being a plant.

—John Harper (1977, p. 27), *Population Biology of Plants*

9.1 When games are not just fun but also necessary

Chapter 8 explored a variety of empirical approaches to develop predictive models of fitness as non-linear functions of traits and environments. Optimization theory has long been a tool of choice for understanding how natural selection favors certain phenotypes over others in a given abiotic environment. If we can assume that the success of a strategy does not necessarily depend on the density of individuals in the population or what other individuals in the community are doing, then we can focus on which traits optimize intrinsic growth rates to define the limits of the fundamental niche of a strategy (Figure 9.1). For many applications, this is a reasonable and practical assumption to make, but plants do not survive and reproduce in isolation. If selection depends on the density of individuals, but not on what others are doing, then maximizing population size can be used as a criterion for evolutionary stability (Roughgarden, 1976). It is important to recognize that observational empirical data cannot easily separate the effects of the environment from the effects of neighbors and neighbor strategies—it has long been known that fitness must also be evaluated within a competitive context to understand the realized niche (Haldane, 1937).

One of the greatest challenges for plant strategy theory is understanding the demographic consequences of strategies in the presence of other strategies. When strategy success depends on what other individuals in the community are doing, then evolutionary game theory becomes a useful (and some would argue, necessary) tool (Maynard Smith, 1982; McNickle and Dybzinski, 2013). Evolutionary game theory (and the related suite of models that go by the name Adaptive Dynamics or Darwinian Dynamics) account for both density and frequency dependence (Figure 9.1) because population growth rates of strategies depend not only on the abiotic environment, but also on the frequencies (i.e., proportions) of other strategies in the game, as well as the density of

Plant Strategies. Daniel C. Laughlin, Oxford University Press. © Daniel C. Laughlin (2023). DOI: 10.1093/oso/9780192867940.003.0009

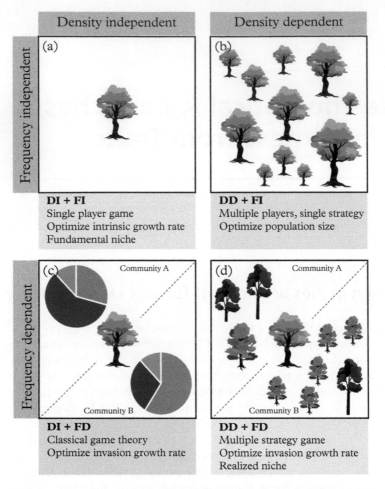

Figure 9.1 *The four combinations of density and frequency independence and dependence. Density is the number of individuals of each species, whereas frequency is the proportional abundance of each species or strategy. Densities and frequencies are of course intertwined concepts because in a multispecies community if one knows the densities of each species then frequencies are also known.*

individuals possessing those strategies (Metz et al., 1992; Diekmann, 2004; McGill and Brown, 2007; Metz, 2012).[1] Recall Table 3.2, which introduced the notation and their definitions. If there is only one player, then the game is the same as a standard optimization problem (it is density and frequency independent), but once there are multiple players, then the game is afoot. This chapter introduces the mechanics of classical games,

---

[1] Density dependence is an intuitive concept. "Frequency dependence" is not. Perhaps "strategy frequency dependence" might be better because the concept is meant to convey that selection depends on the strategies in the community. I stick with frequency dependence in this book for consistency with the literature.

lays out the general framework of modern evolutionary game theory, and explores some key discoveries and novel insights that game theory has contributed to our understanding of plant strategies. In an evolutionary game, the prize of winning is the ability to keep playing, but the cost of losing is extinction.

9.2   Game theory explores the fitness consequences of frequency dependence

Game theory was originally developed as a tool to solve conflicts of interest assuming that the players exhibit rational behavior. Classical games involve players, strategies, payoffs, and rules (Vincent and Brown, 2005). Payoffs are quantifiable values associated with the outcome of a game and depend on the strategies that are chosen. Winning strategies accrue higher payoffs. A rational player's goal is to select the strategy that maximizes payoff, but it is not a simple optimization problem because the payoffs are functions of the individual's strategy as well as the strategies of others. Strategies can be discrete, mixed, pure, or continuous. Discrete strategies are categorical (e.g., fight or flee, herb or tree) and the payoffs can be summarized in matrix form, where the rows and columns are the strategies, and the matrix elements are payoffs. Mixed strategies occur when an individual plays one of the discrete strategies from the matrix game with a probability between 0 and 1, whereas pure strategies occur when the individual always chooses the same discrete strategy. Continuous strategies are any value within a vector of a continuous variable and can represent maximum height, seed mass, specific leaf area, or any other continuous plant trait.

Central to game theory is the evolutionarily stable strategy (ESS), which resists an invader with a different strategy that is introduced at low density. In other words, the ESS is a peak on a fitness landscape, where alternative strategies (often called mutants) have lower fitness and are therefore unsuccessful at invading from low densities. Game theorists define fitness as the per capita population growth rate of a strategy and extend their models for multiple generations to observe how the frequencies of strategies can evolve over time according to the payoffs in terms of fitness. Evolution gets hung up at an ESS because drifting away from it would decrease fitness. Therefore, ESSs are the most likely strategies that should exist in nature. Like the "Nash equilibrium" in classical game theory, an ESS is defined as a "no-regret" strategy where no one can benefit by independently changing their strategy.

The hawk–dove game is a matrix game that serves to make these abstract ideas a little more concrete before we move into modern games that are based on demographic fitness. The hawk–dove game is frequency dependent but is, oddly, also density independent (Figure 9.1c). Maynard Smith and Price (1973) originally introduced game theory to evolutionary biology using examples of animal conflict. Imagine a population of individuals trying to obtain a resource V. When two individuals confront one another, they can react in one of two ways: they can either act like a hawk and fight for the resource by escalating until injured or until their opponent retreats, or they can act like a dove and simply display and retreat immediately if they are attacked. Injuries reduce fitness by a cost C. The rules of this game are relatively simple. The value of the resource and the cost of getting injured are the critical quantities to track, and the rules of the interactions determine the payoffs of each strategy. The rules are used to calculate the payoffs, which are defined as the change in fitness arising from the contest. If two hawks interact, each individual has a 50% chance of obtaining the resource and a 50% chance of getting injured; therefore, the payoff for being a hawk if your opponent is also a hawk equals $(V-C)/2$ (Table 9.1). If a hawk interacts with a dove, the hawk takes the resource and no one is injured, so the payoff for the hawk equals V. If a dove interacts with a hawk, the dove retreats before getting injured and their payoff equals zero, that is, there are no gains in fitness. Finally, if two doves interact, the resource is shared equally and so the payoff equals $V/2$.

The game can be analyzed to determine which strategy is evolutionarily stable. To play the game, we start by assuming that the population size is infinite and that individuals pair off at random and choose to be a hawk or a dove for each interaction. We need to set the rules to determine the value of the resource V and the cost of injury C. For the first scenario, let $V = 2$ and $C = 1$. Table 9.1 shows the payoff matrix for this scenario. In this case, the dove strategy is not an ESS because the payoff for a hawk interacting with a dove is always greater than the payoff for a dove interacting with a dove. A mutant hawk strategy can invade a population of doves and dominate it (Figure 9.2a). If the value of the resource is greater than the cost of injury ($V > C$), that is, if the payoff for a hawk to interact with another hawk is positive, then the hawk strategy is an ESS because it is worth the risk of getting injured to obtain the resource. If it is always worth playing the hawk strategy, then it is a "pure" strategy.

But what happens if the cost of injury is greater than the value of the resource ($V < C$)? If we keep the resource value the same and change the cost of injury to 3 (see second payoff matrix in Table 9.1), then neither hawk nor dove is an ESS. In fact, the population may settle into an equilibrium where individuals randomly switch their strategy at some probability. This is an example of a "mixed" strategy (Figure 9.2b). Note that this classic game is frequency dependent but not density dependent because the emphasis is on the relative *frequencies* of the strategies, not their *absolute values*. It is important to remember that the outcomes are dependent on the rules. Changing the rules (i.e., the model assumptions) results in a different outcome.

Table 9.1 *Payoff matrices for the hawk–dove game using different costs C.*

General payoff structure			Payoffs when $V = 2$ and $C = 1$		Payoffs when $V = 2$ and $C = 3$	
	Hawk	**Dove**	**Hawk**	**Dove**	**Hawk**	**Dove**
Hawk	$(V-C)/2$	V	1/2	2	−1/2	2
Dove	0	$V/2$	0	1	0	1

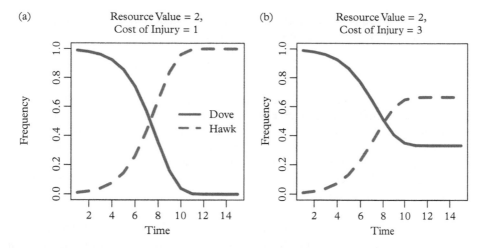

Figure 9.2 *Outcomes of the hawk–dove game using two different costs of injury. See the book's R scripts to reproduce these games.*

Evolutionary biologists extended games like this to determine when a mutant strategy could evolve in a population to better understand drivers of microevolution. While strategies were originally conceptualized in the context of animal behavior, strategies "can be applied equally well to any kind of phenotypic variation, and the word strategy could be replaced by the word 'phenotype', for example a strategy could be the growth form of a plant, or the age at first reproduction" (Maynard Smith, 1982, p. 10). Evolutionary games can be used to model the microevolution of single species or the macroevolution of communities. In plant communities that compete for resources, it is difficult to separate density from frequency dependence, so modern games in plant evolutionary ecology include both types of selection.

9.3   Games for light increased plant height

The hawk–dove game is a matrix game with discrete strategies, but what about continuous traits? After all, the last several chapters have emphasized the continuous variation in phenotypic traits among species. To set game theory in the context of plants, let us now consider a game constructed to simulate individual plants competing for light where strategies vary continuously along plant height (Falster and Westoby, 2003; Kokko, 2007). If a tall individual interacts with a short individual, the tall species obtains the light and the short species gains nothing. It is not difficult to imagine that being tall would be an ESS under those assumptions of fitness payoffs. Light comes from a single direction (from above), so the advantage gained by a slightly taller plant is asymmetrical. Asymmetrical competition refers to the fact that the taller plant acquires the light and deprives its shorter neighbor of that light, so the competitive effect of the taller plant on the shorter plant is disproportionately larger than any small effect that the shorter plant has on the taller plant. Competition for resources belowground, in contrast, is typically considered to be symmetric (recall Figure 3.6). The success of an individual competing for light depends on the height of its neighbors, which makes light competition explicitly a frequency-dependent process.

Competition for light is relative (McNickle and Dybzinski, 2013). The most competitive height is just a little bit taller than your neighbor, but no more than that! The reason for this is because of the costs associated with growing tall. The costs of maintaining a much taller canopy than the shorter neighbor would lead to a loss of fitness. Purely ecophysiological approaches that lack this relative perspective of frequency dependence attempts to explain the evolution of height exclusively from biophysical limits imposed by hydraulic constraints. However, game theory has taught us something altogether different. We tend to think that productive sites are dominated by woody plants because they have the energy and resources required to grow tall. But in the absence of competition, the evolutionarily fittest height, even in a productive site, would be at ground level so that all resources could be allocated to reproduction, rather than allocating resources toward growing tall (Givnish, 1982). Productive sites by themselves were not the driver of increasing maximum heights in trees. Tall species are adapted to productive sites because the consequences of not keeping up with their neighbors is deep shade and eventual death (McNickle and Dybzinski, 2013). After all, "a plant growing by itself would not gain, in seed or pollen production, by having a massive woody trunk" (Maynard Smith, 1982, p. 177). In other words, tall trees did not evolve thick trunks just because they could—they did so because they had to.

Thom Givnish (1982) was one of the first plant ecologists to apply game theory to plant competition for light. He was particularly interested in the adaptive significance of leaf height among forest herbs growing in temperate forests that vary in productivity. Givnish's model balanced the structural costs of supporting leaf height with the photosynthetic benefits. I refer interested readers to a fun step-by-step description of a similar model of plant height that includes a simplified translation into a payoff matrix (see Kokko, 2007). The model suggested that the ESS was a tall herb in productive dense vegetation, but in sparse vegetation there is little to be gained by growing tall, so the ESS is a shorter plant. One of my favorite groups of wildflowers provide a good example of this. *Monarda fistulosa* and *Monarda didyma* are relatively tall herbs that tend to grow in moist, rich soil where competition for light can be fierce. However, their cousin *Monarda punctata* is much shorter and grows in sandy, poor soil where competition for light is much reduced. In 1999, I interviewed for a graduate position at the University of Wisconsin and Thom quizzed me about my understanding of the ecology of these plants over our lunch conversation. As a trained botanist in the Midwest USA, I knew those species well and talked energetically about where I had seen them growing. It was a great conversation that I remember vividly.[2] Givnish (1982) demonstrated that empirical height measurements across a range of different taxa supported the predictions of the model.

---

[2] I didn't get the position.

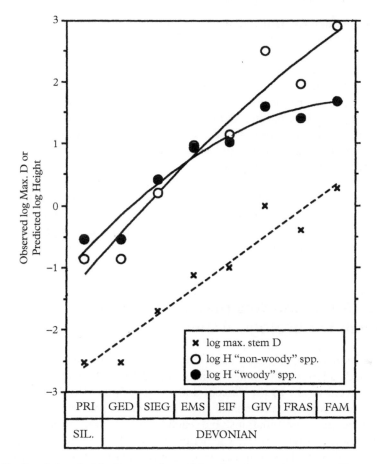

Figure 9.3 *Predicted plant height (log-transformed) based on the maximum stem diameters reported for early Paleozoic plants.*
Reproduced from Niklas (1994) with permission.

Competitive games for light have been played since the origination of the arborescent life form. Niklas (1994) used allometric equations to estimate plant height from fossilized trunk diameters across a range of taxa from the late Silurian and Devonian periods. Figure 9.3 shows that the height of these early land plants increased over time during this period—you can almost see the competitive games playing out over 60 million years of evolution! We cannot be certain whether competition for light drove this progression toward taller lycopsids. Growing tall may have also improved dispersal of their spores, but we don't know if selection on height for dispersal could have driven such strong height differentiation. Bond (2019) argued that these early arborescent plants would not have cast much shade, but according to the insights that game theory has made with respect to competition for light and the inefficiencies of growing tall, it appears that light competition was likely an important evolutionary force in the first forests on Earth.

Many game theoretical models of height evolution converged on a single ESS for height (Falster and Westoby, 2003), implying that a single height would dominate the forest and no other height could invade. But the most conspicuous structural feature of both temperate and tropical forests is the vertical stratification of the forest canopy into layers. Forest ecologists often define three or four distinct vertical strata in closed canopy forests. There is a ground layer, an understory of woody shrubs and treelets, the canopy layer of dominant trees, and sometimes an emergent layer above this canopy. This mismatch between game theory and field observation did not sit well at all with many ecologists. There must be a way to structure the game to allow for multiple height strategies to coexist in a way that agrees with observations in the field. This problem wasn't solved right away. But before we get to the solution, we need to first define the basic structure of modern evolutionary game theory and to see how it unifies population dynamics with plant strategies.

9.4   The *G*-function sets the evolutionary play in the ecological theater

Modern evolutionary game theory made major advances in the 1980s and 1990s. Two strands evolved: adaptive dynamics (Diekmann, 2004; Metz, 2012) and Darwinian dynamics (Vincent et al., 1993; Vincent and Brown, 2005). Each strand applies evolutionary game theory to understand the evolution of continuous strategies, which are taken to be trait combinations, but each uses different terms for similar concepts (McGill and Brown, 2007). I generally follow the notation defined by Vincent and Brown (2005) as I find it the most conceptually attractive, convenient, and tractable.

In evolutionary game theory, there are players, strategies, and payoffs. Players are individual organisms, strategies are heritable phenotypes (traits), and payoffs are fitness expressed as the per capita growth rate of a strategy in a given abiotic and biotic environment. The focus of the game is on strategies that persist through time because the players either live or die but the strategies are passed on through the generations. The ESS remains a central concept, but studies suggest (Vincent and Brown, 2005; McGill and Brown, 2007) that the definition of an ESS needs to be expanded to include convergence stability, which is the condition that a population will evolve to the ESS if it is close enough to it. Maynard Smith's definition of an ESS required that it be resistant to invasion, but it didn't guarantee that it was possible to evolve to it.

Evolutionary game theory defines fitness as the per capita population growth rate, which is equivalent to r or $\log(\lambda)$. Chapter 8 defined per capita population growth rates to be functions of the strategy of the focal species and the environmental conditions (Eq. 8.2), but here we include the strategies and densities of other species. In general, the fitness function expresses per capita population growth rates as a function of three vectors:

$$\frac{1}{N}\frac{dN}{dt} = F(u, x, e) \tag{9.1}$$

where **u** is a vector of strategies, **x** is a vector of population densities of each strategy, and **e** is a vector of environmental variables such as resources, disturbance regimes, and climate (recall Eq. 3.4). Classical game theory models focused on populations that could evolve multiple strategies, but these have been extended to communities of species with different strategies. Thus, at its core, evolutionary game theory models population dynamics as a function of traits and environments. More generally, a fitness generating function (*G*-function) can be analyzed to determine the outcome of any game with S strategies:

$$G(v, u, x, e)|_{v=u_i} = F(u, x, e), \; i = 1, \ldots, S. \tag{9.2}$$

To compute the fitness of any individual with strategy u_i, we evaluate G at $v = u_i$. In other words, setting the argument $v = u_i$ computes the fitness for any focal individual with strategy u_i in the environment that is defined by the other strategies **u**, the population densities **x**, and the environment **e**. The environment **e** could include resources, disturbances, and temperature. This method allows us to calculate whether an invading strategy u_i could invade a set of existing strategies **u** with population sizes **x**. We will see this in action because this is how pairwise invasibility plots are computed.

Sometimes game theorists keep track of not only the population densities **x**, but also the sum of all species population densities $N = \sum x_i$, and the vector of strategy frequencies **p**, where $p_i = \frac{x_i}{N}$ and $\sum p_i = 1$. In this view, the function is expressed as $G(v, u, p, N, e)$. Recall the point of all this: evolutionary game theory is required when the fitness of a strategy is both density dependent and frequency dependent. We can now formally define these terms using the notation above. Given a set of strategies **u**, fitness is density dependent if

$$\frac{\partial G(v, u, p, N, e)}{\partial N} \neq 0, \tag{9.3}$$

and frequency dependent if

$$\frac{\partial G(v, u, p, N, e)}{\partial p_i} \neq 0, \; i = 1, \ldots, S. \tag{9.4}$$

Put plainly, if any non-zero change in fitness occurs with changes in density, then natural selection is density dependent (Eq. 9.3). If any non-zero change in fitness occurs with changes in frequencies of strategies, then natural selection is frequency dependent (Eq. 9.4). In plant communities, natural selection is simultaneously density and frequency dependent, so both equalities will be true.

G-functions can be expressed analytically in three steps: (1) define an ecological model of population dynamics; (2) select a set of strategies **u**; and (3) hypothesize how the individual's strategy v and all other strategies **u** affects the parameters in the population model. Once the demographic parameters become functions of v, **u**, **x**, and **e**, then the population model becomes a game theoretic model. In other words, when population dynamics are expressed as functions of strategies, then the population dynamics lead to strategy dynamics. It is time for an example, so let us consider a game of plant competition.

We start with the Lotka–Volterra competition equations for S separate species expressed in terms of per capita population growth rate:

$$F_1(x) = \frac{r_1}{K_1}\left(K_1 - \sum_{j=1}^{S} \alpha_{1j} x_j\right)$$

$$\vdots \quad = \quad \vdots \qquad\qquad (9.5)$$

$$F_S(x) = \frac{r_S}{K_S}\left(K_S - \sum_{j=1}^{S} \alpha_{Sj} x_j\right)$$

where the demographic parameters include r (intrinsic growth rate when x is near zero), K_i (carrying capacity, the non-zero equilibrium population density when all other species are at zero density), and α_{ij} (the per capita effect of species j on species i). For simplicity, we let r be equivalent across all species, but K and α_{ij} are written as functions of a trait v within the trait set **u**. The *G*-function for a given species i can be written as

$$G(v, u, x) = \frac{r}{K(v)}\left(K(v) - \sum_{j=1}^{S} \alpha(v, u_j) x_j\right) \qquad (9.6)$$

Now express K and α_{ij} as distribution functions of strategy v:

$$K(v) = K_m \exp\left(-\frac{v^2}{2\sigma_k^2}\right) \qquad\qquad (9.7)$$

$$\alpha(v, u_j) = 1 + \exp\left(-\frac{(v - u + \beta)^2}{2\,\sigma_\alpha^2}\right) - \exp\left(-\frac{\beta^2}{2\sigma_\alpha^2}\right) \qquad (9.8)$$

where K_m is the maximum value of K, σ_k represents the range of resources, σ_α represents species niche width, and β introduces asymmetry into the competition coefficients (Vincent and Brown, 2005). Figures 9.4a and 9.4b illustrate these two distribution functions. The outcome of this game strongly depends on the values of all these parameters. Recall the emphasis in Section 9.2 on how the rules determine the outcome of the game. In this example, increasing values of σ_k^2 permits a coalition of strategies to coexist. When $K_m = 100$, $r = 0.25$, $\sigma_\alpha^2 = \sigma_k^2 = \beta^2 = 4$, then a single ESS evolves at $u_1 = 1.21$. However, if σ_k^2 is increased to 12.5 and $S = 2$, then a coalition of two strategies can coexist at $u_1 = 3.12$ and $u_2 = -0.24$ (Figure 9.4).

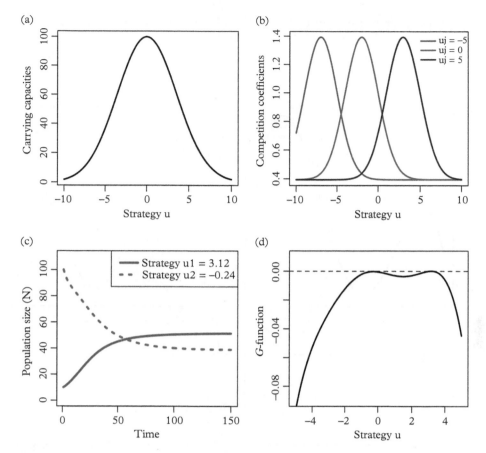

Figure 9.4 *The Lotka–Volterra competition game. (a) Carrying capacities* K *as a function of strategy values* u *where* $K_m = 100$ *and* $\sigma_k^2 = 12.5$. *(b) Competition coefficients* α_{ij} *as a function of strategy values* u *where* $\beta^2 = \sigma_\alpha^2 = 4$ *and* $u_j = -5$, 0, *or 5. (c) A coalition of two strategies coexist no matter the starting values of the two populations. (d) The* G-*function plotted at equilibrium abundances illustrates the two evolutionarily stable strategies (ESS) at their fitness maxima.*

One important feature of adaptive dynamics is that fitness landscapes change shape over time as species evolve toward the ESS (Figure 9.5). So, just as the fitness landscape under density and frequency independence changes shape along environmental gradients (Laughlin and Messier, 2015), the fitness landscape expressed by the *G*-function "heaves and bulges" (Waxman and Gavrilets, 2005) as populations move across it over time. Fitness landscapes are highly complex surfaces that change in response to (1) the environment, (2) the traits of species, and (3) the population sizes of the species. In the first few time steps of the Lotka–Volterra competition game, the species are attracted to the fitness peak to the right and so shift their strategies to the right. Once the first species reaches the ESS at $u_1 = 3.12$, the second species shifts direction and is drawn to the higher peak on the left, eventually settling into the second ESS at $u_2 = -0.24$. Once

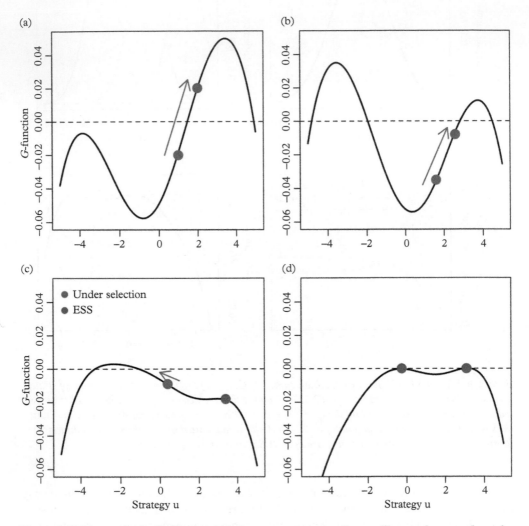

Figure 9.5 *Changes in the G-function over time versus strategy values* u, *illustrated as snapshots of the fitness landscapes that change shape over the course of evolution. (a–b) At first, the two species move to the right by their attraction toward the evolutionarily stable strategy (ESS) at* $u_1 = 3.1294$, *(c) but when one species gets close, a valley separates the two, and then the second species starts moving to the left to climb the peak toward the second ESS at* $u_2 = -0.2397$. *(d) The two evolutionarily stable strategies at equilibrium abundances (same as Figure 9.4d).*

Reproduced from the example in Vincent et al. (1993) with permission.

a species reaches a fitness peak and increases in abundance, the peak is pulled down to G = zero and becomes resistant to invasion. At equilibrium, the per capita population growth rates of both strategies equal zero and all other possible strategies exhibit lower fitness.

This example using the phenomenological Lotka–Volterra equations is as simple as they get (Vincent and Brown, 2005). When interactions involve mechanisms of resource reduction, then the model predictions become much more applicable to plant strategy theory. Unfortunately, the solutions also become more difficult to solve and the papers become less tractable for field ecologists. There do not seem to be easy generalizations about the shape of the drawdown of the fitness landscape around a strategy (Falster et al., 2017), and the shape of trait-based competition functions are unlikely to be simple Gaussian surfaces (Falster et al., 2021b). Many of the key papers in evolutionary game theory are only cited by other game theorists and tend to be ignored by empirical ecologists. It is time for that to change.

9.5 Games of defense resemble a Prisoner's dilemma

The evolution of plant defense involves a trade-off between allocating resources to defense or growth (Herms and Mattson, 1992), where the optimal strategy balances the costs of being eaten versus the cost of defense (Endara and Coley, 2011). In the presence of herbivores, the defended plant will achieve higher fitness than the undefended plant, but in the absence of herbivores the undefended plant will achieve higher fitness because they are allocating more resources to growth. Defenses may both minimize attack rate but also deflect herbivores onto their undefended neighbors (Oksanen, 1990), so plant fitness will depend not only on the rate of herbivory but also on the defense strategies of its neighbors (Grubb, 1992).

Interactions between defended and undefended plants can make the optimal plant defense strategy frequency dependent (Tuomi et al., 1999), where competitive interactions between plants may shift the ESS. One possible reason for this is because undefended plants can gain protection by simply coexisting with defended plants. This situation resembles a Prisoner's dilemma, which is a situation where two players have two options whose outcome depends on the choice made by the other simultaneously. It is typically described in terms of two prisoners separately deciding whether to confess to a crime, and why they might betray each other even if it appears to be in their best interest to cooperate. According to Tuomi and colleagues (1999, p. 80), "defended plants act as helpers by reducing the risk of herbivore attack, while undefended plants are the cheaters which benefit from the reduced attack rate without any cost of defence." Recent game theory models have predicted that compensatory growth responses that allow plants to tolerate moderate levels of herbivory in the absence of defenses can be a single strategy ESS (McNickle and Evans, 2018).

9.6   Pairwise invasibility plots illustrate the potential for new strategies to invade

Demography and fitness is the currency that links the ecological theatre to the evolutionary play (Hutchinson, 1965) and evaluating whether alternative strategies can invade an existing strategy is a key criterion for sympatric microevolution. Pairwise invasibility plots (PIPs) are used to determine whether a strategy is an ESS by plotting fitness in a matrix defined by the resident strategy on the x-axis and the invading strategy on the y-axis. The intersection of the diagonal line $x = y$ and the curve that delineates switches from positive to negative fitness are evaluated to determine whether the ESS exists, whether it is convergent stable, and whether multiple strategies could coexist within each population.

For example, Metcalf and colleagues (2008) combined integral projection models of monocarpic perennial plants (*Carlina vulgaris* and *Carduus nutans*) with adaptive dynamics to test whether flowering strategies for each species aligned with game theory predictions of ESS for this trait. Their analysis focused on the intercept (β_0) of a logistic size-flowering probability relationship, where larger values of β_0 increased the probability of flowering and reduced average size at flowering. They found that ESS existed for flowering strategies for both species and that the observed flowering strategies aligned perfectly with the ESS (Figure 9.6). When the resident flowering strategy for *Carlina* was $\beta_0 = -12$, then no other strategy could invade. Take a vertical slice through Figure 9.6a at $\beta_0 = -12$, and you will see the entire line is found within regions of negative fitness. Therefore, it was an ESS. It was also convergent stable because resident strategies near the ESS will evolve toward it. The same is true for *Carduus*. When the resident flowering strategy for *Carduus* was $\beta_0 = -5$, then no other strategy could invade. Importantly, the predictions of the model align well with the observed values of β_0 in nature shown by the crosshairs in Figure 9.6.

This analysis was one of the first integrations of empirical demographic models with game theoretic models, but more have since been conducted. Similar analyses have shown that vegetative dormancy, once viewed as a nuisance, may actually be an adaptive strategy to minimize mortality when growing in unpredictable and harsh environments (Shefferson et al., 2014), and that flowering probabilities that evolve in response to climatic variability may buffer against population decline (Williams et al., 2015). Methods for using IPMs to study density-dependent evolution are currently being explored and expanded at great depth (Rees and Ellner, 2016).

9.7   Games of establishment drove the evolution of seed size

Now that the foundations of evolutionary game theory have been established, we can explore insights that game theory have contributed to our understanding of

Figure 9.6 *Pairwise invasibility plots (PIPs) for (a)* Carlina vulgaris *(Carline thistle) and (b)* Carduus nutans *(musk thistle), where flowering strategy is represented by the intercept (β_0) of a logistic size–flowering probability relationship. Shaded regions (-) indicate unsuccessful invasion of the resident strategy (x-axis) by an invading strategy (y-axis). Observed flowering intercepts are indicated by the black dot with 95% confidence intervals. (c)* Carlina vulgaris. *(d):* Carduus nutans.

(a–b): Reproduced from Metcalf et al. (2008) with permission. Note that other applications of PIPs plot positive fitness regions in dark and negative fitness regions in white (Geritz et al. 1997). (c) Photo by Evelyn Simak, reproduced here under a Creative Commons Attribution-Share Alike 2.0 license (https://commons. wikimedia.org/wiki). (d) Photo by Max Licher, reproduced here under a Creative Commons Attribution-Share Alike (CC BY-SA) (https://swbiodiversity.org/seinet).

plant strategies. Much attention has been given to traits of seeds, height at maturity, leaf economics, and plant defense. Ellner (1985a,b) explored a density-dependent model of ESS germination strategies in annual plants. His model demonstrated that optimal strategies in density-dependent models of growth and yield differ significantly from density-independent models, indicating that "density-independent theories of life-history evolution in random environments may have limited applicability to density-regulated populations" (Ellner, 1985b, p. 81). Ellner concluded the first paper with a series of hypotheses that could be tested by empirical ecologists: the mean growth rate of the population should strongly correlate with the survivorship of dormant seeds, germination fractions approaching 100% should only occur in populations where the mean growth rate is greater than the survivorship of dormant seeds, and germination fraction should be no larger than 1 minus the frequency of years unsuitable for reproduction. His second paper concluded that germination fraction is predicted to decrease with increased seed survivorship and with increasing variation in total seed yield.

Rees and Westoby (1997) applied evolutionary game theory to ask whether multiple seed mass strategies could coexist in communities. Previous theoretical predictions concluded that a single optimal seed mass would maximize fitness (Smith and Fretwell, 1974), but this prediction was at serious odds with empirical observations that seed mass variation within plant communities was large compared to the small variation between communities and environments (Salisbury, 1942; Rees et al., 1996). Rees and Westoby thought that the discrepancy could be because previous optimization approaches ignored frequency dependence and so did not account for the fact that the shape of fitness landscapes depends on which other strategies are present in the milieu. Their analysis initially assumed that individuals from larger seeds have a larger negative effect on individuals with small seeds, and vice versa. They first developed a model for single species to evaluate whether a single optimal ESS strategy can be invaded by initially rare mutants. The G-functions described how fitness is a function of the seed mass of the invader, the seed mass of other resident species, and population sizes of each species. These functions were coupled with models of annual plant competition (similar to those in Eq. 8.5) (Watkinson, 1980), where the competition coefficients were functions of seed mass, and where seed production was a decreasing function of seed mass (recall the seed size–seed output trade-off from Figure 6.21).

The single-species model concluded that a singular ESS seed mass exists when the competitive advantage of large seeds is unbounded, but if limits to this competitive advantage exist than smaller seeds can invade. This result was strongly dependent on values of the important parameter s, a scale parameter that determined how rapidly the competitive advantage of big seeds accrued over small seeds, which highlights the importance of the assumptions that are embedded in the theoretical models. This result echoed the results of Geritz (1995) who, using a different ESS formulation, found that the large-seed strategy can be invaded by smaller-seed strategies because the more abundant smaller seeds will be able to disperse to safe sites that are not occupied by large seeds. When limits are placed on the competitive dominance of large seeds in the context of a multi-species model, Rees and Westoby discovered that coalitions of multiple strategies with diverse seed masses could stably coexist: the fitness landscapes were rather flat (Figure 9.7a). In other words, a continuous spread of seed masses can coexist, as suggested by theory (Leimar et al., 2013). This result foreshadowed later results of Falster and colleagues (2017), who also emphasized the flatness of fitness landscapes and argued that this neutrality emerged as a product of trait-differentiated competition. Rees and Westoby suggested that any low-dimensional model would likely exhibit flat

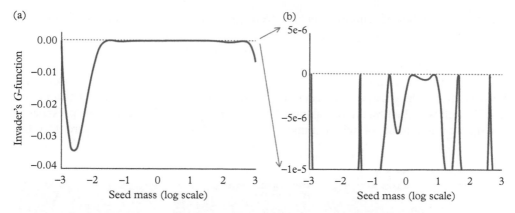

Figure 9.7 *(a) The invader's G-function against log seed mass illustrates a coevolutionary stable state with seven strategies. (b) The vertical scale is reduced to show the fine details of the "flat" fitness landscape.*
Reproduced from Rees and Westoby (1997) with permission.

landscapes because differentiation of species in trait space is more difficult to detect with fewer traits (see Laughlin (2014a) for an in-depth treatment of this fact). If coexistence depends on niche differentiation, then adding more traits to the model would simultaneously enhance species differentiation and the steepness of the selection gradients. The flat fitness landscape also implies that detecting change in performance based on seed mass alone will be difficult with observational data from the field. Their model provided an eco-evolutionary explanation for the wide spread of seed mass within plant communities relative to among communities and explains how this variation is spread between species rather than within species.

9.8 Games for light under a disturbance regime diversified height and leaf economics

It is time to take another look at games for light. Previous games for light either identified a single tall ESS where light competition is a tragedy of the commons, or a two ESS solution that includes just a single shade-tolerant strategy. The trade-off between growth in high light and survival in shade is the key to this diversification, but previous theoretical models could not explain the diversity of woody plant strategies that are commonly observed in forests. To address this challenge, Falster and colleagues (2017) developed a game theoretical model that simulated the eco-evolutionary dynamics of plant height at maturity and leaf mass per area (LMA). The game included size-structured growth, mortality, and reproduction of metacommunities linked through dispersal against a background of disturbance. It included two physiological trade-offs in plant function to set the rules for the game: LMA represents a trade-off between the cost of leaf construction and leaf lifespan, and height at maturity specifies when an individual switches allocation of energy from growth to reproduction. All of this was set in the context of adaptive dynamics where the payoffs were defined in terms of per capita population growth rate and the focus was on the invasibility of new strategies into resident communities.

Falster demonstrated that when both traits evolve together, at least 11 peaks in the bivariate fitness landscape could coexist within the forest in the long-term (Falster et al., 2017). This was the first analytical model that could successfully explain the presence of multiple combinations of height at maturity and LMA within a forest (Figure 9.8). But

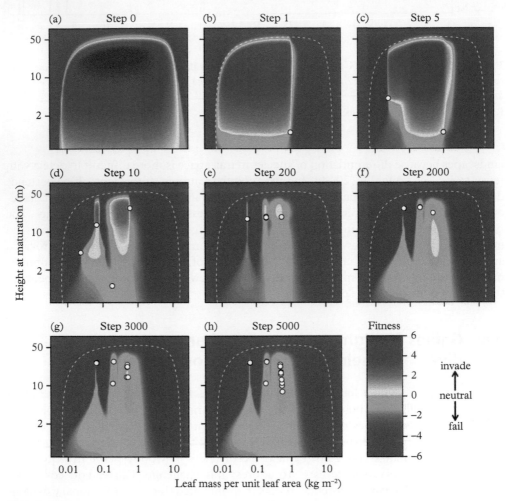

Figure 9.8 *Fitness landscapes at different steps of metacommunity assembly, where the colors represent the invasion fitness of rare species competing with resident species (circles). Invasion is only successful in regions with positive invasion fitness (yellow to red). The ridge-shaped fitness plateau seen in (h) occurs due to the coexistence of a high diversity of shade-tolerant phenotypes.*

Reproduced from Falster et al. (2017) with permission.

they went even further to show the predictions of their model across a range of site productivity and disturbance frequencies that represent values from vegetation types such as arid shrublands and tropical rain forests. This is a major step toward resolving one of the major problems of plant strategies described in Chapter 1: how the same traits can be involved in generating strategies across broad climatic gradients and within communities driven by patch dynamics. Their model demonstrated that less productive sites that are infrequently disturbed support fewer unique shade-tolerant late-successional strategies. This paper demonstrated several key points. First, it demonstrated how more than two continuous and orthogonal traits can evolve to multiple ESS. Second, it included variation in both disturbance frequency and ecosystem productivity, the two main drivers of plant strategy evolution (Grime, 2001). Third, it demonstrated how a diversity of slow-growing shade-tolerant strategies could evolve in forests. Fourth, like the flat fitness landscapes driven by seed mass (Rees and Westoby, 1997), the fitness landscapes for height at maturity alone were also flat, supporting many more strategies than LMA alone. They called these flat fitness landscapes "evolutionarily emergent near-neutrality." If flat fitness landscapes are indeed common, then we need to confront the reality that one optimal strategy does not exist in a given environment. If fitness landscapes are always rugged, empirical models that rely on simple Gaussian surfaces will struggle to find the contours of fitness landscapes.

9.9 Games for soil resources increased root length

Soil resources impose strong constraints on plant production. Competition for soil resources is not asymmetrical because nutrients and water are not supplied from a single direction (Craine and Dybzinski, 2013). Nutrient uptake is limited by the slow diffusion rates of minerals, and diffusion rates become even slower when water is limiting. Plants from semi-arid and nutrient-limited systems like grasslands exhibit higher than expected root length density (root length per volume of soil). The high root length density could be driven by two selection pressures that are not mutually exclusive. First, the slow diffusion rates in dry soil might require plants to proliferate roots throughout the soil to mine mineral resources when they become exhausted at the root surface. Second, competitive interactions may have driven higher root production rates over what is optimal in the absence of competition in the same way that light competition caused increasing allocation to woody stems. One model of nutrient acquisition suggested that sufficient nutrients can be acquired in the absence of competition with much less root length than is typically observed, except when ammonium becomes diffusion-limited in very dry soil (Craine, 2006). When competitors are added to the model, the optimal allocation to root length increases and competition for resources becomes a tragedy of the commons. Models of root trait evolution suggest that roots have become thinner over evolutionary time (recall Figure 6.13) thereby improving their efficiency of soil exploration per unit of carbon invested (Comas et al., 2012; Ma et al., 2018). This provides additional support for the idea that root length per unit carbon has increased in response to increasing competition for soil resources. In a world without competition, we may not have seen an increase in specific root length over time.

Strategies for optimal allocation of biomass in forest ecosystems have been studied for decades, but only recently have these questions been asked from a game theory perspective. And "just as the physical properties of a moving fluid depend on the characteristics and interactions of individual atoms, the dynamics of the world's forests depend on the characteristics and interactions of individual trees" (Dybzinski et al., 2011, p. 153). Evolutionary game theory was used to find ESS for allocations to leaves, wood, and fine roots for trees in old-growth forests across a gradient in nitrogen availability. In Dybzinski's model, invading strategies would be deemed successful if their populations could increase from low-density and the ESS is the resident type that resists invasion by all invaders. They concluded that with increasing nitrogen availability, ESS allocation to foliage and wood increases, but allocation to fine roots decreases (Figure 9.9). These ESS strategies do not maximize growth rates in monoculture because the most competitive strategies are not necessarily the best but are rather those that create conditions for others that are worse. Their model also clarified that the trade-off between root-to-shoot allocation is likely to be driven by a trade-off between fine roots and wood, not between roots and leaves. Dybzinski and colleagues (2015) extended this model to explore how

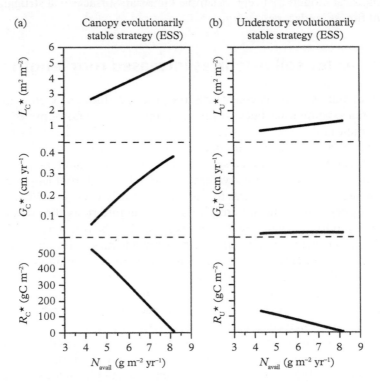

Figure 9.9 *The evolutionarily stable strategy (ESS) of leaf area index ($L_x^\star$), stem growth rate ($G_x^\star$), and fine root mass ($R_x^\star$) across a gradient in nitrogen availability (N_{avail}) for individuals in (a) the forest canopy and (b) the forest understory.*

Reproduced from Dybzinski et al. (2011) with permission.

elevated concentrations of CO_2 in the atmosphere would change the dynamics, and were surprised to conclude that carbon storage under elevated CO_2 is independent from nitrogen supply rate because competition for nitrogen drove increased fine root biomass and competition for light drove increased allocation to wood.

One of the main conclusions from adaptive dynamics models is that density and frequency dependence promotes substantially taller plants and greater root production than would be physiologically optimal in the absence of competition. These conclusions were made by analyzing the effects of light and nutrient limitation, but what about competition for water? Farrior and colleagues (2013a) developed a game theoretic model for both water and light competition. In a scenario where the soil is always water saturated, the ESS is to allocate little to root production (Figure 9.10). In a scenario where the soil is always water limited, then a tragedy of the commons develops, and the ESS would allocate so much to fine roots to acquire the water that no carbon would be left for growth and reproduction, making the solution unfeasible for closed-canopy forests. These paradoxical results mirrored other studies (Zea-Cabrera et al., 2006). However, Farrior discovered a resolution to the paradox by varying the total portion of the growing season in water saturation (q) and the rate of constant rainfall during water limitation (R_{dry}). Competitively dominant carbon allocation strategies depend on the timing of the rainfall during the growing season. If water limitation occurred for less than 70% of the growing season, then a feasible fine root allocation strategy exists (Figure 9.10). In other words, if additional rainfall arrives during a period of water limitation, then the surprising ESS solution is to increase allocation to fine roots but not to leaves to maintain a competitive edge.

Farrior and colleagues (2013b) tested these predictions in a field experiment involving nutrient and water additions and showed that root production did in fact increase with increasing water, but the response depended on whether nitrogen was limiting. They developed a game theoretic model of competition for both water and nitrogen to help explain these results. If leaf construction is constrained by nitrogen uptake, then leaf production will increase with nitrogen but not water addition. Adding water enhances competition for light and exacerbates nutrient limitation. They concluded that water additions cause an increase in fine root production because of the enhanced nitrogen limitation, but if nitrogen is added with water to ameliorate nutrient limitation, then fine root production is decreased. In other words, the enhanced root production with increasing water availability is most pronounced when nitrogen is limiting.

This chapter has demonstrated that plants produce more leaves, stems, and roots than what is optimal for the most efficient harvesting of required resources for growth because excess tissue production provides a competitive advantage by pre-empting the resources before others get to them (McNickle and Dybzinski, 2013), although interactions between resources can affect the intensity and direction of selection (Farrior et al., 2013b). McNickle and colleagues (2016) developed this observation further to see how these game theoretical dynamics play out over global gradients in carbon and nitrogen availability where plant fitness depends on its own leaf, stem, and root production as well as the production of these tissues by others. They developed a G-function that states that fitness of the focal plant is a weighted product of net carbon and nitrogen

Always water saturated _____

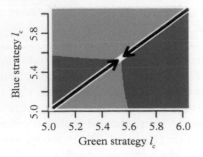
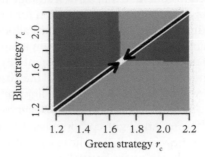

Always water limited _____

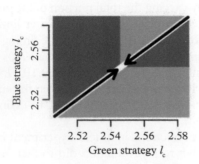
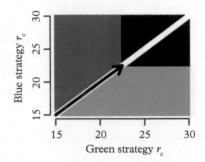

Water limited with periods of water saturation _____

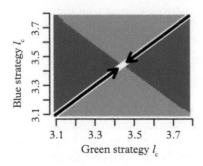
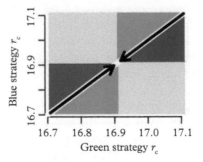

Figure 9.10 *Pairwise invasibility plots (PIPs) for strategies of leaf area index (l_c) on the left and root area index (r_c) on the right. Green colors indicate the green strategy dominates, blue indicates that the blue strategy dominates, yellow indicates that neither is dominant. A tragedy of the commons occurs for water use in plants with water limitation (middle row), where the evolutionarily stable strategy (ESS) would allocate so much to fine roots that no carbon would be left for growth and reproduction (black region).*

Reproduced from Farrior et al. (2013a) with permission.

harvested. Carbon harvest is a function of photosynthetic rate, aboveground competition and costs of tissue construction and respiration. Nitrogen harvest is a function of nitrogen uptake, belowground competition, and the costs of tissue construction. Surprisingly, their game theoretic model accurately predicted net primary production, plant functional types and biomes, and the allocation of total biomass into leaves, stems, and roots across gradients in carbon and nitrogen availability (Figure 9.11). In their words:

> As a game theory treatment of plant growth and allocation, we argue that [the model] is successful because plant investment into roots, leaves, and stems may be a triple tragedy of the commons as plants amplify production of these tissues to pre-empt each other's access to light, water and nutrients and maximize their competitive ability (McNickle et al., 2016, p. 7).

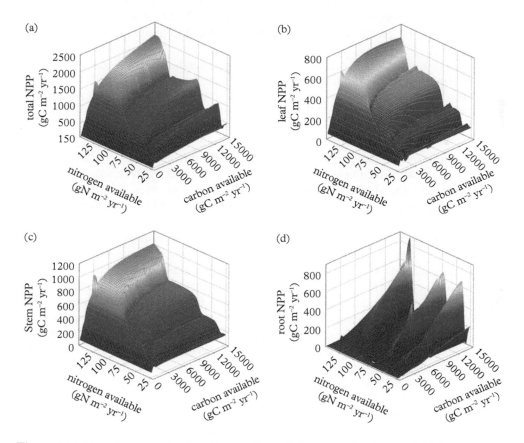

Figure 9.11 *Net primary production along gradients of nitrogen and carbon available for (a) total, (b) leaf, (c) stem, and (d) root production.*
Reproduced from McNickle et al. (2016) with permission.

9.10   Testing game theory predictions with empirical models

It has been argued that the ubiquity of density—and frequency-dependent natural selection calls for a game theoretical approach to understanding plant strategies (Brown and Vincent, 1987; Vincent et al., 1993; Diekmann, 2004; Metz, 2012). Game theory has provided us with important insights into the evolution of plant strategies, and applications of game theory to plant strategy theory are really starting to pick up steam.

Is game theory the only way to learn about plant strategies under density and frequency dependence? Are we doomed to playing theoretical games with no hope of evaluating the predictions of these games with empirical data? This question has been asked before. Brown and Vincent (1987, p. 78) noted that their "theory is testable, but a comprehensive experimental approach is required." Waxman and Gavrilets (2005, p. 1152) asked for a greater emphasis on empirical tests and wrote:

> we believe that workers in Adaptive Dynamics need to come up with testable predictions of their approach to evolution, that are novel, in the sense that they cannot be easily achieved by any other approaches. It would be especially interesting to see a comprehensive comparison of empirical data and the corresponding predictions of Adaptive Dynamics.

To their credit, many of the recent game theoretical models of plant competition have included empirical tests, albeit some more direct than others (Dybzinski et al., 2011; Farrior et al., 2013b; McNickle and Evans, 2018). However, even these theorists cautioned about their ability to conduct strong empirical validations of game theoretical results when the models are built on parameters that are not directly measurable in the field. For example, Farrior and colleagues (2013a, p. 324) admitted, "empirical tests of the predicted relationship between tree structure and rainfall are complicated by the idealized rainfall in the model. For example, what are the values of R_{dry} and q in any location?" This is exactly what McGill and Brown (2007, p. 427) referred to in their annual review: models must be developed in collaboration with field ecologists to make them "less heuristic and more amenable to empirical test." One recent theoretical model of coexistence generated empirically testable predictions in water-limited Mediterranean habitats (Levine et al., 2022). It may be impossible to design the decisive experiment that orthogonally crosses abiotic factors with biotic interactions (Reich et al., 2003), but what is impossible for one generation may be the next major empirical advance by the next, and I am optimistic.

The rules of the game and the fitness payoffs determine whether a strategy is an ESS. Parameters are often fine tuned so that more than one ESS is possible (Vincent et al., 1993; Rees and Westoby, 1997). Remember, "assuming the mathematics are correct, the important features of a model are its assumptions and its predictions" (Tilman, 1988, p. 64). In game theory, the assumptions are found within the fitness payoffs. In the hawk–dove game, the relative value of the resource and the cost of injury determined whether a strategy was an ESS. Changing the costs and benefits changes the model predictions. Game theory is a powerful tool to understanding the consequences of the assumptions that we make about competitive interactions between plants, but we must know when to admit when the theoretical approach reaches its limitations for predicting outcomes on the ground. Perhaps reframing what we measure in field experiments in terms of payoffs and costs, rather than phenomenological competition coefficients, is what is really needed.

Goldberg (1996) articulated how competition for resources must be viewed from two perspectives to understand the whole process: competitive responses and competitive effects. These ideas are strongly related to frequency dependence because the optimal trait value of the focal individual will depend on the trait value of individuals in the local neighborhood. Individual competitive ability includes the response of the focal individual to resource limitation caused by individuals in their local neighborhood and the effect that an individual has on the suppression of other individuals in the neighborhood (Goldberg, 1990; Goldberg and Landa, 1991). From the perspective of trait-based competitive abilities, responses can be quantified as the change in fitness (or fitness component) of a focal individual to different abundances of neighbors as a function of its traits, and effects can be quantified as the change in fitness (or fitness component) of a focal individual to different abundances of neighbors as a function of their traits. These cannot be analyzed in the context of an equilibrium solution, but they can be analyzed using long-term datasets of growth, survival, and reproduction where the local competitive neighborhoods are known. There are two empirical approaches worth exploring.

First, if the effects of density and frequency dependence can be reduced to impacts on resource availability at the scale of the focal plant, then in some ways these effects are already included in optimization models that only involve resources. After all, the effects of increasing density are often manifested as reductions in resource availability. For example, given that density of tall neighbors is the only way to reduce light availability, quantifying the fundamental niches of species with respect to light gradients will give us ample data for predicting how traits influence population performance in recently disturbed high-light environments versus long-established vegetation. Goldberg recommended that seedling response to shade can be experimentally quantified by growing seedlings in established forests, but they could also be planted in experimental shade.

Second, empirical analyses of long-term observational data could also be leveraged to better understand fitness responses to neighborhood density as a function of the focal plant's traits and the surrounding neighborhood's traits. This is analogous to an individual "playing the field" (Maynard Smith, 1982), where the competitive environment is described by the mean and variance of traits in the neighborhood. Westoby hoped that his LHS plant strategy scheme would be adopted as a framework for re-analyzing the thousands of field experiments and ecophysiological studies that have accumulated more rapidly "than have been satisfactorily digested and interpreted" (Westoby, 1998, p. 215). But alas, this synthesis of competition experiments never materialized. However, it may not actually be a fruitful exercise given that short-term competition experiments only capture the "transient dynamics" of competition and not the long-term outcomes of competition that are seen over longer time spans (Tilman, 1988).

Despite the lack of a formal meta-analysis of the literature, Kunstler and colleagues (2016) conducted a global meta-analysis of the traits that underpin tree competition and demonstrated the widespread relevance of traits across multiple forested biomes. This study decomposed the effects of traits on growth rate into several quantities, but I focus on the following three effects: (1) the effect of traits on maximum growth rates in the absence of competition; (2) the effect of traits on tolerance of competition (i.e., competitive response), where the growth of the focal individual possessing the trait is less affected by neighborhood density; and (3) the competitive effect, where neighborhoods possess traits that suppress the growth rate of the focal individual. They found that maximum growth rates were highest in tree species with low wood density and high SLA (Figure 9.12). Species with low wood density were also intolerant of competition and had weak effects on neighbors and species with high SLA had low competitive effects. These results demonstrated that the classic trade-off in forests that underpins patch dynamics is trait based: traits generate the trade-off between growth rate in the presence of competition versus growth rate without competition.

The life of a plant is greatly modified by its neighbors, and projections of vegetation response to global change must contextualize physiological responses in the context of competition. The study by Kunstler and colleagues (2016) provides a "trail map" (Levine, 2016) for how to use empirical data to test the effect of traits on fitness in the context of density and frequency dependence. Will empirical ecologists increasingly come to view their valuable field observations in the context of game theoretical predictions? A reckoning of empirical approaches (Chapter 8) and game theoretical approaches (this chapter) will provide the most promising pathway toward the maturation of plant strategy theory.

9.11   Chapter summary

- When the success of a strategy depends on what other individuals in the community are doing, then evolutionary game theory becomes a useful (and some would argue, necessary) tool.

- The evolutionarily stable strategy (ESS) is a strategy that resists an invader with a different strategy that is introduced at low density. In other words, the ESS is a peak on a fitness landscape, where alternative strategies (often called mutants) have lower fitness and are therefore unsuccessful at invading from low densities.

- Productive sites by themselves are not the driver of increasing maximum heights in trees. Tall species are adapted to productive sites because the consequences of not keeping up with their neighbors is deep shade and eventual death. In other words, tall trees do not produce thick trunks just because they can—they do it because they must.

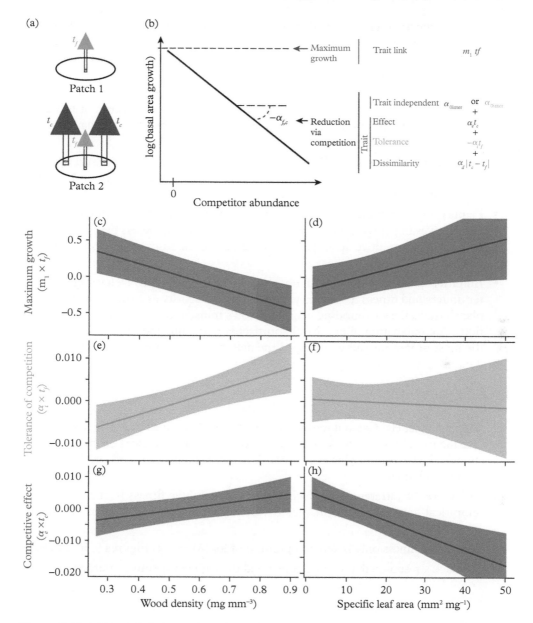

Figure 9.12 *(a) Empirical observations of growth rate on focal individuals f vary in the abundances of competitors of species c. (b) The model estimates how trait values of the focal individual (t$_f$) and abundances and trait values of competitors (t$_c$) influence growth rates of a focal tree. Maximum growth (red) was a function of the traits of the focal tree. Reduction in growth rate per unit basal area of competitors (black) was modeled as the sum of trait independent effects (blue), the effect of competitor traits (t$_c$) on their competitor effect (purple), the effect of the focal tree's traits (t$_f$) on its tolerance of competition (green), and the effect of trait dissimilarity (orange). (c–h) Empirical results of the hierarchical models illustrating variation in maximum growth (red), competitive tolerance (green), and competitive effects (purple) as a function of wood density and specific leaf area.*
Reproduced from Kunstler et al. (2016) with permission.

- In evolutionary game theory, there are players, strategies, and payoffs. Players are individual organisms, strategies are heritable phenotypes (traits), and the payoff is fitness expressed as the per capita growth rate of a strategy in a given abiotic and biotic environment.

- Once the demographic parameters of a population model become functions of the focal individual's trait, the traits in the community, the densities of those that possess those traits, and the environmental conditions, then the population model becomes a game theoretic model.

- Pairwise invasibility plots (PIPs) are used to determine whether a strategy is an ESS by plotting fitness in a matrix defined by the resident strategy on the x-axis and the invading strategy on the y-axis.

- Plants produce more leaves, stems, and roots than what is optimal for the most efficient harvesting of required resources for growth because excess tissue production provides a competitive advantage by pre-empting the resources before others get to them.

- Empirical analyses of long-term observational data could also be leveraged to better understand fitness responses to neighborhood density as a function of the focal plant's traits, the surrounding neighborhood's traits, and the environmental conditions. A combination of empirical approaches and game theoretical approaches is the most promising pathway toward the maturation of plant strategy theory.

9.12   Questions

1. Explain what Hal Caswell meant when he tweeted "models based on assumptions are the most powerful tool for understanding the consequences of those assumptions" (Hal Caswell, March 26, 2020). How does this relate to results of game theoretical models?

2. What are the differences and similarities between how fitness is quantified using empirical approaches in Chapter 8 versus game theoretical approaches in this chapter?

3. Why does fitness only reach an optimum of $\log(\lambda) = 0$ in Figures 9.5d and 9.7?

4. Is space a resource that can be consumed through competitive interactions?

10

Applying Plant Strategies in Conservation and Restoration

The push to incorporate trait ecology into conservation science aims to establish evidence-based approaches for linking widely available and/or easily measured traits to the complex processes that drive extinction risk.

—Rachael Gallagher and colleagues (2021, p. 928) "A guide to using species trait data in conservation," in *One Earth*

We need acts of restoration, not only for polluted waters and degraded lands, but also for our relationship to the world. We need to restore honor to the way we live, so that when we walk through the world we don't have to avert our eyes with shame, so that we can hold our heads up high and receive the respectful acknowledgment of the rest of the earth's beings.

—Robin Kimmerer (2013, p. 195), *Braiding Sweetgrass*

10.1 Conservation and restoration are fields grounded in hope and reciprocity

Billions of years of evolution have produced a fabric of life across the planet that teems with diversity (Díaz, 2022). The air we breathe, the water we drink, the food we eat, the shelter we dwell in—all are a direct result of this biological diversity (Daily, 1997; de Groot et al., 2002; Lavorel, 2013). Yet, within a geological blink of an eye, human society has wreaked havoc by unraveling this fabric of life at its seams. The average abundance of native species has declined by 20%, 75% of the land surface has been altered by human actions, and 1,000,000 plant and animal species are threatened with extinction (Brondizio et al., 2019). Powerful forces have pursued policies of unabated expansion of the human footprint across ever more vast regions of the planet, even though "growth for the sake of growth is the ideology of the cancer cell" (Abbey, 1977).

My relationship with plants was kindled by Dave Warners, my first mentor and plant ecology professor. We wandered through the forests, meadows, dunes, savannas, wetlands, and prairies of Michigan to survey rare habitats and collect wild seed. It was a revelation that we could collect seeds and grow them—the act of harvesting seed in

Plant Strategies. Daniel C. Laughlin, Oxford University Press. © Daniel C. Laughlin (2023). DOI: 10.1093/oso/9780192867940.003.0010

nature transformed my concept of what seeds were—they became corporeal beings that can be gathered and grown to repair degraded landscapes. We grew seedlings in the college greenhouses and planted them in the earth. We replaced manicured lawns with species-rich wet meadows and dry prairies (Figure 10.1). Restoring degraded ecosystems through the reassembly of lost habitats was a tangible way of doing good. Restoring ecosystems lit a wildness inside me that "is perennially within us, dormant as a hard-shelled seed, awaiting the fire or flood that awakes it again" (Snyder, 2010, p. 14). To this day, few things bring me deeper joy than watching native species that I have planted in my yard reproduce and disperse throughout my neighborhood. Ecological restoration is a creative and defiant movement that oozes with optimism (Palmer, 2019).

Conservation biology focuses on the preservation of biodiversity, whereas restoration ecology focuses on the reassembly of communities. They are intertwined and complementary pursuits. These sciences are both grounded in hope and reciprocity and are a mutually beneficial means of developing meaningful relationships with the plants that sustain society (Jordan, 2003). Robin Kimmerer (2013, p. 229) noted, "In some Native languages the term for plants translates to 'those who take care of us'." Not everybody has a positive relationship with plants. In the movie *The Goblet of Fire*, Harry Potter deeply disappointed me when he told his good friend Neville Longbottom, "No offense but I really don't care about plants." In the end, Neville the herbalist saved the day by giving Harry some gillyweed to eat so that he could breathe underwater and survive the second deadly task of the Triwizard Tournament.[1] This scene is a rather good representation of a condition commonly called plant blindness (Wandersee and Schussler, 1999; Balding and Williams, 2016). It is perhaps more appropriately called "plant awareness disparity" (Parsley, 2020), which is the fact that most people do not notice plants around them and therefore cannot appreciate their importance. Even when overlooking wild landscapes in scenic areas, many people simply see a sea of green things, rather than individual plants: each species looks the same, diversity does not exist to them (Marder, 2013).

I recently encountered this kind of plant awareness disparity close to home. I take daily walks with my pup along Jacoby Ridge, the same prairie where the story of plant strategies began in the Prologue, and one morning I noticed early activities of construction plans to bury a water pipeline. Naturally, I was concerned that such a disturbance would interfere with the abundant local wildlife and cause an influx of invasive species. I contacted the people in charge, offered my assistance, and asked to see their vegetation management plan. They were super nice but they thought I was worried that the pipeline would be an eyesore so they assured me that "no one will even notice that it's there once it's buried." They thought that I wouldn't notice the 1-km long and 10-m wide disturbance across the prairie. As Aldo Leopold (1949, p. 197) wrote:

---

[1] For the record, the movie diverged from the book, in which it was Dobby, the freed house-elf, who gave Harry the gillyweed.

Figure 10.1 *Examples of restoration projects across a variety of ecosystems. (a) Wet meadow restoration at Calvin College, Michigan in 1999. (b) Xeric prairie restoration at Pennsylvania State University in 2001. (c) Ponderosa pine forest on the North Rim of Grand Canyon National Park, Arizona in 2007. (d) Mixed conifer forest on the North Rim of Grand Canyon National Park, Arizona in 2009. (e) Temperate rain forest restoration project in New Zealand in 2015. (f) Sagebrush steppe restoration project in Grand Teton National Park, Wyoming in 2021.*

All photos (except in e) by the author. (e): Photo provided by Catherine Kirby with permission.

> One of the penalties of an ecological education is that one lives alone in a world of wounds. Much of the damage inflicted on land is quite invisible to laymen. An ecologist must either harden his shell and make believe that the consequences of science are none of his business, or he must be the doctor who sees the marks of death in a community that believes itself well and does not want to be told otherwise.

These words have haunted me for decades. So, I engaged in conversation with the grounds people about the unintended consequences of disturbing an intact prairie and discussed the positive things we can do to reassemble a productive prairie. It took some persistence, but I am very happy that they hired a local restoration company that drill seeded my native seed mix into the disturbed ground. The species were selected based on their traits to resist weed invasion. Seedlings are emerging and now I have a new monitoring project for my Vegetation Ecology class.

In my classes I try to compel my students to imagine a world bereft of plants and impress upon them what a terrible place that would be. I am unsure that this exercise ever really sinks in, maybe because it is impossible for a human brain to genuinely imagine a world without plants—even when blind to them, our subconscious knows that we could not exist without them. The science of conservation and restoration faces a steep uphill battle, but it is one of the most important battles we will ever face as a species. If we do not stand up for the plants that need saving, no one will.[2]

This chapter explores how plant strategies are of central importance to the vital applied sciences of biodiversity conservation and ecological restoration. According to Phil Grime (2001, p. xxviii) "The primary purpose in recognizing [plant strategies] is to understand the assembly of plant communities and ecosystems and to interpret their responses to environmental change and management." In my experience, however, it is all too common to question the relevance of plant traits in conservation and restoration practice. There is a curious perspective that the traits of plants are irrelevant because decisions are made on the basis of species, not their traits. On the one hand, it is absolutely true that managers cannot manage traits per se, but are experts at managing species (Laughlin, 2014b). On the other hand, managers are keenly interested in traits, including both life history and functional traits. I have worked with managers who want to know the lifespan of weeds, the dispersal mechanisms of rare plants, and the palatability of leaves to grazers. Traits are of fundamental interest to managers; they are not simply theoretical constructs that have no bearing on the day-to-day realities of managing the land. In fact, incorporating traits into restoration practice will more effectively reach both traditional and functional restoration targets (Merchant et al., 2022).

The chapter is organized into four parts. First, it begins with a discussion of applying traits to conservation in relation to demographic rates, functional rarity, grazing management, and planning for climate change. Second, it discusses applying traits in restoration to understand and predict restoration trajectories and design compositions to achieve

---

[2] Take it from the Lorax: "Unless someone like you cares a whole awful lot, nothing is going to get better. It's not" (Seuss, 1971).

functional outcomes. Third, we look at how traits are used to predict species invasiveness and the resistance of native plant communities to invasion. Fourth, we look toward the future to apply plant strategies in assisted migration.

10.2 Plant strategies inform decisions in conservation biology

Conservation biology is the science of rarity and is applied to prevent species from going extinct. There are four ways in which plant strategies are used in conservation biology. First, demographic analyses are needed to identify the demographic rates that are most vulnerable to extinction threats. Second, the concept of species rarity has been generalized to the new concept of functional rarity. Third, traits are being used to generalize across taxa to understand how plants respond to management actions, especially grazing. Fourth, traits are used to assess the vulnerability of species and to make decisions about conservation actions.

The most conspicuous applications of plant strategies in conservation are the demographic tools for diagnosing the cause of population declines (Brigham and Schwartz, 2003; Crone et al., 2011; Crone et al., 2013; Zipkin and Saunders, 2018). Demographic models are necessary for assessing population viability of species that are at risk of extinction by estimating how long it may take for a declining population to reach an extinction threshold (Morris and Doak, 2002). Elasticities describe the proportional change in population growth rate from proportional changes in each population matrix element holding all other elements constant and can be thought of as the relative importance of each matrix element (recall Chapter 4). Such analyses determine the specific life history transitions or vital rates that have the biggest effect on population growth rate in order to guide management actions (Caswell, 1989), and population models can be used to prioritize which landscapes and environments should be targeted for species reintroduction projects (Caughlin et al., 2019).

Some excellent examples of using population models in conservation come from studies on desert cactus. First, *Echinocactus platyacanthus* (candy barrel cactus) is an endangered species in Mexico and models showed that stasis of adults had the highest relative importance for population growth rate (Jiménez-Sierra et al., 2007). Second, populations of the endangered *Astorphytum capricorne* in the northern Chihuahuan Desert in Mexico were studied because one of the main causes of population declines is illegal harvesting by collectors that exclusively dig up large adult plants. Population models suggested that management actions should focus on enhancing seedling recruitment and reducing adult mortality (Mandujano et al., 2015). Third, *Coryphantha robbinsorum* from southeastern Arizona has been the topic of much discussion in the literature. Demographers have concluded that increasing the survival of established plants, especially adults, would be the most effective conservation strategy, not necessarily increasing the rate of recruitment (Schmalzel et al., 1995; Fox and Gurevitch, 2000). The survival and growth of large established plants is a critical life history stage for long-lived individuals, yet the focus of empirical research is often on reproduction and seedling dynamics

(Bruna et al., 2009). We need to study the entire trajectory of the life history, not just its component parts. There continue to be discussions among demographers about the use of asymptotic versus stochastic approaches to understanding the sensitivity of vital rates, but one thing is clear: knowledge of demographic rates are necessary to understand the population dynamics of species with different life history strategies (Silvertown et al., 1996).

The traditional paradigm in conservation has focused on understanding what causes the rarity of species. Deborah Rabinowitz (1981) suggested that there were seven forms of rarity, generated by variation in three factors: geographic range (restricted vs. widespread), habitat specificity (specialist vs. generalist), and local population size (scarce vs. abundant). If a species was widespread, a habitat generalist, and locally abundant then it is common. However, the other seven possible combinations of these factors describe differences in rarity.

Violle and colleagues (2017) proposed that functional rarity is another key aspect of biodiversity that should be conserved to maintain critical ecosystem functions. This approach has been referred to as the "ecology of outliers." Functional rarity increases with increasing distinction of the traits (i.e., the relative distance in trait space to another species) and with increasing scarcity of the organism that possesses the most distinctive traits (Figure 10.2). Building on Rabinowitz's (1981) foundational framework for species rarity, Violle and colleagues (2017) described 12 forms of functional rarity driven by variation in four factors: geographic range (restricted vs. widespread), local

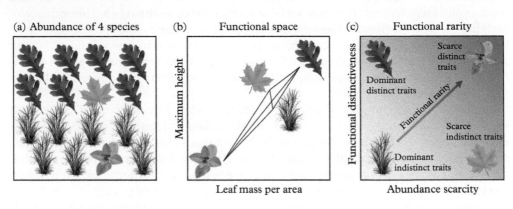

Figure 10.2 *Conceptual illustration of the concept of functional rarity using four species and two traits adapted for plants (a) The "classical" view of taxonomic rarity ranks species by their abundances in a community. The oak and grass species are more abundant than the maple and the dicot herb. (b) The "modern" view of trait rarity ranks trait distinctiveness based on distances in trait space. The lines that form the diamond shape indicate distances in trait space. In this example, the bright green dicot herb in the lower left is the most functionally distinct species. (c) The "integrated" view of functional rarity combines abundance scarcity and functional distinctiveness, where the scarcest and most functionally distinct species are the most functionally rare.*
From Violle et al. (2017) with permission.

population size (scarce vs. abundant), regional functional distinctiveness (unique vs. shared), and local functional distinctiveness (distinct vs. redundant).

Bagousse-Pinguet and colleagues (2021) showed that functional rarity and evenness were important predictors of ecosystem multifunctionality, demonstrating that this concept has direct relevance to the maintenance of ecosystem services. The framework also has direct relevance to conservation and management of rare habitats. For example, few ancient woodlands still remain within the densely populated landscapes of Europe and are under increasing pressure from land use change. Kimberley and colleagues (2013) tested whether species that were indicators of ancient woodlands were functionally distinct from generalist woodland species. They found that ancient woodland indicator species were short perennial species with a large seed mass, suggesting that poor dispersal ability may be partly responsible for their rarity.

Plant traits are now being considered as responses to a variety of different land uses and management actions, especially grazing by domestic and wild ungulates (Vesk and Westoby, 2001; Garnier et al., 2007). Traits have been shown to respond differently to grazing and other forms of biomass removal (Louault et al., 2005). In calcareous grasslands in Germany, life form, growth form, vegetative spread, lateral spreading distance, seed mass, and germination season responded to management treatments that included grazing, mowing, mulching, and burning (Kahmen et al., 2002). Grazing encouraged woody life forms by removing species with small seeds and persistent seed banks, mulching increased dominance of ground-layer species, and burning increased species with nutrient storage strategies (Kahmen and Poschlod, 2008). In Scottish grasslands, intense grazing led to an increase in annual species, early flowering and seed dispersal, a rosette habit, and higher requirements for light (Pakeman, 2004). In New Zealand grasslands, heavy grazing led to a decrease in plant height and leaf dry matter content (LDMC; Laliberté et al., 2012). In grasslands in both Argentina and Israel, grazing resistant species were shorter and exhibited smaller leaves with higher specific leaf area (SLA; Díaz et al., 2001). Similarly, in rangelands in the southwestern USA, protection from heavy grazing led to a decrease in SLA and leaf nitrogen (Strahan et al., 2015). A global-scale meta-analysis indicated that grazing favored shorter lifespans, shorter heights, prostrate habit, and both stoloniferous and rosette architectures over tussock forms (Díaz et al., 2007). However, in Australian shrublands and woodlands, traits could not predict responses to grazing (Vesk et al., 2004), suggesting that the effects of traits on grazing response vary along gradients of productivity and evolutionary response to grazing (Milchunas et al., 1988).

Traits also provide important information to guide conservation planning because they represent the capacity of species to respond to a changing climate (Andrew et al., 2022; Harper et al., 2022). The vulnerability of a species to climate change is a composite of the exposure of the species to environmental change, the sensitivity of the species to environmental change, and the capacity to adapt to change (Willis et al., 2015; Foden et al., 2019; Thurman et al., 2020). Butt and Gallagher (2018) developed a framework for using plant traits to guide conservation actions under climate change and argued that long-lived species with limited dispersal ability are inherently more at risk than species with short generation times and long-distance dispersal capacity. By combining traits related to capacity to reproduce and disperse with species attributes that include niche breadth and range size, they were able to rank species overall risk. For example, *Indigofera colutea* was ranked as low risk because it exhibits high reproductive rate, a broad abiotic niche, and a wide geographic range (Figure 10.3). In contrast, *Acacia axillaris* was ranked as high risk because it exhibits low reproductive rate, a restricted abiotic niche, and a narrow geographic range. The authors cross-referenced the species risk scores with in-situ and ex-situ management plans and concluded that many species are not adequately included in conservation action plans.

Gallagher and colleagues (2021) proposed a three-step framework for using traits to improve conservation outcomes and provided an example of assessing the recovery potential of 17,197 species following the unprecedented bush fires of the 2019–2020 austral summer. The first step is to identify the information needed to support the conservation goal. In this example, the objective was to identify the species in most need of recovery actions. Given the large number of species that needed to be assessed, they needed to identify species at most risk of poor recovery after the fires, but demographic data on population status was not available across all species being assessed. The second step is to identify the most informative traits to support the conservation goal. Chapter 7 discussed how several traits predict response to fire, including resprouting, obligate seeding, and growth form (Gill, 1975; Collins, 2020). Species that have weak resprouting and seeding responses to fire would be ranked as most at risk, especially if they have restricted geographic ranges. The third step is to state the limitations of the traits as proxies for the information needed to support the conservation goal. In this example, trait data was available for 8,000 species that were being assessed (Falster et al., 2021a). Meaningful inference could be drawn for those species, but the biggest limitation was the lack of trait data for the other 9,197 species. Modern algorithms that combine trait and phylogenetic relationships could be used to fill the data gaps to make reasonably informed decisions across the remaining species. The framework laid out by Gallagher and colleagues (2021) provide a roadmap for rigorous integration of plant strategies into the critical conservation work of managing threatened species.

10.3   Applying plant strategies to enhance success in ecological restoration

Ecological restoration is "the process of assisting the recovery of an ecosystem that has been degraded, damaged, or destroyed" (SERI, 2004). This includes the removal of abiotic and biotic stressors that prevent natural recovery of degraded ecosystems as well as direct actions to initiate the reassembly of communities. Restoration is about managing community assembly because it manipulates abiotic conditions, the dispersal of seeds from the species pool, and the density of non-native competitors. Given the centrality of traits to our understanding of community assembly (McGill et al., 2006; Keddy and Laughlin, 2022), plant strategies are central to restoration. This section is organized into the following general principles: (1) traits can inform predictions of the trajectory of community dynamics and ecosystem functioning over time (Funk et al., 2008; Laughlin,

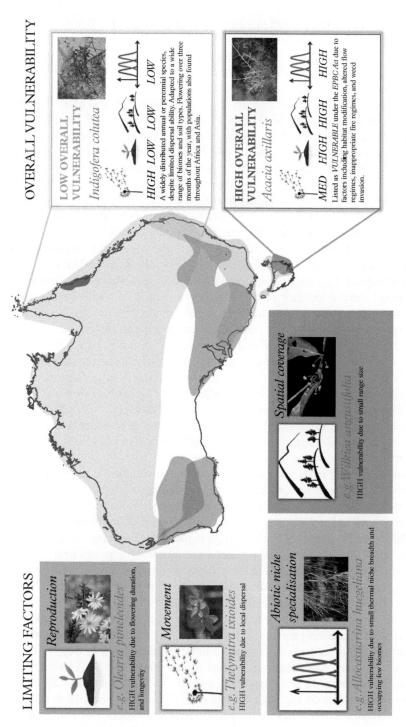

Figure 10.3 *Species traits related to reproduction and movement (dispersal) can be combined with abiotic niche specialization and spatial coverage to evaluate vulnerability of species to climate change. Two species are provided as examples of both low-risk and high-risk species based on their rankings for the four trait factors.*

Reproduced from Butt and Gallagher (2018) with permission.

2014b; Perring et al., 2015; Brudvig et al., 2017; Laughlin et al., 2017c); (2) traits can influence establishment success in response to different environmental conditions and restoration treatments (Balazs et al., 2020; Zirbel and Brudvig, 2020); and (3) traits can inform species selection to optimize functional outcomes (Laughlin et al., 2017b; Zirbel et al., 2017; Laughlin et al., 2018a; Fiedler et al., 2021).

Ecologists are engaged in an important debate about how to set and achieve restoration goals in a rapidly changing world (Hobbs et al., 2009; Higgs et al., 2014; Murcia et al., 2014). Historical reference conditions have long been the gold standard on which to base restoration goals in many ecosystems (White and Walker, 1997; Landres et al., 1999; Moore et al., 1999; Swetnam et al., 1999) because the goal is often to restore the ecosystems to their pre-disturbance state. However, future environmental conditions under climate change will be nothing like the historical climatic conditions (Harris et al., 2006; Choi et al., 2008; Hobbs et al., 2009). Rather than restoring historical assemblages that may not survive the changing environmental conditions, a more general approach has emerged that emphasizes achieving functional objectives to restore resilient ecosystems (Hobbs and Cramer, 2008; Seastedt et al., 2008; Jackson and Hobbs, 2009).

To achieve these goals I propose that restoration ecologists should no longer simply ask *what* species were dominant historically, but rather, we should be also asking *why* those species were dominant historically (Laughlin et al., 2017b). Presumably, the answer to this question is that the species that were dominant historically exhibited traits that gave them a fitness advantage (Keddy and Laughlin, 2022). As the environment continues to change, the trait values that optimize fitness will change as well. If we understand which traits confer fitness in the new environmental conditions (Aitken et al., 2008), then we can manage for species with more favorable trait combinations to reduce mortality risk and enhance restoration success (Funk et al., 2008; Laughlin, 2014b; Cadotte et al., 2015; Ostertag et al., 2015). Restoration projects can then serve as "acid tests" of ecological theory (Bradshaw, 1987b) for learning how traits affect demographic fitness across environmental gradients.

Let us explore how traits have been used to evaluate restoration success. Traits can explain species performance in restoration projects by distinguishing between strategies that promote early colonization and strategies that promote long-term persistence in dense vegetation (Pywell et al., 2003). Across ten bunchgrass prairie restoration projects, plant traits explained as much variability in change in plant cover as species identity, supporting the feasibility of using traits as a general framework for predicting responses to restoration treatments (Clark et al., 2012). Restoration practitioners can enhance establishment success by promoting plant strategies that fit the specific conditions of the project (Ostertag et al., 2015; Balazs et al., 2020; Leger et al., 2021).

There are now numerous examples across a range of ecosystems that have tested the power of traits for explaining restoration success. Plant traits determined establishment success in tallgrass prairie restorations in Michigan. Prairies that were tilled were dominated by taller species with larger seeds and higher SLA, but SLA declined with increasing age-since-restoration, fire frequency, and dry soil (Zirbel et al., 2017). Low leaf nitrogen species established slightly better under higher herbivory, high root mass fraction species established better in wetter soil, and low-SLA species established better under high light conditions (Figure 10.4) (Zirbel and Brudvig, 2020). In California

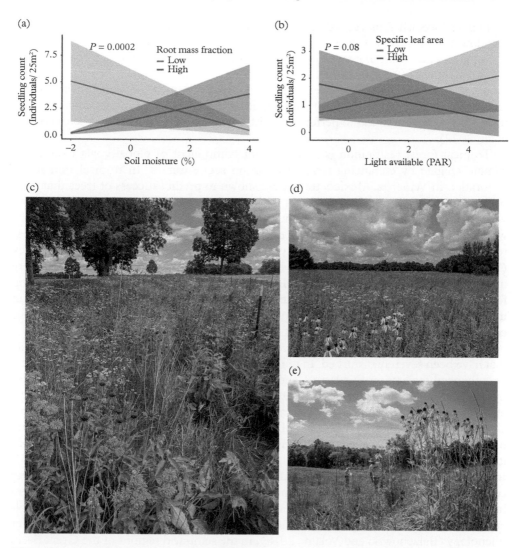

Figure 10.4 *Seedling counts the first year after seeding in a tallgrass prairie restoration experiment were a function of trait-by-environment interactions. Seedling counts were higher for species with (a) higher root mass fraction in wetter soil and (b) higher SLA in darker environments. (c–e) Restored prairies at the W.K. Kellogg Biological Station in southwestern Michigan, USA taken 4–6 years after seeding.*

(a–b): Reproduced from Zirbel and Brudvig (2020) with permission. (c–d) Photos by Lars Brudvig, and (e) photo by Chad Zirbel; reproduced here with permission.

grasslands, carbon addition treatments benefitted shorter, larger-seeded, and nitrogen-fixing species, whereas mowing treatments tended to favor shorter species with high SLA (Sandel et al., 2011). Within urban grasslands in Berlin, Germany, tall height and low SLA characterized successful species introductions (Fischer et al., 2013). Tree cutting in Swedish fens caused an increase of shade-sensitive vascular plant species with lower SLA (Hedberg et al., 2013), and tree cutting to restore calcareous grasslands in Belgium caused an increase of species with lower seed mass and shorter lifespans (Piqueray et al., 2015). In the cold deserts of the Colorado Plateau shrublands ecoregion, taller species with larger seeds performed better in more seasonal environments (Figure 10.5) (Balazs et al., 2020), and species with higher LDMC and lower root dry matter content were found to perform consistently better across a range of sites (Balazs et al., 2022).

Tropical forest restoration is an important component of global goals to mitigate climate change by enhancing terrestrial carbon sequestration. In tropical rain forest restoration in Veracruz, Mexico, traits were shown to predict success of trees that were planted in pastures to enhance natural recovery of forest cover. Growth and survival were positively related to crown size and negatively related to seed mass, but leaf traits were uninformative (Martínez-Garza et al., 2013). Similarly, in tropical dry forest restoration in Costa Rica, wood density, photosynthetic rate, and water-use efficiency (WUE) explained variation in growth and survival, but morphological leaf traits were uninformative (Werden et al., 2018). Restoration targets go beyond just establishing forest cover and can include targets to enhance habitat for pollinators. For example, in the Atlantic rain forests of Brazil, restoration targets of high floral trait diversity of trees could be met, but this proved more difficult for non-tree species (Garcia et al., 2015).

Traits are also being used to improve establishment success in harsh environments that have been severely disturbed. In the restoration of mines, functional trait dissimilarity enhanced facilitation (Navarro-Cano et al., 2019), and in harsh limestone quarries traits influenced whether species colonized cliffs, embankments, or platforms (Gilardelli et al., 2015). Traits influenced establishment success in semi-arid revegetation programs along roads in East Spain, where high SLA species were more successful in north-facing productive habitats and species with seed mucilage secretion and low seed susceptibility to erosion were more successful in harsher sites (Bochet and García-Fayos, 2015). Intraspecific trait variation among populations within the same species can also inform ecotype selection to improve establishment success. Populations of *Elymus glauca* from harsher sites in the Mediterranean climate zone of California exhibited dehydration tolerance and summer dormancy through lower SLA, lower specific root length, and earlier phenology (Balachowski and Volaire, 2018). In the semi-arid regions of the Great Basin in Nevada, the most successful ecotypes of *Elymus elymoides* were smaller plants with smaller seeds and earlier phenology (Kulpa and Leger, 2013). Counterexamples do exist, however, and caution is warranted—sometimes traits are better predictors of performance of fully established plants than for recently planted seedlings that are struggling to survive an extreme drought (Gardiner et al., 2019).

These examples illustrate that traits can inform predictions of establishment success and the resulting vegetation dynamics after restoration. Traits can therefore be used to select species for restoration to maintain ecosystem functions and services (Giannini et al., 2017; Carlucci et al., 2020). Several quantitative tools have been developed to select species for restoration based on functional traits (Laughlin et al., 2018a; Wang et al., 2020; Wang et al., 2021; Ladouceur et al., 2022; Santala et al., 2022). These tools

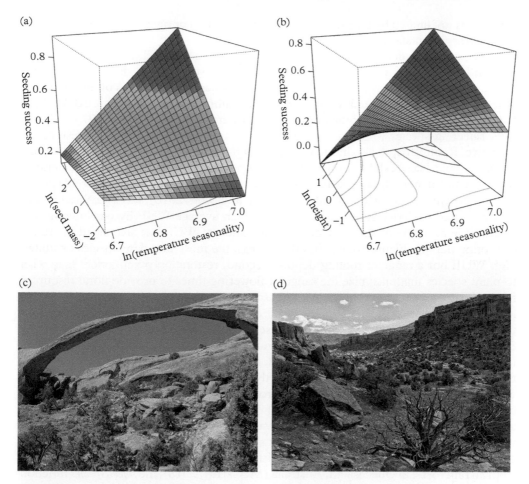

Figure 10.5 *Seeding success in restoration projects in the Colorado Plateau, USA was a function of trait-by-environment interactions. (a) Larger seeds and (b) taller plants enhanced seeding success in highly seasonal climates. Representative photos of the Colorado Plateau shrublands ecoregion: (c) Landscape Arch in Arches National Park, Utah, USA, and (d) Big Dominguez Canyon in the Dominguez-Escalante National Conservation Area in western Colorado, USA.*
(a–b): Reproduced from Balazs et al. (2020) with permission. (c–d): Photos by author.

can be used to derive species assemblages that exhibit traits to achieve specific functions, such as optimizing pollinator habitat (M'Gonigle et al., 2016), invasion resistance (Yannelli et al., 2018), or drought resistance and fire tolerance (Laughlin et al., 2017b). Practitioners may want to restore a community that converges on one trait but diverges on another. The *selectSpecies* R function (Laughlin and Chalmandrier, 2018; Laughlin et al., 2018a) can be used to derive an assemblage of species that is constrained to the average value of traits, while maximizing the diversity of other traits. The *selectSpecies* function is not limited by the number of traits to constrain or diversify, despite claims to the contrary (Santala et al., 2022).

Three examples demonstrate how to use this function with real data to design a seed or planting mix for a restoration project by selecting species with the highest relative abundances. First, restoring ecosystems that are resilient to drought is often an important management goal (Funk et al., 2015). Drought-tolerant plants can exhibit high WUE, the rate of carbon assimilation per unit of water used (Noy-Meir, 1973). Therefore, selecting species with traits that converge on high WUE can be one restoration objective. Rooting depth also influences drought tolerance, but a drought-resilient community would likely exhibit a diversity of rooting depths to optimize complementary water use throughout the soil profile (Hooper et al., 2005). Using a dataset of 48 species from a serpentine grassland in California (Funk and Wolf, 2016), the *selectSpecies* function was used to derive an assemblage with a high average WUE by constraining the assemblage to the 67th percentile of the distribution of WUE but diversified the range of rooting depths (Figure 10.6a). In other words, the most abundant species exhibited high WUE but a range of rooting depths. Second, restoration practitioners may wish to plant species that maximize the range of flowering times to provide floral resources for pollinators throughout the growing season, while simultaneously constraining the list to species that can tolerate infertile soil conditions. The *selectSpecies* function was used to derive an assemblage with a diverse range of flowering times by diversifying flowering date, but constraining leaf C:N ratio to the 67th percentile of the distribution of leaf C:N ratio (Figure 10.6b). Third, restoration practitioners who are restoring rain forests by planting tree seedlings directly into clearings may wish to plant species with high SLA to promote fast seedling growth and rapid canopy closure. However, they may also want the canopy to stratify after reaching the sapling stage. Species with greater carbon allocation to dense wood tissue will exhibit slower growth than species with low wood density. Therefore, canopy stratification may be achieved by planting species with a diversity of wood densities. Using a dataset of 41 tree species from a subtropical rain forest in Queensland, Australia (McCarthy, 2018), the *selectSpecies* function was used to derive an assemblage with a diverse range of wood densities, but a high average SLA (Figure 10.6c). As I write this book, experiments are being conducted around the world to test the efficacy of traits in ecological restoration, and the results from these experiments will inform the next generation of restoration science.

10.4   Plant strategy theory is directly relevant to invasive species management

Understanding the factors that make a species invasive has been a topic of great interest among conservation and restoration scientists. And for good reason. It has been estimated that > 13,000 plant species have become naturalized in regions they have been introduced (van Kleunen et al., 2015). Invasive species are recognized as one of the greatest threats to the integrity of indigenous ecosystems and it is rare to find a habitat on earth that is not invaded by a non-native species.

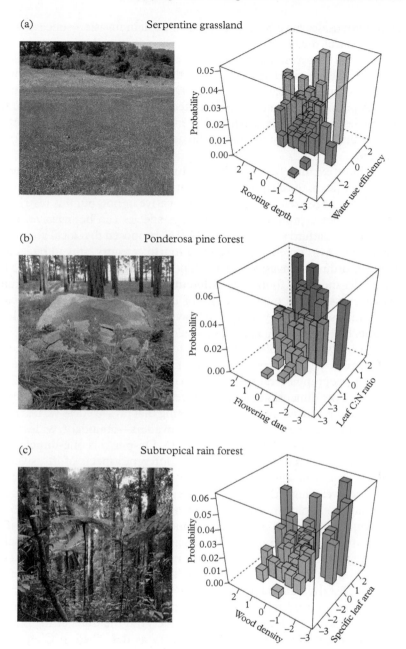

Figure 10.6 *Examples of species assemblages (i.e., discrete probability distributions) derived from the* selectSpecies *R function that simultaneously constrain one trait and diversify another to restore (a) serpentine grassland in California, USA, (b) ponderosa pine forest in Arizona, USA, and (c) subtropical rain forest in Queensland, Australia. Each bar represents one species and its location within the 2-dimensional trait space.*

Reproduced from Laughlin et al. (2018a) with permission. (a): Photo provided by Jennifer Funk. (b): Photo by the author. (c): Photo provided by Brandon Clark.

Dozens of hypotheses have been proposed to explain invasion success, and they all align with the central themes of this book. Invasion is hypothesized to be a function of human-mediated dispersal and propagule pressure, the biotic characteristics of the invading species, the abiotic characteristics of the environment, and the biotic characteristics of the community (Catford et al., 2009; Drenovsky et al., 2012; van Kleunen et al., 2018). A predictive theory of plant strategies should therefore be able to predict which phenotypes could invade a particular environment (Chapter 8) contingent on the phenotypes that are already present in the native community (Chapter 9). This is a challenging task.

One of the most exciting frameworks for studying invasion draws on the model of rarity discussed earlier in the chapter (Rabinowitz, 1981), but flipped on its head. Traditional approaches categorized species as either invasive or not, but this restrictive binary categorization did not capture the range of ways species can be invasive. Building on Rabinowitz (1981), Catford and colleagues (2016) proposed that local abundance, geographic range size, environmental range, and a fourth category, spread rate, can be used to categorize (or ordinate) invasive species. The approach was applied to 251 herbs in Victoria, Australia and it was found that different traits were linked with different dimensions of invasiveness (Palma et al., 2021). For example, fast-spreading species tended to be tall with small seeds or they produced heavy animal-dispersed propagules. Locally abundant plants had lower SLA and heavy seeds.

This framework was applied more broadly to invasive species in Europe using vegetation plots, species distributions, traits, and historical origins to determine drivers of local abundance, geographical range size, and habitat breadth (Fristoe et al., 2021). The authors first tested the dimensionality of the three-dimensional model and found that all three variables were positively correlated, indicating that invasiveness varies from species that are poor invaders to species that are "super invaders—abundant, widespread aliens, that invade diverse habitats" (Fristoe et al., 2021). Despite this one-dimensional variation in invasiveness, different factors explained different aspects of invasiveness. They showed that earlier introduction dates were positively associated with all three dimensions. The species were aligned along two aboveground axes of plant form and function (size and the leaf economics spectrum) following Díaz and colleagues (2016). Invasive species that originated outside of Europe attained higher local abundances if they had acquisitive leaf traits, and geographical range size was greatest for invaders with acquisitive leaf traits regardless of origin. There were many complex patterns discussed in this study, and it was noted that traits explained relatively little variation compared to date of introduction. Clearly, we still have much to learn about what makes a plant invasive.

The vast majority of research that sought links between plant traits and invasiveness have focused exclusively on the traits of the invader, but the traits of the resident community are just as important. Plant strategy theory as described in Chapters 8 and 9 integrates the biotic characteristics of the invading species, the abiotic characteristics of the environment, and the biotic characteristics of the community. Two theoretical frameworks hold promise for increasing our understanding of invasion dynamics. First, evolutionary game theory explicitly calculates the fitness of an invader in the context of the traits of the residents, and was proposed as an important framework for studying biological invasions (Pintor et al., 2011). Second, modern coexistence theory can be applied to calculate the fitness of an invader in the context of the resident community (Hallett et al., 2019). Both theories explicitly acknowledge that some resident communities may be more resistant to invasion than others. For example, traits have been shown to predict which species could colonize and produce high biomass in recipient communities that varied in diversity (Catford et al., 2020). Managers can exploit this feature by enhancing traits in the native community that resist invasion and restoration ecologists can reassemble communities that optimize invasion resistance (Laughlin, 2014b; Laughlin et al., 2018a).

10.5   Plant strategy theory can inform assisted migration

Plant species are currently experiencing extreme weather events and as the climate continues to change plants will need to either respond plastically in their phenology and physiology, adapt via natural selection, migrate to stay within their climate niches, or go extinct (Davis and Shaw, 2001; Kress and Krupnick, 2022). The projected pace of climate change indicates that species will need to disperse faster than their natural abilities allow them to if they are to maintain populations within their current climate niches (Davis and Shaw, 2001; Aitken et al., 2008). Long-lived species such as trees will lag behind short-lived species in their ability to track suitable climatic conditions (Jump and Penuelas, 2005; Lenoir et al., 2008). Risk of extinction will be greatest for species with slow dispersal rates, small population sizes, long generation times, high habitat specificity, and in areas that lack corridors and connectivity (Loss et al., 2011; Gallagher et al., 2015; Wang et al., 2019).

Assisted migration[3] is the intentional translocation of a species outside their current range to facilitate or mimic natural range expansion into a more suitable environment because they cannot disperse to these sites on their own. Multiple arguments in favor of and opposed to assisted migration have made this one of the most contentious topics in conservation biology (Vitt et al., 2010; Hewitt et al., 2011), even though humans have affected the distribution of plant species since prehistoric times (Rossetto et al., 2017). There is also perhaps no other topic that better integrates all the concepts discussed in this chapter because (1) it concerns the conservation of rare and declining species; (2) restoration projects are the most common venues for species introductions; and (3) there is concern that a species could become invasive in their new location (Mueller and Hellmann, 2008; Ricciardi and Simberloff, 2009). Experimental tests of assisted migration use common garden settings to compare demographic rates across environmental gradients (Tiscar et al., 2018; Wang et al., 2019). Traits and plant functional types have proven to be useful generalizations for predicting plant species responses

---

[3] Assisted migration has a variety of synonyms in the literature, including assisted colonization, facilitated migration, assisted range expansion, species translocation, and managed relocation (Hewitt et al., 2011).

to climate change (Harrison et al., 2015; Parmesan and Hanley, 2015). Plant strategy theory should be able to inform where to introduce declining species by matching their traits to a new climatic context but should simultaneously consider other abiotic and biotic contexts. Therefore, experimental tests of assisted migration in common garden settings that use traits to predict demographic performance across gradients could represent some of the strongest tests of plant strategy theory.

Tests of the viability of assisted migration have yielded some promising results. One study tested the assisted migration of *Liatris ligulistylis* (meadow blazing star), a vulnerable species in Canada, by establishing common gardens within and beyond its current range (Wang et al., 2019). *Liatris ligulistylis* thrived 500 km north of its current range with enhanced growth, survival, and flowering rates, suggesting that the species is in climate disequilibrium due to lags in its migration rate. The translocation of threatened orchids to higher elevations in subtropical China demonstrated that the transplants could survive extreme weather events in their new location (Liu et al., 2012). Positive growth and survival responses were observed in translocated provenances of *Populus tremuloides* (quaking aspen) 1,600 km to the northwest, indicating that the benefits of enhanced productivity outweighed the risks (Schreiber et al., 2013). *Pinus albicaulis* (whitebark pine) and *Abies religiosa* (sacred fir) performed well when translocated beyond their current ranges into sites where the climate was predicted to be more suitable in the future (Sáenz-Romero et al., 2021). Southern ecotypes of *Quercus rubra* (red oak), and to some extent *Quercus macrocarpa* (bur oak), exhibited higher survival, lower SLA, faster height and diameter growth, and longer leaf phenology than northern ecotypes when planted in the northern part of the range, suggesting that southern seed sources should be permitted to be planted across seed zones boundaries to maintain forest productivity (Etterson et al., 2020).

Other studies were less positive but have yielded important insights. One study reported that almost all populations of eight plant species that were experimentally planted in a new site for conservation purposes disappeared after 15 years (Drayton and Primack, 2012). Another study demonstrated that lack of local adaptation limits the potential of assisting the migration of drought-tolerant *Pinus nigra* (black pine) seedlings (Tíscar et al., 2018). A common garden study of genotypes of *Populus fremontii* (Fremont cottonwood) concluded that if plants were translocated 6°C they would perform poorly at this time and the authors suggested a more modest transplant distance of < 3°C would avoid problems of maladaptation. Transfer distances that approximate the long-term projections of climate warming constitute a "bridge too far" (Grady et al., 2015) and that incremental steps would be needed to facilitate the migration of *Populus fremontii*.

Climate is only one of many factors that will limit the success of assisted migration. One review article emphasized that the local microenvironment, herbivory, intraspecific competition, and interspecific competition can outweigh the effects of macroclimate in determining the success of species establishment, growth, survival, and continued recruitment (Park and Talbot, 2018). This study suggested that managers considering assisted migration should move beyond single-species models to study how community composition (i.e., density and frequency dependence) will influence demographic success of introducing species beyond their range to assist their migration to more suitable macroclimates. The presence of pollinators and symbiotic fungi will also limit establishment success.

Assisted migration remains controversial, but consensus will remain an impossibility without large-scale common gardens to test its viability (Bucharova, 2017). However, federal regulations that were designed to protect biodiversity are restricting our ability to conduct strong tests. For example, some federal and state laws in the USA restrict the introduction of species outside their current ranges into federally protected reserves and managed forests. These laws were enacted to protect biodiversity from non-native invasive species, but the same law handcuffs our ability to test the viability of assisted migration of species from different climate zones (Williams and Dumroese, 2013; Laughlin et al., 2017b; Etterson et al., 2020). It has been estimated that < 1% of species become invasive when introduced into a new range (Williamson and Fitter, 1996), suggesting low risk, especially since most translocations are relatively short distances rather than intercontinental introductions. However, predicting which species out of one hundred will become invasive in the context of assisted migration remains a research priority.

10.6   Chapter summary

- Conservation biology and ecological restoration are creative and defiant sciences that express optimism in our ability to right the wrongs that humans have inflicted on biodiversity. Plant strategies are integral to successful conservation and restoration.

- Conservation is the science of rarity and is applied to prevent species and ecological functions from going extinct. Demographic analyses are needed to identify the demographic rates that are most vulnerable to extinction threats. The concept of species rarity has been generalized to the new concept of functional rarity. Traits can be used to generalize across taxa to assess responses to grazing management as well as assess vulnerability of species to global change drivers to inform conservation actions.

- Restoration is community reassembly because it manipulates abiotic conditions, the species pool, and the density of non-native competitors. Plant strategies inform predictions of community reassembly and the trajectory of community dynamics and ecosystem functioning over time. Traits influence establishment success in response to different environmental conditions and restoration treatments, and they might be useful to inform species selection to optimize restoration outcomes.

- Plant strategy theory is directly relevant to invasion ecology because invasion success is hypothesized to be a function of human-mediated dispersal and propagule pressure, the biotic characteristics of the invading species, the abiotic characteristics of the environment, and the biotic characteristics of the community. Both evolutionary game theory and modern coexistence theory can be applied to inform invasion ecology.

- Assisted migration is the intentional translocation of a species outside their current range to facilitate or mimic natural range expansion into a more suitable environment because they cannot disperse to these sites on their own. This topic integrates all the concepts discussed in this chapter because it concerns the conservation of rare and declining species, restoration projects are the most common venues for species introductions, and there is concern that a species could become invasive in their new location. Experimental tests of assisted migration in common garden settings to compare demographic rates across environmental gradients could represent some of the strongest tests of plant strategy theory.

10.7   Questions

1. Can you think of other applications in the context of land management where trait-based demographic approaches can be used?
2. How can traits be used to inform grazing management regimes?
3. How can the classical approach to setting targets for ecological restoration inform the selection of the species pool within a trait-based approach to ecological restoration?
4. How can plant strategies inform the restoration of ecosystem functioning?
5. What are the limitations of trait-based theory to restoration and management?

Part V

The Effect of Traits on Demographic Rates

If you have made it this far into the book, then the primary thesis has been laid out in full. The first four parts would be sufficient material for a graduate seminar on plant strategies. However, those of you who seek to advance a more mechanistic theory may be interested in exploring what is still unknown about how traits influence individual demographic rates and underpin demographic trade-offs.

Part V explores an advanced approach to predicting population fitness by using traits to explain variation in growth, survival, and reproduction. These chapters are written for the eager ecologist who wants to not only explore a general theory of plant strategies but for those who want to push the field further by discovering the mechanisms underlying how traits drive demographic trade-offs and fitness across complex environmental gradients.

Chapter 11 reviews the large literature on traits that promote growth, traits that enhance survival, the trade-off between growth and survival rates, and how growth and survival vary with plant size.

Chapter 12 demonstrates the current difficulties of using traits to predict recruitment rates and proposes how to make progress toward understanding recruitment, the trickiest of all the demographic rates.

11

Plant Traits That Promote Growth and Enhance Survival

Extinction is the rule. Survival is the exception.

—Carl Sagan (2006, p. 76), *The Varieties of Scientific Experience*

No species ever became extinct that did not first survive. The true exception is not survival, but continued survival, something that requires a species to show ongoing pluck while enjoying continual and timely good luck.

—Adrian Burton (2016, p. 580), "Staving off extinction," in *Frontiers in Ecology and the Environment*

11.1 Demographic rates alone can be misleading proxies for fitness

A vigorous oak seedling shooting up out of the soil may look like a new recruit, but it could be a sprout from an ancient belowground root stock. The same can be said for many pursuits in ecology. Sometimes popular research directions look fresh and may appear novel and exciting, but they are often just new perspectives on older questions, rather than actual shifts in paradigm, which are exceedingly rare (Kuhn, 1970). Pursuing predictions of demographic rates as a function of traits has ancient roots. However, new methodological advances allow us to integrate predictions across all demographic rates into estimates of demographic fitness, which is invigorating the search for answers to these age-old questions. Population models can be used to quantify demographic fitness, which is determined by stage-specific survival and reproduction rates (Arnold, 1983; Violle et al., 2007). But the world is a big place, with hundreds of thousands of species, and we cannot hope to rigorously quantify demographic rates on every plant species on Earth. Can we use traits to infer demographic rates on species we have not studied demographically? One central goal of ecology is to develop theory that is generally applicable. One of the main goals of this book is to determine the sets of traits that predict variation in fitness, underlain by variation in growth, survival, and reproduction, across the primary environmental gradients on the planet.

Plant Strategies. Daniel C. Laughlin, Oxford University Press. © Daniel C. Laughlin (2023). DOI: 10.1093/oso/9780192867940.003.0011

This book has emphasized modeling fitness directly to test the faculty of functional traits as predictors of fitness (Laughlin et al., 2020b). It is the most direct approach but is less mechanistic than one built on the demographic rates themselves. We spent several chapters emphasizing the importance of growth, survival, and reproduction for estimating demographic fitness—but shouldn't we also determine how traits affect the demographic rates individually to learn how trait effects on fitness are indirectly mediated through the demographic rates? Absolutely! However, first a word of caution.

The alluring prospect that functional traits can explain variation in species performance has invigorated comparative functional ecology in the last few decades (Pugnaire and Valladares, 2007). Inspired by classic evolutionary theory that linked morphology to performance and fitness (Arnold, 1983), ecologists have recently intensified their search for relationships between functional traits and demographic rates (Poorter and Bongers, 2006; Ordonez et al., 2010; Wright et al., 2010; Adler et al., 2014; Kunstler et al., 2016; Visser et al., 2016; Blonder et al., 2018; Laughlin et al., 2018b; Worthy et al., 2020; Chalmandrier et al., 2021). However, they have generally avoided the more challenging links to fitness by focusing on one or two demographic rates in isolation. Analyzing components of fitness in isolation is an important step but testing relationships between traits and demographic rates to infer effects on fitness can be misleading without considering demographic trade-offs (Silvertown et al., 1993; Franco and Silvertown, 2004; Villellas et al., 2015). For example, species with fast individual growth rates may exhibit low individual survival rates. If the growth–survival trade-off can generate co-variation in individual growth rates and individual survival rates that yield equal fitness, all else being equal (e.g., equal reproduction rates), then individual growth rates tell us little about fitness (Mangel and Stamps, 2001; Steiner and Tuljapurkar 2012). Similarly, populations with high survival and low reproduction could have the same fitness as populations with low survival and high reproduction (Silvertown et al., 1993; Sheth and Angert, 2018). Consequently, if a trait is negatively related to survival, then it may be positively related to individual growth or reproduction (Stearns, 1992; McGraw and Caswell, 1996; Metcalf et al., 2006; Rüger et al., 2018). For example, wood density negatively effects individual growth rates but positively effects survival rates because faster tree diameter growth can be achieved by constructing low-density wood, but this comes with a higher risk of damage and death (Wright et al., 2010; Kunstler et al., 2016; Visser et al., 2016). The classic study of teasel (*Dipsacus sylvestris*) demonstrated quite clearly that individual growth rates can be unrelated to population growth rate (Werner and Caswell, 1977), so any attempt to infer something about fitness by considering individual growth rates alone would fail.

In these final two chapters, I review the tremendous amount of work done to unravel the traits that affect growth, survival, and reproduction. Recall that plant strategies are phenotypes resulting from natural selection that enable a population to persist in a given environment. This means that we cannot, and should not, think that a trait can enhance demographic rates in all environments. That could only occur in the fabled Darwinian demon (Laughlin, 2018). However, there do appear to be a few traits that are directly linked to growth, survival, and reproduction, regardless of the environment. For example, rankings of relative growth rates (RGRs) among seedlings appear to be stable across productivity gradients. This would suggest that the main effect of a trait on growth rates does not interact with site productivity. In a regression context, this would imply that some traits have strong "main effects" on demographic rates. It is my contention that these traits are relatively few, and that many other traits will exhibit effects on demographic rates that strongly depend on the environmental context.

11.2　Inherent variation in growth rates varies tenfold

The speed of progression from early to later stages in the life cycle is negatively associated with age at maturity, indicating that individual growth rate is an important trait contributing to the fast–slow continuum (Salguero-Gómez et al., 2016). Functional ecologists have long considered RGR to be a central trait underlying plant strategies (Grime and Hunt, 1975; Grime, 1979; Chapin, 1980). Grime and Hunt (1975) demonstrated that species from a range of habitats exhibited strong interspecific differences in growth rate when grown under standardized, "optimum" conditions (i.e., ample light, water, and nutrients). Species from nutrient-poor sites grew slower than species from productive sites. They called this trait "potential RGR" or "maximum RGR" (RGR_{max}). This study established the standard for interspecific comparisons of traits and set the stage for a grand screening of the British flora that occurred over the 1980s and 1990s (Grime et al., 1997b).

There are many sources of variation in growth rates: growth rates vary among individuals within a species, they vary across stages in the life cycle, and they tend to increase with increasing resource availability. Given all these sources of variation, you might be wondering if it is even possible to detect strong differences in growth rate between species. Since Grime and Hunt's landmark study, ecophysiologists tackled the challenge of determining the traits that underlie the variation in growth rates among a range of herbaceous and woody species (Poorter and Remkes, 1990; Cornelissen et al., 1998; Poorter and Van der Werf, 1998; Veneklaas and Poorter, 1998). Some wondered whether growth rates of species measured on seedlings would retain the same rank order when measured on adults, and it was shown that potential seedling RGR was positively correlated with growth rates in the field and growth rates of adults (Reich et al., 1992; Cornelissen et al., 1998). These results generated confidence in the community of ecologists tackling this problem that it was not only a defensible approach, but the best approach, for quantifying variation in growth rates across species.

Our current understanding of plant growth can be traced back to the work of Blackman (1919), who drew attention to the parallel between compound growth rates of bank accounts and the growth rate of plants, where the rate of increase in plant biomass is proportional to the amount of biomass already present in the plant. This discovery prompted botanists to consider the *relative* growth rate of plants, not just the *absolute* growth rate (Poorter and Garnier, 2007). Blackman proposed that the biomass of a plant (M_2) can be predicted with knowledge of the previous biomass (M_1) and its RGR over time using the following expression:

$$M_2 = M_1 e^{RGR \cdot t}. \tag{11.1}$$

Equation 11.1 implies that plant growth is exponential. You may have experienced this fact as a gardener. One week you see that weeds have started germinating. At first, you aren't that bothered because they appear manageable. Then you leave town for a short holiday only to return to a tall and tangled mass of weeds that exploded seemingly overnight.

RGR quantifies the rate of biomass increase between two time periods and is defined relative to the biomass already present. Therefore, we can write RGR as

$$RGR = \frac{1}{M}\frac{dM}{dt}, \tag{11.2}$$

where dM/dt is the absolute rate of change in biomass and $1/M$ emphasizes that this rate is expressed on a per existing mass basis. The units of RGR are typically expressed as mg g^{-1} day^{-1}; in other words, the amount of increase in mg, per existing plant mass in g, per day. But Eqs. 11.1 and 11.2 do not make it obvious how one would measure and calculate this quantity in practice. If we let $t = t_2 - t_1$, take natural logarithms of Eq. 11.1, and rearrange to solve for RGR, we can see how RGR can be measured experimentally:

$$RGR = \frac{\log \dot{M}_2 - \log M_1}{t_2 - t_1}, \tag{11.3}$$

where log is the natural logarithm, and M_2 and M_1 are biomasses measured at time t_2 and t_1, respectively. In other words, if we measure plant biomass over a period of time, then we can calculate RGR in a straightforward manner. RGR has been shown to vary tenfold among plant species. The typical range of potential RGR in the absence of competition varies between 40 to 400 mg g^{-1} day^{-1}. Herbaceous plants are generally the fastest growing and range between 100 to 400 mg g^{-1} day^{-1}. Woody plants exhibit a range of 10 to 150 mg g^{-1} day^{-1} (Poorter and Garnier, 2007). What drives this variation?

11.3   Specific leaf area and unit leaf rate drive relative growth rate

Tilman's (1988) theory of plant strategies was developed from first principles with an interest in determining drivers of population growth rate (r) among species. He built his model using resource ratio theory, where population abundances can be solved at equilibrium if one knows the consumption and supply rate of different resources that limit population growth (Tilman, 1985). To Tilman, light and soil nutrients were the most important resources limiting plant growth, and plants that could survive at the lowest levels of a resource would be the ultimate winners in competition for that resource. Tilman argued that RGR determined whether a species would be a successful competitor for light or nutrients, and that the allocation of biomass into aboveground versus below-ground organs (i.e., the root-to-shoot ratio) drove variation in RGR. Growth rate, he argued, is determined by the difference between rate of photosynthesis and respiration. Given that root and stem tissues respire but do not photosynthesize, species that allocate more biomass into these organs will have larger respiration rates and slower growth, whereas species that allocate more to leaves will grow the fastest.

Tilman likened algae to a plant that is 100% leaf material because it allocates no resources to growth and maintenance of stems or roots. Recall the argument from Chapter 9 that if light was not limiting, then the most efficient plant would lie flat along the surface of the ground. Single-celled algae have been shown to be able to double their biomass twice per day, which is equivalent to an RGR of 1400 mg g^{-1} day^{-1} (Tilman, 1988). This is much faster than any growth rate of a vascular plant.[1] He compared these values to RGR of species measured by Grime and Hunt (1975): *Poa annua* individuals exhibit maximal growth rates of 380 mg g^{-1} day^{-1}, whereas *Picea sitchensis* grow slowly at a rate of 30 mg g^{-1} day^{-1}. The herbaceous plant *Poa annua* allocates less to stems and roots, and therefore exhibits intermediate growth rate. The tree *Picea sitchensis* grows slowest because of the heavy respiratory cost of stems and roots. To Tilman, the evolution of higher plants from single-celled algae is thus a story about increasing investment in non-photosynthetic material in response to resource limitation. Tilman's model predicts the following trade-off: at the high end of the soil fertility gradient, plant species that allocate relatively more carbon aboveground than belowground are predicted to be better competitors for light; at the low end of the soil fertility gradient, plant species that allocate relatively more carbon belowground than aboveground are predicted to be better competitors for soil nutrients. Stimulated by such a theory, many ecophysiologists set out to test the drivers of RGR.

RGR can be decomposed into a set of component parts. Poorter (1989) discussed several different approaches to decomposing RGR, including those based on leaf area partitioning and nitrogen concentrations, but the most common approach emphasizes the carbon economy (Evans, 1972). RGR can be factorized into a "physiological" component, the unit leaf rate (ULR),[2] and a "morphological" component, the leaf area ratio (LAR):

$$RGR = ULR \cdot LAR \qquad (11.4)$$

ULR is defined as the increase in plant mass per unit leaf area ($1/m^2 \; dM/dt$) and is expressed in the units g m^{-2} day^{-1} (see Table 11.1 to see a list of terms and their definitions). It can be interpreted as the balance between carbon gain through photosynthesis

---

[1] Algal reproduction (population growth) is not directly comparable to individual plant growth. In fact, this distinction between population and individual level parameters led to some of the confusion in the debates over plant strategy theories (see Grace (1990)).

[2] It has been argued that ULR is not just a "physiological component" of RGR because ULR can vary among species due to size differences, in which case other factors like self-shading, tissue turnover, allocation, and morphology can be important (Rees et al., 2010).

Table 11.1 *Traits, abbreviations, and definitions in plant growth analysis, using updated terminology as recommended by Hendrik Poorter to improve understanding.*

Trait	Abbreviation	Definition	Comments
Relative growth rate	RGR	Rate of change in biomass expressed on a per existing mass basis (mg g^{-1} day^{-1}); the increase in mass per existing plant mass per day.	
Unit leaf rate	ULR	Increase in plant mass per unit leaf area ($1/m^2$ dM/dt) and is expressed in the units g m^{-2} day^{-1}.	Traditionally referred to as "net assimilation rate" (NAR)
Leaf area ratio	LAR	Leaf area per unit plant mass expressed in the units m^2 kg^{-1}. Also, the product of SLA and LMF.	
Specific leaf area	SLA	One-sided projected leaf area per unit leaf mass expressed as m^2 kg^{-1}.	The inverse of leaf mass per area (LMA), i.e., SLA = 1/LMA
Leaf mass fraction	LMF	Fraction of leaf mass to total plant mass expressed as g g^{-1}.	Traditionally referred to as "leaf mass ratio" (LMR)

and carbon loss due to shoot respiration, root respiration, exudation, and volatilization. ULR also differs among herbs and woody plants due to inherent differences in tissue carbon concentration. LAR is defined as the leaf area per unit plant mass expressed in the units m^2 kg^{-1}, but LAR can be further factorized into its products specific leaf area (SLA) and leaf mass fraction (LMF):

$$LAR = SLA \cdot LMF \tag{11.5}$$

SLA is the leaf area per unit leaf mass expressed as m^2 kg^{-1}. LMF is the fraction of leaf mass to total plant mass expressed as g g^{-1}. Therefore, RGR is composed of three traits

$$RGR = ULR \cdot SLA \cdot LMF. \tag{11.6}$$

This approach has improved our understanding of the mechanistic basis of growth across species and genotypes (Poorter, 1989). Note that there are elements in the equation for RGR that are directly related to the root-to-shoot ratio. ULR should decline with greater investment in organs like stems and roots with high respiratory demand and LMF is inherently negatively correlated with root-to-shoot ratio.

Do all three traits contribute equally to variation in RGR, or does variation in one trait matter more than variation in the others? The answers to these questions, like so many in plant strategy theory, differ for herbaceous and woody species. Variation in RGR among herbaceous plants is clearly driven by the morphological component of RGR, the LAR (Poorter and Van der Werf, 1998): there was no correlation between ULR and RGR ($r^2 = 0.03$), but LAR was strongly and positively correlated with RGR ($r^2 = 0.69$) (Figure 11.1). Given that LAR is a composite trait of SLA and LMF (Eq. 11.5), it is important to determine which of these morphological traits is more important. Both traits were positively correlated with RGR, but SLA explained double the amount of variation in RGR than LMF ($r^2 = 0.53$ vs. $r^2 = 0.27$, respectively), indicating that SLA plays a central role in determining the inherent variation in RGR among herbaceous plants. A large meta-analysis using a standardized metric of importance further confirmed the ranking of importance of these traits: LAR was more important than ULR,

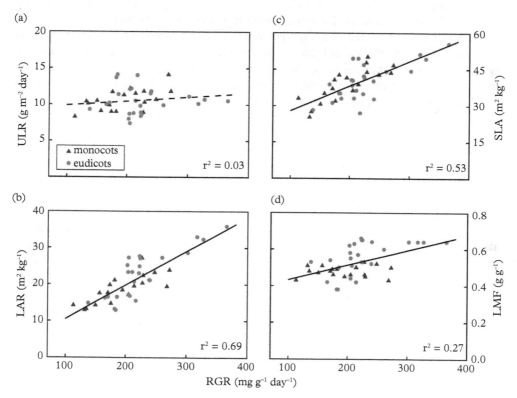

Figure 11.1 *Unit leaf rate (ULR) (traditionally called net assimilation rate or NAR) versus relative growth rate (RGR), where monocots and eudicots are indicated by different colors and symbols. (b) Leaf area ratio (LAR) versus RGR. (c) Specific leaf area (SLA) versus RGR. (d) Leaf mass fraction (LMF, traditionally called leaf mass ratio or LMR) versus RGR.*

Adapted from Poorter and Remkes (1990) with permission, and generously provided by Hendrik Poorter.

and SLA was more important than LMF (Poorter and Van der Werf, 1998). These results elevated SLA in the minds and hearts of functional ecologists.

But what about woody plants? Is SLA the principal driver of growth in shrubs and trees? The allocation of biomass into secondary thickening of the xylem to grow tall is a tremendous strain on growth and maintenance respiration costs, which should elevate the importance of ULR. As woody seedlings grow, they tend to invest more in stems at the expense of leaves (and roots in larger trees) and SLA tends to decline. Both of these ontogenetic changes led to reduced LAR and ULR, which reduced RGR as tree seedlings grew in size (Veneklaas and Poorter, 1998). However, studies of temperate tree species (Cornelissen et al., 1996; Hunt and Cornelissen, 1997a) and Australian woody plants (Saverimuttu and Westoby, 1996) demonstrated that LAR, and its component SLA, were the strongest drivers of RGR. Nevertheless, it does seem that LMF becomes more important in woody species (Cornelissen et al., 1998).

Shipley (2006) attempted to resolve the importance of the determinants of RGR by conducting a meta-analysis of RGR across a range of species and life forms. This analysis included a range of sizes, in contrast to previous controlled experiments that standardized sizes (Poorter and Van der Werf, 1998). Shipley concluded that LMF was never related to RGR, and that ULR was the best predictor of RGR. This was in stark contrast to other evidence that SLA was the primary driver of RGR. Shipley included light irradiance in his analysis and speculated that, for woody species, ULR was always the most important variable, but for herbaceous species the importance of ULR increased and SLA decreased with increasing daily quantum input. However, Poorter and Van der Werf (1998) found no evidence for such a role of light in their analyses.

Tilman's plant strategy theory proposed that species with high root-to-shoot ratios are predicted to grow slowly because of allocation to belowground organs. But LMF, the closest correlate to root-to-shoot ratio, has consistently been shown to be unrelated to RGR. Hunt and Cornelissen (1997b) demonstrated mathematically that Tilman's model was similar to the ecophysiological approaches just discussed, but they concluded that SLA, rather than whole-plant allocation per se, is the vehicle through which plants have evolutionarily adapted their RGR. The merits of Tilman's prediction have been vigorously debated over the last few decades. For example, Shipley and Peters (1990) found no evidence for a relationship between root-to-shoot ratio and RGR in wetland plants, and Elberse and Berendse (1993) found no inherent difference in biomass allocation between grasses from nutrient-rich and nutrient-poor sites. Tilman made many tests of this theory in old fields and prairies in Minnesota and there always seemed to be one or two species that bucked the trend. These discrepancies between his experimental results and the model predictions caused him to adjust the theory, which is a good sign of scientific progress. Craine (2009) reviewed the empirical evidence and came to the strong conclusion that there is little evidence to support Tilman's resource-ratio theory of plant strategies. In short, Craine argued that the theory is flawed because it assumes that nutrient concentrations are well mixed in soil solution (they are not), and that root-to-shoot ratio is the mechanism by which plants compete for soil nutrients (root length proliferation is likely the more relevant root trait, as discussed in Chapter 6). I refer you to pages 91–118 of Craine (2009), and to Craine (2005) for a more detailed summary of the evidence. Overall, Tilman's model placed great importance on RGR at the expense of survival and reproduction, the other components of fitness. It is not growth alone that is under selection.

11.4 Effects of traits on growth rate change with plant size

Many other empirical studies have reported weak correlations between traits and individual growth rate causing some to question the usefulness of traits for understanding growth (Wright et al., 2010; Paine et al., 2015). However, it has become clear with time that the classic growth equations are confounded by plant size. Recall that the equation for RGR assumes that plant growth is exponential. While this may be true for the early stages of growth in herbaceous plants (like the weeds in your garden), it is not a good descriptor for most plants across their life cycle (Paine et al., 2012). In fact, plants tend to experience reduced growth rates as they approach their maximum size (Figure 11.2), especially if they are growing with neighbors. Plants are less efficient at larger sizes because they start shading their own lower canopies, their leaves may start to age or senescence, and they need to allocate more resources to non-photosynthetic tissues like stems and roots and to reproduction.

Rees and colleagues (2010) demonstrated that interpretations of RGR decompositions are strongly impacted by differences in plant size across studies. Leaf area and leaf mass are both size dependent. They discovered that when sizes are not corrected, then ULR is overestimated as the most important variable, a result that agrees with Shipley's (2006) meta-analysis that did not control for plant size. ULR was no longer correlated

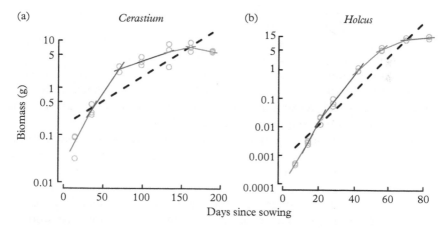

Figure 11.2 *Changes in biomass over days since sowing in (a)* Cerastium diffusum *and (b)* Holcus lanatus. *The solid-colored lines indicate the slopes that would be inferred were these plants to grow exponentially during every census interval, whereas the dashed lines indicate the constant relative growth rate that would be inferred after fitting the traditional model of exponential growth to all the data.*

Reproduced from Paine et al. (2012) with permission.

with RGR when standardized by plant size, suggesting that its affects are driven by differences in size (Rees et al., 2010). SLA became the most strongly correlated variable with RGR when sizes were controlled. Moreover, a ranking of the importance of variables demonstrated that SLA was the most important, followed closely by ULR (Rees et al., 2010). Controlling for plant size is a challenge and cannot be perfectly achieved—after all, comparing a mature *Arabidopsis* to a one-day-old *Quercus* seedling is difficult to justify.

RGR can also be decomposed into traits related to the nitrogen economy in plants (Ingestad, 1979; Poorter, 1989). When taking that approach, then RGR is positively related to the concentration of nitrogen in the plant (Berendse and Elberse, 1990). A common explanation for this positive relationship is because nitrogen is a major component of the enzyme Rubisco, which drives carboxylation in photosynthesis. The importance of Rubisco is clear: it is the most abundant protein on the planet (Raven, 2013)! Nitrogen is used elsewhere in the plant beyond the photosynthetic machinery, but given that nitrogen concentration by mass is positively correlated with SLA (Reich et al., 1997; Wright et al., 2004), nitrogen is clearly a part of a suite of traits that contribute to fast RGR.

Herbaceous plants exhibit faster rates of growth than woody plants (Grime and Hunt, 1975). Woodiness comes up repeatedly as an important functional distinction between species (Theophrastus, 287 BCE; Díaz et al., 2016). The binary classification into woody and non-woody categories may be a useful first approximation to divide species into groups, but continuous variation in the density of the stem tissue plays a prominent role. Wood density should be correlated with RGR because trees that invest more carbon per unit volume into denser wood cannot produce the same volume of wood as a species that invests less carbon per unit volume of wood (Roderick, 2000; Chave et al., 2009). Trees that invest less carbon into their stems per unit volume can allocate more toward leaves to grow even faster. For example, the tropical Balsa tree (*Ochroma pyramidale*) in the mallow family (Malvaceae) is prized for its extremely fast growth rates (up to 5 m of height growth per year!), and also has one of the lowest density woods on the planet (0.18 g cm^{-3}). You may be familiar with this wood if you ever played with small wooden airplanes as a child—they are usually made from balsa. Species with dense wood have a lower cross sectional area of conduits and thicker outer walls, which makes them less vulnerable to drought and freezing-induced embolisms (Hacke et al., 2001). This may limit transpiration and photosynthesis in trees with dense wood.

Many studies have demonstrated that wood density is negatively related to tree growth rates (Enquist et al., 1999; Muller-Landau, 2004; King et al., 2006; Poorter et al., 2010; Wright et al., 2010; Rüger et al., 2012). There is some evidence to the contrary, at least when the fastest-growing species are removed from the analysis (Nascimento et al., 2005). One study detected a stronger effect of vessel diameter (which is related to wood density) on tree growth rates (Fan et al., 2012). One study found that wood density was strongly and negatively correlated with RGR in greenhouse-grown tree seedlings, but not in wild-grown seedlings or mature trees (recall Figure 7.4), suggesting that the effect of wood density on RGR may change over ontogeny and be dependent on the local environmental conditions (Laughlin et al., 2017a). In summary, fast growth in woody plants is associated with high SLA, high leaf nitrogen concentration, and low wood density.

Using a theoretical model of plant growth (Falster et al., 2011; Falster et al., 2016; Falster et al., 2018), Gibert and colleagues (2016) suggested that changes in the strength and even direction of trait effects on growth should actually be expected. Through a meta-analysis of the literature, they found that the relationship between SLA and growth rate shifted from positive in seedlings to non-significant in adult plants (Figure 11.3). Wood density exhibited consistently negative effects on growth rate across ontogeny. The positive relationship between maximum height only became significant in adult

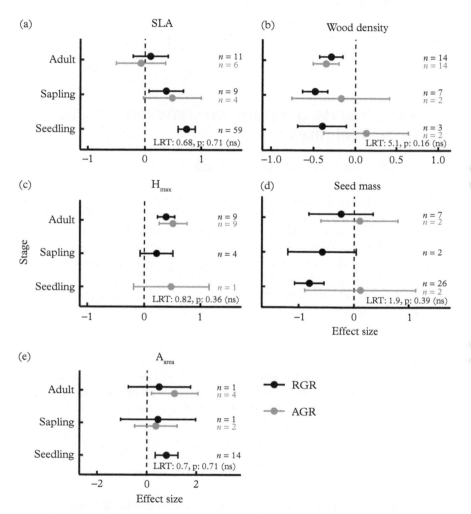

Figure 11.3 *Meta-analysis results of the effects of traits on relative growth rate (RGR) and absolute growth rate (AGR) on woody plants at three ontogenetic stages (seedlings, saplings, and adults). Effect size was calculated as z-transformed cross-species Pearson's product–moment correlation coefficients (r) between the growth rate and the trait, weighted by the number of species in the sample.*

LRT = likelihood ratio test; n = the number of correlations contributing to each effect size.
Reproduced from Gibert et al. (2016) with permission.

size classes. The negative effects of seed mass on growth rate in seedlings became unde-tectable in saplings and adults. The main reason that trait effects on growth change with plant size is the decrease in the ratio of leaf area to biomass for structural support, so RGR tends to decline with plant size (Givnish, 1995; Enquist et al., 2007). Moreover, these increasing costs of supporting tissue devalue the benefits of inexpensive leaf con-struction costs of low SLA (Gibert et al., 2016). This work has now made it clear that the classical trade-off between slow-to-fast species requires a nuanced incorporation of size into the framework—something that metabolic scaling theory cannot easily accommo-date (Enquist et al., 2007; Falster et al., 2018). This undoubtedly complicates the search for linear plant strategy dimensions. Fortunately, incorporating size into demographic rate models is exactly what structured population models like integral projection models (IPMs) have been designed to do.

11.5   Traits generate the trade-off between growth and survival rates

Growth and survival are intimately linked. Recall from Chapter 4 that the growth–survival kernel in an IPM recognizes that growth from one year to the next can only happen if a plant survives. But survival does not guarantee positive growth. Plants can stay the same size, they can shrink, or they can even go dormant if they don't feel like getting out of bed after a long winter or dry season (Volaire and Norton, 2006). In fact, there is an inherent trade-off between growth and survival.

The growth–survival trade-off is one of the most well-recognized trade-offs in ecol-ogy and is driven by variation in functional traits (Grime, 2001; Martínez-Vilalta et al., 2010; Reich, 2014; Rüger et al., 2018). It is thought to be most clearly expressed as a trade-off between growth in optimal conditions versus survival when resources are limiting. In other words, species that grow fast when resources are abundant die when resources are scarce, and species that survive when resources are scarce grow slowly when resources are abundant. In the tropical forest of Barro Colorado Island, Panama, the trade-off was strongest among species when growth rates of the fastest-growing saplings were related to mortality rates of the slowest-growing saplings (Figure 11.4) (Wright et al., 2010). In closed-canopy forests, light drives this trade-off. Shade-tolerant species with the lowest whole-plant light compensation points (LCP) exhibit the high-est survival rates in low light but the lowest height growth rates in high light (Lusk and Jorgensen, 2013). But limitation of water and mineral nutrients could also set up a trade-off between fast growth in wet and rich soil versus survival in dry and poor soil (Reich, 2014).

Recall from Chapter 6 that the growth–survival trade-off among tropical forest trees was most strongly correlated with wood density and SLA (Rüger et al., 2018), but the degree to which functional trait variation is the mechanism that drives this demographic trade-off is unclear. Siefert and Laughlin (2023) built one integral projection model

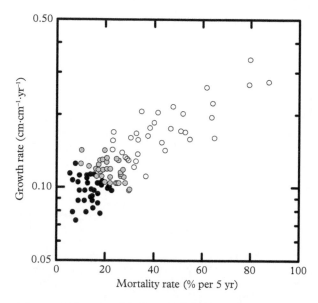

Figure 11.4 *The growth–survival trade-off is illustrated here as a positive correlation between growth rate and mortality rate (the opposite of survival rate) across tropical tree species on Barro Colorado Island, Panama. The trade-off is strongest when growth rates of the fastest growing saplings are related to mortality rates of the slowest growing saplings.*
Reproduced from Wright et al. (2010) with permission.

for temperate tree species in the Eastern USA where growth, survival, and reproduction were functions of individual size, species-level average trait values, environmental conditions, and local neighborhood density. We used this model to test whether the growth–survival and stature–recruitment trade-offs (Rüger et al., 2018) emerged from the effects of traits on demographic rates in our models. We found that wood density was the underlying driver of the growth–survival trade-off among saplings in temperate forests (Figure 11.5). Saplings of species with low wood density grew quickly but had low survival rates, whereas saplings of species with high wood density had high survival rates but grew slowly. This growth–survival trade-off naturally emerged from the opposing effects of wood density on growth and survival in our models. We also found evidence of a stature–recruitment trade-off mediated by maximum height and wood density. Species with low maximum height and low wood density had high recruitment at small sizes but low growth and survival at larger sizes, whereas species with tall maximum height and dense wood had high growth and survival as large trees but produced few recruits when they were small. Interestingly, we did not find evidence of a growth–survival trade-off or stature–recruitment trade-off driven by specific leaf area.

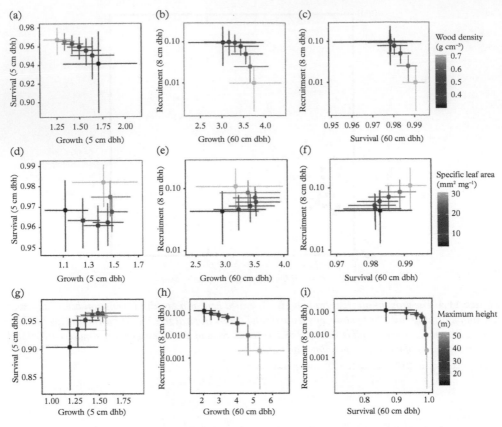

Figure 11.5 *Trait-mediated demographic trade-offs in temperate forest tree species. Points represent predicted demographic rates (error bars show 90% credible intervals) for an average temperate tree in the Eastern USA with (a–c) varying wood density, (d–f) specific leaf area, or (g–i) maximum height. A negative relationship between demographic rates across values of a trait indicates a trait-mediated demographic trade-off. The first column (a,d,g) shows the relationship between predicted growth and survival at 5-cm diameter, diagnostic of a growth-survival trade-off in saplings. The second (b,e,h) and third (c,f,i) columns show relationships between growth and survival, respectively, at 60-cm diameter and recruitment at 8-cm diameter, diagnostic of a trade-off between recruitment early in life and growth and survival later in life (i.e., the stature–recruitment tradeoff).*
Reproduced from Siefert and Laughlin (2023) with permission.

11.6 Causes of death differ between semelparous and iteroparous species

Just as growth is size dependent, survival rates also change with ontogeny. Large plants survive at much greater rates than small plants. Survival may decline at the largest of sizes due to senescence, but this may not be a widespread phenomenon in perennial plants (Chapter 5). Annual plants, by definition, do not survive to the following year: flowering is fatal. In semelparous perennials, demographers have shown that survival to the flowering stage is also size dependent (Metcalf et al., 2003; Rees et al., 2006; Metcalf et al., 2008). However, death in semelparous species is driven by internal factors that drive senescence (Munné-Bosch, 2007) as well as external factors. Internal factors cannot be modelled as functions of traits in response to the environment, so it is important to separate trait-based predictions of survival rates for semelparous and iteroparous species.

Survival rates are intimately connected to lifespan. Higher survival rates inevitably lead to longer lifespans. Put another way, long-lived plants have intrinsically higher rates of survival than short-lived species. Because of this, it is possible that some traits determine the probabilities of achieving a certain lifespan, and others determine the probability of surviving the vagaries of environmental factors. Death in iteroparous perennial species is driven mostly by external factors, adverse environmental conditions such as drought or freezing temperatures. Human survival rates in cold climates increased with the rise of technological advances such as clothing, shelters, and heating systems. The ability to protect ourselves from the elements improved our quality of life and hence, our survival and lifespan increased. Likewise, perennial plants protect themselves from adverse conditions, especially those that need to over winter in temperate and arctic climates or survive the dry season. But survival after disturbances like fire also require the plant to protect themselves. This uncanny ability to survive, and perhaps achieve immortality, is because plants exhibit modular growth.

11.7   A clonal herb is a tree on its side that is potentially immortal

Recall that modules are the basic architectural building blocks of a plant and are defined as construction units produced by apical meristems that consist of a node, an internode, a leaf, and an axillary bud (Klimešová et al., 2019). The plant body is therefore a repeating series of such modules. If one module is lost, then others can take over and the individual plant can still survive. Death in modular organisms only occurs when every single module dies (Rohde and Bhalerao, 2007; Munné-Bosch, 2014). For example, a tree can continue to live as long as one apical shoot meristem and one apical root meristem are connected by a vascular system. Indeed, it can be very difficult to know in field surveys whether a tree is actually dead or not (Anderegg et al., 2012)—it sometimes takes repeated surveys. The fate of the meristem defines the ability of a plant to persist past the annual and into perennial stages—the perennial plant by definition retains the shoot apical meristem in an indeterminate state (Lundgren and Des Marais, 2020).

Modularity permits trees to get taller and herbs to get wider. Harper (1981) quipped that a clonal herb is just a tree lying on its side (Silvertown and Charlesworth, 2001). Plants are generally viewed as stationary, sedentary, and sessile organisms that spend their entire lives in the same location, and the only time they can travel across a landscape to immigrate is in the form of a dispersing seed. Two-thirds of all plant species exhibit clonal behavior and many of these have the uncanny ability to spread laterally across the ground (Herben and Klimešová, 2020). In the recent past, plant species may have been distinguished as being either clonal or non-clonal in trait databases, but it is increasingly recognized that there is broad continuous variation in a plant's ability to propagate vegetatively (Klimešová et al., 2018). In fact, a clonal plant with short rhizomes may be functionally more similar to a non-clonal plant than to a clonal plant with long rhizomes. Clonal growth organs include leaves (bulbs), stems (rhizomes), and roots (sprouting).

Clonality is often called vegetative reproduction, but John Harper railed against the view that vegetative spread is reproduction: "If a tree spreads vertically we call it growth, but if a clover spreads laterally we call it reproduction—nonsense," said Harper, quoted in Turkington (2010). Sexual reproduction produces new genetic individuals that develop from a zygote. As a Darwinian demographer, Harper felt strongly that demographers should be interested in genetic individuals. Genets are genetically identical individuals that can vegetatively propagate new individual ramets that may or may not become physiologically distinct. Ramets often have stems that emerge from the ground, so they appear to be individuals, but they were produced through vegetative growth. Quaking aspen (*Populus tremuloides*) stands provide a clear example. Aspen trees look demonstrably individual—we can measure their diameter at breast height just like any other tree—but each aspen ramet belongs to a larger clonal genet. This can be seen in dramatic fashion when aspen turn color in the fall,[3] where individual clones can be distinguished from far away because each clone turns color at a different rate causing the colors to be patchily distributed across the mountain forests. Clonality makes counting genets very difficult. It is impossible to track genets over successive years using standard field methods—genetic tools may be required, but these are costly. This methodological constraint has limited progress in our understanding of the demography of clonal plants. Demographers disagree on whether to include clonality within the fecundity function or in the growth and survival function. One solution is to account for clonality separately in its own matrix, which can be added to the **A** matrix in a matrix population model (Salguero-Gómez et al., 2015; Janovský et al., 2017). There are multiple solutions for accounting for clonality, but not all of them lead to identical results and impact the calculation of generation time.

Clonal growth has important consequences for survival rates and lifespan. The capacity of clonal plants to continually generate new ramets may permit the genet from escaping the vagaries of senescence (Silvertown and Charlesworth, 2001) and also tends to increase generation time. Some have argued that the modularity in plants gives them the ability to escape physiological deterioration, making them potentially immortal

---

[3] Leaves turn color because of the breakdown of chlorophyll and the uncovering of other pigments like anthocyanins and carotenoids. Leaf senescence is an important nutrient conservation process in perennials. Catherine Smart (1994) called it "the last will and testament of a leaf, in which bequests of nutrients are made to the rest of the plant."

(Peñuelas and Munné-Bosch, 2010). Accurate aging of organisms that live for millennia is tricky, and molecular techniques are often relied on (de Witte and Stöcklin, 2010). Assuming these estimates are close to the truth, we can conclude that the oldest organisms on the planet are clonal plants (recall Table 5.1). For example, clones of creosote bush (*Larrea tridentata*) in the Mojave Desert have been estimated to be 11,000 years old (Vasek, 1980), and clones of King's lomatia (*Lomatia tasmanica*) in Tasmania have been estimated to be 43,600 years old (Lynch et al., 1998)! There are reports that Pando (Latin for "I spread"), a clone of aspen in Utah, is more than 80,000 years old (Sussman, 2014), but even if this is hyperbolic, microsatellite mutation rates suggest that some aspen stands in Canada were 10,000 years old (Ally et al., 2008). These numbers are hard to comprehend.

11.8   Bud banks and belowground storage organs extend lifespans

Plants have evolved a bewildering variety of solutions to the challenges of growing and surviving. Herbaceous plant solutions are different from woody plant solutions. Understanding plant modularity and its relation to demography has been a long-standing problem in plant ecology (Raunkiaer, 1934; White, 1979), but fortunately a standardized set of terms and protocols for measuring modular traits in plants has been proposed (Klimešová et al., 2019). Despite the diversity of belowground storage organs, it was proposed that a few general continuous traits can be measured on virtually all plants: lateral spreading distance, multiplication rate, persistence of connection, bud bank size, and bud bank depth (recall Figure 6.12). These modular growth traits are directly linked to plant persistence and rates of survival. The likelihood of surviving to the next year is a function of many traits and depends on environmental factors. For now, we are concerned primarily with traits that directly enhance perennation, and hence, survival rates and lifespan. All else being equal, species with large bud banks and large belowground storage units exhibit higher survival rates and live the longest lives (Munné-Bosch, 2014).

All vascular plants can resprout from buds. Bud banks promote perennation, survival, and longevity and consist of "all viable axillary and adventitious buds which are present on a plant and are available for regrowth, branching and replacement of shoots through a season or for vegetation regeneration after an injury" (Klimešová and Klimeš, 2007, p. 124). Recall that seed banks are stashes of dispersed seeds in the soil that have the potential of germinating and contributing to recruitment rates. Bud banks, on the other hand, are stashes of dormant meristems that can be found either above or belowground. In contrast to seed banks, bud banks are directly attached to living plants and interest in bud banks has been reinvigorated in the last two decades. Species differ considerably in the number of dormant buds that are present on individual plants, which lead to differences in survival rates both over unfavorable seasons (e.g., winter, dry season) and following disturbances (Bellingham and Sparrow, 2000; Klimešová and Klimeš, 2007; Fidelis et al., 2014). For example, tuber-forming orchids produce a single belowground bud per year, whereas weedy perennial species like *Agropyron repens* can produce tens of thousands (Klimešová and Klimeš, 2007).

Harper (1977) introduced the term bud bank to represent the hidden population of meristems, but interest in the location of meristems goes back to Christen Raunkiaer, who recognized the critical role that meristems played in global species distributions. Raunkiaer (1934) evaluated how plants protect meristems and developed a life form classification scheme to distinguish among the many ways plants solved the problem of perennation (recall Figure 2.1) and proposed five main categories. Phanerophytes produce their perennating meristems high above ground. We know them more casually as trees and shrubs. Raunkiaer divided these into four separate categories according to height, but we will not consider these details. Chamaephytes are also woody plants, but their perennating meristems are produced close to the ground. In cold climates the meristems are buried beneath the snowpack. Hemicryptophytes produce their meristems at the soil surface. Grasses and sedges are a prime example of this group. Cryptophytes produce the meristems beneath the soil surface or under water. Therophytes are not perennials at all; they are annuals that survive unfavorable seasons as embryos within a seed. At least one meristem is required to survive. Therophytes do not produce a single lasting meristem after flowering, and so, the whole plant dies. The second thing to note is that Raunkiaer focused exclusively on meristem location. We now know that herbaceous and woody plants can have similar lifespans extending several hundreds of years. Therefore, the number of perennating buds may be more indicative of lifespan, rather than mere location.

The bud bank consists of all dormant meristems both belowground and aboveground on stems, roots, rhizomes, corms, bulbs, bulbils, and tubers. The diversity of forms that have evolved to promote above- and belowground survival is enormous (Bonser and Aarssen, 1996; Klimešová and Klimeš, 2007). One of the key differences between bud banks and seed banks is that seeds can persist in the soil for decades and centuries in the absence of the parent plants. Bud banks, in contrast, only persist in the presence of parent plants because, given the modular nature of plants, the plant coroner needs to declare that all meristems are dead before they can declare the whole plant to be dead. Likewise, buds cannot readily disperse like seeds. Growth of bud banks is primarily controlled by apical dominance through the hormone auxin. The bud bank typically remains dormant until apical dominance is broken, either through the senescence of the leading branch or disturbance to the stem.

Trees (phanerophytes) are good examples of species with large aboveground bud banks, given that all meristems are protected beneath bud scales over winter. But some species of trees and shrubs stand out because they contain meristems on the trunk that can survive crown fires and resprout (Burrows, 2013; Pausas and Keeley, 2017). This phenomenon is called epicormic resprouting. A few species such as messmate (*Eucalyptus obiqua*) and shining gum (*Eucalyptus nitens*) exhibited this remarkable recovery after the devastating bush fires near Marysville, Victoria in Australia. Belowground bud banks are also starting to get the attention they deserve. Consider the remarkable underground trees called geoxylic suffrutices (Meller et al., 2022). Geoxyles propagate their stems laterally underground and send up shoots and flowering branches that do not grow tall.

Rather, individuals attain horizonal diameters > 50 m across. *Jacaranda decurrens* is an example of one species in the Brazilian Cerrado that can live for millennia (Alves et al., 2013; Bond, 2019)! The agent of selection for this unique growth form has been repeated fire. Resprouting can arise from belowground, basal, or aerial meristems. According to a model developed in Australia (Clarke et al., 2013), the probability of surviving a fire increases in the following sequences: basal collar resprouters, aerial epicormic resprouters, basal lignotuberous resprouters, belowground geophytes and xylopodia resprouters, aerial terminal resprouters, and finally to belowground resprouters from rhizomes and roots (recall Figure 7.12).

The densities of bud banks can be impressively high in open grasslands and savannas around the world (Fidelis et al., 2014; Ott et al., 2019; Bombo et al., 2022). For example, bud bank densities in the South African Lowveld Savanna Bioregion could get higher than 2,000 buds per square meter in sites with high fire frequency and the presence of herbivores (Bombo et al., 2022). Bud bank densities in a tallgrass prairie in Kansas, USA ranged from 5,000 to 20,000 buds per square meter, where most of the buds were found on the grasses and densities increased with increasing biomass of rhizomes (Figure 11.6) (VanderWeide and Hartnett, 2015). It is easy to see why these perennial species can attain such long lifespans when they invest so heavily in buds.

Most of the research on belowground traits has focused almost exclusively on resource acquisition by fine roots (Reich, 2014; Laliberté, 2017; Bergmann et al., 2020). However, belowground organs are useful to the plant beyond resource acquisition and play a direct role in survival. Within this belowground compartment, Klimešová and colleagues (2018) distinguish between the structural sub-compartment that plays a role in persistence, anchorage, resprouting ability, and resource transport in clonal plants, and the nonstructural sub-compartment that considers storage compounds like carbohydrates. Thus, the relative allocation of carbon to each of these compartments directly affects perennation. The root-to-shoot ratio was proposed to be a good indicator of investment in belowground resource acquisition. However, most biomass that is belowground is not acquiring resources, but rather transporting or storing resources (Klimešová et al., 2018). We saw how the root-to-shoot ratio was not a good predictor of RGR, and it is also not a good predictor of survival rate because it combines belowground biomass from multiple compartments that contribute differently to plant function.

The absolute biomass of structural and non-structural belowground sub-compartments are likely strong predictors of perennial survival because resprouting depends on two resources: buds and carbohydrates (Ottaviani et al., 2017; Lundgren and Des Marais, 2020). Turnover of the structural compartment is relatively slow in perennial herbs, suggesting that buds will not be depleted quickly through natural senescence, and can therefore be activated any time. It may be that the bud bank is strongly positively correlated with the biomass of the structural sub-compartment, which would make the traits redundant and would not contribute unique information toward our goal of predicting survival (Laughlin, 2014a). In that case, then bud bank density and storage sub-compartments would possibly be complimentary predictors of longevity, where the former represents potential for resprouting, and the latter represents likelihood of resprout success. Species with large storage of carbohydrates

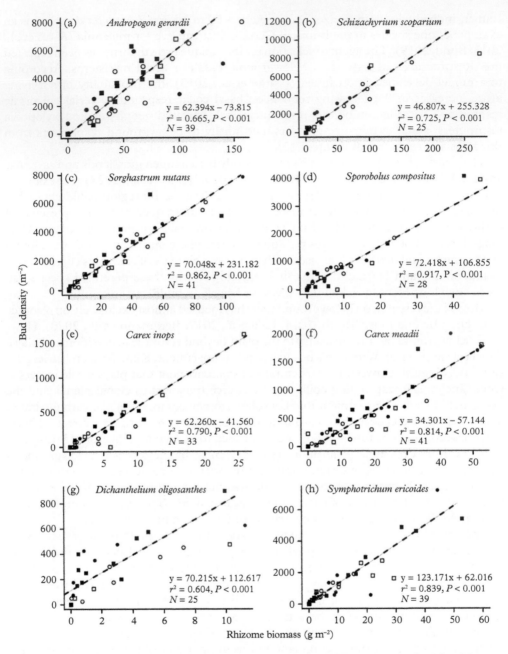

Figure 11.6 *Positive relationships between bud bank densities and rhizome biomass in eight of the most frequently encountered species in a tallgrass prairie. An increase in 1 g/m² of rhizome biomass leads to between 34 to 123 buds per square meter, based on the regression relationships across these species.* Reproduced from VanderWeide and Hartnett (2015) with permission.

can deploy the energy to the shoot to enable a strong resprout responses following disturbance or the unfavorable season (Klimešová et al., 2018). Any species with a storage organ and high density of buds can potentially achieve long lives, and therefore have high rates of survival.

This chapter considered which traits promote growth and enhance survival. We saw that SLA and wood density have clear effects on growth rates, but these rates are dependent on size and ontogeny. The traits that influence survival are linked to the distribution of dormant meristems throughout the plant body, and larger perennial plants will likely contain larger bud banks. Building accurate demographic models for growth and survival across the range of plant phenotypes is an active area of research, especially in this era of increasing drought-induced mortality. Chapter 12 considers the traits that bolster recruitment, a notoriously difficult demographic rate to estimate in plants.

11.9   Chapter summary

- The classical approach to measuring plant growth rate likened it to compound growth rates of bank accounts, where the rate of increase in plant biomass is proportional to the amount of biomass already present in the plant, implying that plant growth is exponential.

- It was determined that variation in RGR among herbaceous plants was driven by the LAR, rather than ULR. Between the two traits that contribute to LAR, SLA explained more variation in RGR than leaf mass fraction, indicating that SLA plays a central role in determining the inherent variation in RGR among herbaceous plants.

- The classic growth equations are confounded by plant size. While exponential growth may be true for the early stages of growth in herbaceous plants, plants tend to experience reduced RGRs as they approach their maximum size, especially when growing with neighbors.

- The relationship between SLA and growth rate shifts from positive in seedlings to non-significant in adult plants. Wood density exhibits consistently negative effects on growth rates across ontogeny. The positive relationship between maximum height and growth rate is most significant in adult size classes. The negative effects of seed mass on growth rate in seedlings is undetectable in saplings and adults.

- The growth–survival trade-off is one of the most well-recognized trade-offs in ecology. Species that grow fast when resources are abundant die when resources are scarce, and species that survive when resources are scarce grow slowly when resources are abundant. This growth–survival trade-off naturally emerged from models of tree demographic rates by the opposing effects of wood density on growth and survival.

- Bud banks promote perennation, survival, and longevity and consist of "all viable axillary and adventitious buds which are present on a plant and are available for regrowth, branching and replacement of shoots through a season or for vegetation regeneration after an injury." The plant coroner needs to declare that all meristems on an individual plant are dead before they can declare that the plant is dead—a rather tall order.

11.10   Questions

1. What traits drive RGR, and does the answer depend on whether the plant is woody or herbaceous?

2. What traits determine survival, and does the answer depend on whether the plant is woody or herbaceous?

3. Do you expect there to be a relationship between traits that influence RGR and traits that influence survival?

12

Plant Traits That Bolster Recruitment

A plant is only the means by which a seed produces more seeds.

— John Harper (1977, p. 29), *Population Biology of Plants*

I have great faith in a seed ... Convince me that you have a seed there, and I am prepared to expect wonders.

— Henry David Thoreau (1913, p. 151), *Excursions*

12.1 Plant recruitment is the most difficult demographic rate to accurately predict

Modeling plant recruitment is hard. Recruitment is often the weakest link in plant population models (Nguyen et al., 2019) and has been ignored in plant strategy models that have emphasized relative growth rate and survival (e.g., Craine, 2009). Quality data on seedling recruitment can be rare in monitoring datasets and seedling recruitment can exhibit enormous spatial variability (Puhlick et al., 2012; Puhlick et al., 2021; Shackelford et al., 2021), often requiring clever modeling techniques to estimate recruitment rates (Chu and Adler, 2015; Shriver et al., 2021). Successfully integrating traits into our understanding of plant recruitment will fill a critical gap in plant strategy theory.

Innumerable stars must align in order for recruitment from seed to occur (Grubb, 1977). Recall from Chapter 4 that the fecundity kernel for a typical perennial plant includes the product of the probability that the plant flowers, the number of seeds produced if it does flower, the establishment probability of each seed produced, and the size distribution of recruits. All these parameters define the probability that an individual plant in time t produces a recruit in time $t + 1$. Now consider this incomplete list of additional difficulties.

First, seed production can be limited by pollinators in outcrossing species, so predicting seed set requires knowledge of pollinator activities (Willson, 1983). Second, it depends on dispersal to a safe site typically removed from the mother plant (Janzen, 1970; Harper, 1977), and dispersal occurs through a variety of different agents, so predicting dispersal requires knowledge of the activities of the dispersal agents (Ridley, 1930; Van der Pijl, 1982). Third, germination requirements are highly variable (Fenner, 1985; Baskin and Baskin, 1998) and establishment probabilities are often exceptionally low. Fourth, recruitment is the product of germination, emergence, and establishment,

Plant Strategies. Daniel C. Laughlin, Oxford University Press. © Daniel C. Laughlin (2023). DOI: 10.1093/oso/9780192867940.003.0012

which are distinct processes that are under unique selection pressures (James et al., 2011). Fifth, reproduction strategies are not aligned with strategies of the established phase (Grime et al., 1988; Shipley et al., 1989). Sixth, recruitment of genets versus ramets is hard to adequately quantify in clonal plants. Seventh, plants can store seeds for later recruitment in soil seed banks (Doak et al., 2002; Adams et al., 2005; Paniw et al., 2017; Nguyen et al., 2019). All of these processes make it hard to find simple generalizations about the regeneration niche of a phenotype (Grubb, 1977).

Recall that our task is to link fitness to traits along environmental gradients. Fitness depends on both the growth and survival matrix and the fecundity matrix in population models. It is therefore impossible to fully understand plant strategies by only considering the traits of established plants. Can we use traits as predictors of recruitment rates in the same way we did with growth and survival in Chapter 11? Before we examine traits directly, let us review how ecologists have approached the regeneration phase.

12.2   Traits of established and regenerative phases are decoupled

One of the difficulties of incorporating regeneration into plant strategy theory is that it is well established that traits of juveniles and traits of mature plants are decoupled (Grime et al., 1987; Grime et al., 1988; Shipley et al., 1989; Leishman and Westoby, 1992). In other words, we generally cannot predict the behavior of seedlings based on the traits of the adults. Grime measured traits on juveniles and adults on hundreds of British plant species and found that the two sets of traits were decoupled (Grime et al., 1987).

Given their lack of consistency with the strategies of the established adult phase, Grime (1989) proposed the existence of five reproductive strategies that overlap with his three strategies of the established phase (recall Table 2.3):

(1) The "vegetative expansion" (V) strategy can be found in lightly disturbed habitats. Clonal plants expand vegetatively through rhizomes and stolons and eventually fragment into self-supporting ramets. These propagated offspring enjoy low risks of mortality compared to true seedlings. This strategy overlaps most with competitors and stress-tolerators.

(2) The "seasonal regeneration" (S) strategy can be found in habitats that are predictably disturbed each year through drought, flooding, trampling, or grazing, and are recolonized by gap colonizers. Short-lived ruderals, competitive ruderals, and stress-tolerant ruderals typify this regeneration strategy.

(3) The "persistent seed bank" (Bs) strategy can be found in habitats that are not frequently disturbed. Plants that use this strategy make long-lived seeds that accumulate in situ year after year, waiting for a small disturbance to stimulate them to germinate. This strategy overlaps with competitors and ruderals.

(4) The "numerous small wind-dispersed seeds or spores" (W) strategy can be found in habitats with unpredictable disturbance or inaccessible sites (e.g., cliff faces). Plants that produce copious amounts of seeds that readily disperse are included in this group.

(5) The "persistent juveniles" (Bsd) strategy can be found in unproductive habitats in plants that grow slowly as juveniles and are most common among stress-tolerant plants.

Shipley and colleagues (1989) measured 20 traits, including seed mass, germination rates, and growth rates of seedlings and clonal traits, height, and leaf thickness on 25 species of marsh plants (recall Figure 2.10). The traits of the juvenile plants were interpreted as adaptations to regenerate in gaps (small seeds with rapid germination and fast RGR) or regenerate in established vegetation (large seeds with slow germination and slow RGR), whereas the strategies of established plants were differentiated by their propensity to occupy space vertically (height) or horizontally (clonal spread). Leishman and Westoby (1992) confirmed the lack of correlation between regenerative and established traits on an even larger analysis of 43 traits measured across 300 species from semi-arid woodlands in New South Wales, Australia.

I have always maintained that uncorrelated traits are far more useful than correlated traits when building predictive models of community assembly because uncorrelated traits provide unique information about how plants function (Laughlin, 2014a). Orthogonality of traits can help us to understand how so many species can coexist in species-rich communities. Using the example in the previous paragraph, Shipley and colleagues (1989) organized the leading dimensions of juveniles and adults into a two-by-two orthogonal table with four possible strategies. One strategy was composed of stress-tolerant juveniles and adults, and they were adapted to infertile undisturbed sites. Another strategy was composed of stress-tolerant juveniles but competitive adults, and they were adapted to fertile soil and recruit into small gaps in the vegetation. A third group was composed of fugitive juveniles and competitive adults that are adapted to fertile soil, recruit in large canopy gaps, with infrequent gap formation. The last group was composed of fugitive juveniles and adults that were adapted to fertile soil, and frequent and large gap formation for recruitment.

The lack of association between regenerative and established traits may be a blessing rather than a curse of higher dimensionality. Perhaps we can build models where a set of traits on mature plants can predict variation in growth and survival, and a set of traits on young seedlings can predict variation in recruitment (Larson et al., 2016; Garbowski et al., 2021; Leger et al., 2021). This would prove to be particularly powerful in the context of constructing fecundity kernels for trait-based population models. Let us consider the traits of seeds that have received the most attention.

12.3   Plants produce either a few large or many small seeds

As discussed in Chapter 5, reproductive traits are under strong selection at young life history stages. Species that reach reproductive maturity quickly should have higher net reproductive rates because more individuals in the population are reproducing before they die. Moreover, species with higher reproductive effort invest more resources into reproduction during each reproductive event. But both of these life history choices come at a cost of growth and survival (de Jong and Klinkhamer, 2005). In one study, unusually heavy investment in reproduction in *Betula alleghaniensis* and *Betula papyrifera* trees led to reduced growth, die back of terminal buds and branches, and reduced subsequent flowering (Gross, 1972). Annual, biennial, and short-lived monocarpic plants reach maturity early, invest heavily into reproduction, yet live shorter lives than perennials. These life histories are reflected in their fecundity kernels where the relative importance of reproduction (measured by vital rate elasticity) is large. These are some of the few traits that would consistently increase your reproductive rate, but successful recruitment of species with these traits will be highly contingent on the environmental context. For example, annuals are common in frequently disturbed habitats or in arid regions with brief rainy seasons; otherwise, these traits are selected against and are not dominant features of the vegetation.

The most well-studied plant trait that has been proposed to be related to recruitment is seed size. Offspring size is an important contributor to fitness (Roff, 1993), but offspring size exhibits a trade-off with offspring number. The number of seeds produced by a plant is somewhat analogous to clutch size in animals (Lack, 1947), and consequently seed size has important links to establishment probabilities, which are critical parameters that are difficult to estimate in population models. Seeds in angiosperms are comprised of three types of tissue: the embryo is a fusion of maternal and paternal gametes, the endosperm, and the surrounding tissues are from purely maternal origin. However, each species exhibits different ratios of these tissues, and each component affects different aspects of the roles that seeds play for the regeneration of a population. Seeds need to travel, find a safe site, remain dormant or germinate, then support a seedling until it can photosynthesize on its own. The relative abundance of each of these three tissues and the variety of shapes and sizes of seeds could reflect the many functions and sometimes conflicting selection pressures on seeds (Willson, 1983).

There is enormous variation in seed size across species. Orchids, such as the coral-root orchid (*Corallorhiza maculata*) hold the world record for smallest seeds, appropriately called dust seeds, that weigh in at 0.000001 g. Large plants like oak trees and palms can easily support a massive seed, but they will produce relatively fewer of them on a per area basis. The largest known seed is produced by the Coco-de-mer palm (*Lodoicea maldivica*), weighing in at 10,000 g. Like Mary Willson, I have not been able to confirm whether this is the weight of the seed or the fruit, but regardless, the seed is still massive. If you count the number of zeros separating the masses of these two species you will see that seed mass can vary by 10 orders of magnitude globally! There is more variation in seed size than in any other plant trait, suggesting that natural selection had a lot of potential variation to work with. This massive variation is no doubt part of the reason that understanding variation in the causes and consequences of seed size has captured the attention and imagination of ecologists across the world.

Seed size is also remarkably stable within species conserving strong interspecific differences (Harper et al., 1970), suggesting that seed size is under strong stabilizing selection (Westoby et al., 1997). It may be under strong stabilizing selection because a gain in fitness from increasing offspring number by reducing size is offset by the loss of fitness because of reduced success of each offspring (Stearns, 1992). There are clear examples of species that exhibit variation in seed size (Willson, 1983), but taking everything into consideration, seed mass exhibits more variation between species than within species, making it a great trait for comparative ecology.

Plants face a tremendous trade-off in how they can allocate their limited resources into reproduction. The allocation of resources into vegetative growth limits seed production (Vallejo-Marín et al., 2010; Herben et al., 2015). Investment in clonal growth exhibits trade-offs with investment in seed. Sutherland and Vickery (1988) observed that ramet production across five *Mimulus* species was negatively correlated with fruit set and seed production. Bullock and colleagues (1995) observed grass species dynamics in a grassland and observed a trade-off between species that colonized gaps in vegetation by seed versus those that grew into gaps through vegetative spread.

The number of seeds produced by a plant is constrained by the size of each seed. A plant can either produce hundreds of small seeds or a handful of large seeds. Seed mass is closely related to life form—small herbaceous plants, on average, produce smaller seeds than large trees (Salisbury, 1942; Fenner, 1985; Moles et al., 2007). Small plants cannot support very large seeds due to biomechanical constraints, but they can produce many lightweight, small seeds (Rees, 1997b). After controlling for the effects of plant size and phylogeny, seed mass was clearly negatively correlated with seed number (Jakobsson and Eriksson, 2000).

Based on these correlations, would creating more small seeds or few large seeds increase your probabilities of recruitment? Recall from Chapter 11 that specific leaf area tends to increase RGR across species, and rankings of growth rates across species is relatively consistent across a gradient of productivity. Similarly, bud banks and belowground storage are consistently related to survival and longevity across environments. Is there an analogous trait for recruitment? Clearly, generating hundreds of small seeds is necessary for recruitment in short-lived annuals, but is this true in all environments? Much work has gone into answering this question, and the answer is that the environmental context is extremely important (de Jong and Klinkhamer, 2005). We cannot generalize about consistent main effects of traits on recruitment rates in plants.

Several studies have linked large seeds to greater establishment probabilities across a range of hazardous conditions, suggesting that seed size could be a useful predictor of establishment probabilities in such conditions. Large seeds tend to exhibit higher rates of emergence and produce larger seedlings, which can be helpful for surviving the hazards of the establishment phase (Jakobsson and Eriksson, 2000). For example, large-seeded species performed better under grazing pressure (Armstrong and Westoby, 1993), in nutrient-poor soils (Jurado and Westoby, 1992), when planted deep in the soil or under litter (Westoby et al., 1997), and exhibited less dependence on mycorrhiza for establishment (Allsopp and Stock, 1995). Some studies also found that large-seeded species perform better when growing in established vegetation (Burke and Grime, 1996), in deep shade (Salisbury, 1942; Grime and Jeffrey, 1965; Leishman and Westoby, 1994a), or in dry environments (Baker, 1972; Leishman and Westoby, 1994b), but note that several counter examples can be found (Westoby et al., 1997). The mechanism underlying enhanced performance could either be because large seeds produce larger seedlings, a reserve effect where endosperm reserves can support carbon deficits, or a metabolic effect since larger seeds have lower relative growth rates. Westoby and colleagues (1997) concluded that the reserve effect is most likely the cause of enhanced performance of large seeded species in hazardous environments.

Moles and colleagues (2007) undertook a highly ambitious global scale analyses of seed size and showed that seed size tends to decrease with increasing latitude, but differences in growth form and habitat type explained the majority of seed size variation. Seeds are largest in rain forests and evergreen sclerophyll vegetation, and smallest in grasslands. Not only are differences in seed size large across vegetation types and life forms, but also there can be up to four orders of magnitude range in seed size within a single flora (Salisbury, 1942; Leishman et al., 1995).

Recall from Chapter 9 that classic optimization models of seed size within a species assume that seed size trades off with seed number and that seedling success increases with seed size (Smith and Fretwell, 1974). This approach predicts a single optimal seed size, that plants with more resources should produce more seeds instead of larger ones, and that large seeds would be advantageous in more hazardous environments (de Jong and Klinkhamer, 2005). Game-theoretical models that incorporate seedling competition suggest that a range of seed sizes should be present within a single habitat (Geritz, 1995; Rees and Westoby, 1997), and that is indeed what is observed in nature (Moles et al., 2007). Habitats that experience disturbance-driven gap dynamics where openings are frequently made within closed canopies would allow for spatial and temporal variation in environmental conditions—this would permit the coexistence of species exhibiting a range of seed sizes.

12.4   There is a need to expand the search from seed mass to other regeneration traits

It is now clear that efforts must cast a wider net in the search for traits that predict recruitment by considering dispersal and establishment, persistence, and germination timing (Saatkamp et al., 2019). Larson and Funk (2016) articulated a trait–filter framework for identifying traits that are important at each stage of the recruitment process under selection by both abiotic and biotic filters. The stages include seed production, dispersal, germination, emergence, seedling establishment, and clonality (Figure 12.1). The authors acknowledge that the potential list of traits is not exhaustive, yet their framework is an important step toward deeper integration of traits into the complex processes involved in recruitment.

Seed production could be affected by flower morphology and phenology, leaf and stem economics, seed mass, and vegetative height (Henery and Westoby, 2001; Adler et al., 2014). Dispersal could be affected by dispersal agent, height, seed mass, terminal velocity, granivory, and frugivory (Schupp et al., 2010; Thomson et al., 2011; Tamme et al., 2014). Germination could be affected by hydrothermal time traits, dormancy, seed coat hardiness, and seed mass (Baskin and Baskin, 1998; Bradford, 2002; Larson et al., 2016). Emergence, which ranks among the least studied of these processes, could be affected by germination phenology, embryonic leaf and radical traits, shoot width, shoot strength, and seed mass (Sydes and Grime, 1981; James et al., 2011;

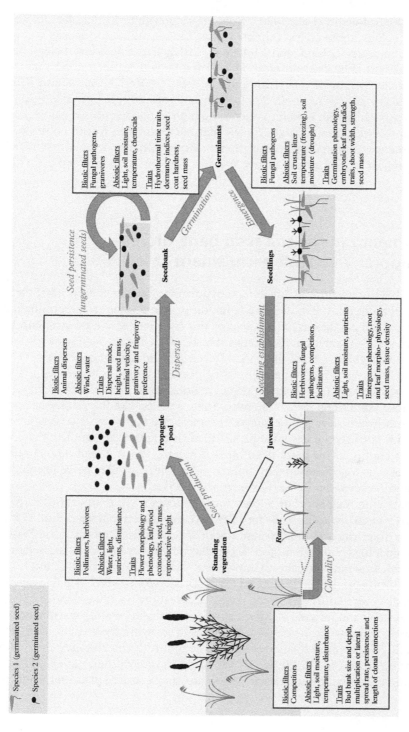

Figure 12.1 *A trait-filter framework to identify the traits that influence all phases of recruitment, including seed production, dispersal, germination, emergence, seedling establishment, and clonality.* Reproduced from Larson and Funk (2016) with permission.

Larson et al., 2016; Larson et al., 2020). Seedling establishment could be affected by emergence phenology, root and leaf economics, seed mass, and tissue density (Alvarez-Clare and Kitajima, 2007; Funk and McDaniel, 2010; Butterfield and Briggs, 2011; Larson et al., 2015). Finally, belowground bud banks and lateral spread rate affect the capacity of clonal growth (Dalgleish and Hartnett, 2006; Klimešová and Klimeš, 2007; Fidelis et al., 2014; Ott et al., 2019; Bombo et al., 2022). Leaf mass per area tends to increase with age for seedlings due to the high initial resource requirements of "start-up" plants (Westoby et al., 2022). We still have much to learn about the predictive performance of traits for each of these demographic processes that together determine the regeneration niche of plants. We now explore dormancy and dispersal in more detail.

12.5 Dormancy in the soil seed bank drives temporally variable recruitment rates

Seeds have evolved an uncanny ability to remain dormant, further complicating the prediction of recruitment rates. This clever adaptation prevents them from germinating at the wrong time when conditions are too cold or dry for successful establishment. Germination is triggered by environmental cues. Horticulturists and seed ecologists have spent many years trying to crack the code of seed germination requirements (Deno, 1993; Baskin and Baskin, 1998). Some species respond to simply being soaked in water. Other species, particularly from cold temperate regions, must be cold stratified by chilling them in moist conditions for 2–3 months. Some species, like the *Penstemon* species of western North America, require a complex temperature cycle to induce germination. Many seeds with hard seed coats must be scarified or chipped, to simulate the conditions that seeds endure in the gizzards of birds. Amazingly, some seeds are stimulated by wildfire smoke, as was discovered in some uncommon Australian herbs like *Calytrix leschenaultia* and *Lechenaultia biloba* (Dixon et al., 1995).

Seed lifespan is nothing short of astonishing. Students revived an extinct variety of squash (*Cucurbita maxima*) after growing an 800-year-old seed that was found in a clay plot in a Native American archaeological site on the Menominee Reservation in Wisconsin, USA. Sacred lotus (*Nelumbo nucifera*) seeds that were recovered from an ancient lake bed at Pulantien, Liaoning Province, China were successfully germinated after 1,300 years (Shen-Miller et al., 1995). Date palm (*Phoenix dactylifera*) seeds dated to be 2,000 years old were discovered during an archeological excavation in a stone jar at King Herod the Great's Palace at the fortress of Masada—these were successfully germinated and researchers are now growing the palms with hopes of growing the ancient date variety (Sallon et al., 2020). Seeds of the narrow-leaved campion (*Silene stenophylla)* remained viable after being frozen for 32,000 years according to radiocarbon dating. While the scientists could not get the frozen seeds to germinate, they cultured placental tissue from the fruit and successfully raised a plant, the plant flowered and produced seeds, and the seeds germinated at high rates (Yashina et al., 2012).

Seed dormancy has been shown to be negatively correlated with lifespan and vegetative propagation (Rees, 1997b). Short-lived species tend to produce seeds that can remain dormant for many years, whereas long-lived species, in particular clonal plants, tend to produce short-lived seeds, indicating a trade-off between sexual reproduction and vegetative propagation (Silvertown and Charlesworth, 2001). Seeds that disperse long distances also tend to exhibit short residence times in the soil (Cook, 1980).

Given the difficulty of experimentally determining the longevity of seeds, it would be particularly useful if traits could predict seed longevity. Thompson and colleagues (1993) determined that seed size and shape together predict the persistence of seeds in soil in northwestern Europe. Seed shape is an index measured as the variance between length and width. Round seeds have low variance in their dimensions because they have the same length and width, whereas long and linear seeds have high variance in their measurements of length and width. Small and round seeds persist in the soil for longer than large and more linear seeds (Thompson et al., 1993). Small seeds may be more difficult to find by predators and large seeds may be more exposed to pathogens. However, the generality of these findings has been called into question given the lack of support in other flora around the world (Leishman et al., 2000).

The phytochrome system of seeds prevents germination by perceiving the ratio of red-to-far-red wavelengths of light (Pons, 2000). The visible spectrum of radiation ranges from 400 to 750 nm in wavelength. Red light is approximately 660 nm and far-red light is approximately 730 nm. The red-to-far-red (r:fr) ratio is defined as the ratio of photon flux density at 660 nm and 730 nm. This ratio changes throughout the depth of a forest canopy. The r:fr ratio above the canopy is 1.2, but it declines to 0.18 below the canopy because of the absorption of red wavelengths by the leaves. Some seeds exhibit increasing germination rates in response to the r:fr ratio, indicating that they preferentially germinate in open conditions (Yirdaw and Leinonen, 2002). One study linked this response to seed size in forest herbs, where small-seeded species exhibited a strong germination response to the r:fr ratio, whereas the vast majority of large-seeded species exhibited no response (Jankowska-Blaszczuk and Daws, 2007). This suggests that small-seeded forest plants only germinate in open sites with high r:fr ratio, which are temporally rare and only occur when gaps in the canopy open due to tree falls, necessitating the persistence of the small seeds in the soil seed bank. Large-seeded plants have a shorter lifespan in the seed bank, and germination is not affected by the r:fr ratio.

The dynamics of soil seed banks are finally starting to receive the attention they deserve in plant population models (Paniw et al., 2017; Nguyen et al., 2019; Logofet et al., 2020), but their importance to understanding population viability has been known for over two decades (Doak et al., 2002). Persistence in the soil seed bank is an important and difficult parameter to estimate. Much more comparative work across species is needed to determine the traits that predict seed dormancy for population models.

12.6 Masting and dispersal drive recruitment through time and space

Many trees produce enormous crops of seed at irregular intervals while producing few or no seeds in between. This phenomenon is known as "masting," named after many of the trees that produce "mast" (i.e., nuts or large seeds). Janzen (1971) puzzled over the temporal and spatial dynamics of plant recruitment and proposed that seed predation was such a strong selective agent that plants irregularly mast to satiate seed predators. Plants that produce a few seeds every year run the risk of their seeds being eaten, thereby limiting recruitment. Plants that produce millions of seeds, sometimes synchronously across the whole population, overwhelm the predators, which cannot possibly eat all the seeds. A few seeds then survive, germinate, and establish. If you can produce more seeds than seed predators can consume, you increase your chances of successful recruitment.

Trees are not the only species that mast. The long-lived tussock grass in New Zealand, *Chionochloa pallens*, produces mast seeding events that satiate their primary seed predators—one moth and two flies (Kelly and Sullivan, 1997). Kelly and colleagues (2000) found that extreme and synchronized mast seeding occurred in 11 species of *Chionochloa* in response to warm summer temperatures in the previous summer. Seed predation was proportionally constant despite hundredfold variation in seed production, leading the authors to speculate that seed predators in this system are hard to satiate.

Just as masting has been proposed to be a mechanism of temporal escape from predation, dispersal has been proposed to be a mechanism of spatial escape from density-dependent factors of mortality. There are many reasons why a plant should disperse. Unless the species is shade tolerant, the seed will need to find an opening in the canopy rather than try to grow beneath the canopy of its mother. Janzen (1970) and Connell and colleagues (1971) proposed that the seed also needs to escape the specialist predators and pathogens that have been attracted or supported by the parent. These density-dependent effects are typically attributed to Janzen and Connell, but they appear to have been proposed much earlier by Henry Ridley (1930), who wrote:

> In almost every plant the greatest number of its seeds fall too near the mother plant to be successful, and soon perish. Only the seeds which are removed to a distance are those that reproduce the species. Where too many plants of one species are grown together, they are apt to be attacked by some pest, insect, or fungus.

These effects have been shown to be quite strong in grasslands as well as forests, and perhaps are a mechanism of species coexistence (Petermann et al., 2008). The theory continues to be immensely popular to test empirically, but the ubiquity of its predictions remain controversial (Chisholm and Fung, 2020; Song et al., 2021). Some species, like *Miliusa horsfieldii* (Caughlin et al., 2015), exhibit quite clear improvements in germination and seedling growth rates as distance from adult conspecifics increases. The decline in number of diaspores that can germinate will be a function of distance from source and the initial number of diaspores produced (Eriksson and Kiviniemi, 1999), but recruitment rate will initially increase with distance from source due to conspecific negative density dependence.

The diversity of dispersal syndromes in plants is testament to the selection pressures that plants face in their quest for safe sites for establishment (Van der Pijl, 1982). Some plants explosively eject the seed from fruit—for example, the tropical tree *Hura crepitans* fruit shatters and scatters seeds up to 14 meters! Seeds that disperse the farthest via wind are small, have low terminal velocities, and are often equipped with a plume to aid transport through the atmosphere. Some seeds have flotation devices that aid in dispersal via water. Many seeds, especially weedy species with names like "sticktight" and "stickseed," are equipped with spines, bristles, barbs, or hooks and attach to the fur of mammals (or the clothes of humans) and are dispersed on animal exteriors. Other seeds are equipped with attractive oily rewards like elaiosomes, in the case of ant-dispersed seeds, or fleshy fruits. Colorful fruits are attractive to birds that can see in color but have no sense of smell (except for a few species such as vultures), whereas smelly fruits that lack color tend to attract mammals that lack color vision but have acute senses of smell. Many trees are dispersed by birds who cache large numbers of seeds to eat later, but inevitably a few seeds are forgotten or lost and establish a new tree.

12.7 The spatial scales of reproduction require methodological considerations

Dispersal in space away from the mother plant requires demographers to consider spatial scale when quantifying reproductive rates. This poses unique challenges. Demographers typically set up permanent plots or transects to tag individuals so that their growth and survival can be continually remeasured over time. They also monitor these plots for new recruits. These plots are often rather small—they can be a few meters squared for herbaceous plants, and perhaps a hectare for trees. But, for reasons just discussed, seeds need to travel away from their mother to succeed, so it might be difficult to definitively link a mother to their progeny to accurately measure recruitment rates because the actual recruits are not found in the permanent plot. Recruits may be coming from mothers that are outside the plot. Moreover, a light-demanding species will need to recruit into a gap in the canopy, which might not be available until a gap opens, which could take decades in closed-canopy vegetation. So how do demographers link individual reproduction to new recruits?

The standard approach is to (1) estimate seed production on the plot across all the individuals in the plot in year t; (2) count the number of recruits in the plot in year $t + 1$; and then (3) estimate an establishment probability as the ratio of recruits in year $t + 1$ to seeds produced in year t. The common approach used in long-term grassland studies has been to model recruitment at the scale of the plot. Adler and colleagues (2007) analyzed permanent chart quadrats where the locations of individual plants are mapped each year (recall Figure 5.2). These quadrats were originally promoted by Frederick Clements (1905), and were applied throughout rangelands of the western

United States.[1] Adler worked hard to rediscover many of these quadrats from sites in Kansas, Idaho, Colorado, and New Mexico, USA, and used the old maps to analyze the demography of long-lived grassland species (Adler et al., 2004; Adler et al., 2006; Adler et al., 2007; Adler et al., 2010; Adler et al., 2018a). Moore and colleagues (2022) also relocated many of these same chart quadrats throughout Arizona, USA and revived their mapping. The basal area of established plants is mapped every year on these chart quadrats, so it is relatively straightforward to monitor survival and growth of individual plants. Bur recruitment has been a challenging demographic rate to model. New recruits into the quadrats are detected by the mapping methods, but it is impossible to know from which mother the recruit came. A perhaps more serious problem is that these maps do not include information on number of seeds produced per flowering plant, so the standard way to estimate establishment probability cannot be used. To solve this problem, Adler has modeled these recruits at the plot-scale, which predicts the number of recruits based on plot-level basal cover of each species. To try to account for the fact that the recruits could be coming from plants outside the plot, he used a mixing fraction of basal cover in the plot to the basal cover on nearby quadrats. This method relies on having a sample of plots within a landscape to accurately model this mixing fraction. This method accounts for high reproduction of a species in a quadrat even when most of the flowering individuals are located outside the quadrat.

Because of differences in how herbaceous plants and trees are measured, there is another difficulty in comparing tree recruitment to other species: tree demographic modelers often consider recruits to be trees that enter the sapling phase (> 2.54 cm diameter at breast height), rather than seedlings that emerge from the ground. This introduces a lag into how individuals produce tree recruits in tree demographic models and might bias comparisons of establishment probabilities across herbs and woody plants. You may recall from the Preface that this book began with a discussion of the low probabilities of plant establishment. We have now come full circle.

The estimated values of establishment probabilities likely have high uncertainties in their estimates and also vary strongly from year to year and site to site. Modeling these dynamics is one of the great challenges facing plant ecologists in the coming century, for recruitment is a key process that will enable plants to shift their distributions in response to climate change. Average establishment probabilities have been estimated to be as low as 0.0140 yr^{-1} for *Bouteloua gracilis* (Lauenroth et al., 1994) and as high as 0.295 yr^{-1} for *Bouteloua eriopoda* (Lauenroth et al., 1997). I extracted 18 estimates of establishment probabilities from four studies (Hesse et al., 2008; Merow et al., 2014a; Sheth and Angert, 2018; Levin et al., 2019) contained within PADRINO, the open access database of integral projection models (Levin et al., 2022), and the values ranged from 0.00 to 0.99, with a median establishment probability of 0.0001076. The peak of this distribution will be very close to zero with a tail that extends all the way to one. Understanding seedling establishment success as a function of traits within specific environmental contexts will not only improve ecological theory, but also will enhance our

---

[1] You can read a thorough review of the deployment of these quadrats in White (1985a).

ability to successfully restore degraded ecosystems (Larson et al., 2021; Shackelford et al., 2021).

All in all, plant recruitment is a spatially and temporally dynamic process that poses serious challenges for linking traits to demography. However, if we continue to focus most of our research effort into understanding the traits that govern growth and survival at the expense of recruitment, we are missing half of the story. We need to get the fecundity function right.

12.8 Chapter summary

- Modeling plant recruitment is hard. Recruitment is often the weakest link in plant population models and has been ignored in plant strategy models that have emphasized relative growth rate and survival. If we can successfully integrate traits into our understanding of plant fecundity, then we can fill a critical gap in plant strategy theory.

- One of the difficulties of incorporating regeneration into plant strategy theory is that it is well established that traits of juveniles and traits of mature plants are decoupled. In other words, we generally cannot predict the behavior of seedlings based on the traits of the adults.

- Seed mass can vary by 10 orders of magnitude globally. There is more variation in seed size than in any other plant trait, and this massive variation is no doubt part of the reason that understanding variation in the causes and consequences of seed size has captured the attention and imagination of ecologists across the world. A plant can either produce hundreds of small seeds or a handful of large seeds at any given reproductive event. Small herbaceous plants cannot support very large seeds due to biomechanical constraints, but they can produce many lightweight, small seeds.

- There is empirical support that large-seeded species performed better under grazing pressure, in nutrient-poor soils, and when planted deep in the soil or under litter. Some studies also found that large-seeded species perform better when growing in established vegetation, in deep shade, or in dry environments.

- It is now clear that we must cast a wider net in the search for traits that predict recruitment by considering dispersal and establishment, persistence, and germination timing. A trait–filter framework was proposed for identifying traits that are important at each stage of the recruitment process, including seed production, dispersal, germination, emergence, seedling establishment, and clonality.

- Seed dormancy drives variable recruitment through time. Seed dormancy has been shown to be negatively correlated with lifespan and vegetative propagation. Short-lived species tend to produce seeds that can remain dormant for many years whereas long-lived species, in particular clonal plants, tend to produce short-lived seeds, indicating a possible trade-off between sexual reproduction and vegetative propagation.

- Small-seeded understory plants only germinate in open sites with high r:fr ratios, which are temporally rare and only occur when gaps in the canopy open due to tree falls, necessitating the persistence of the small seeds in the soil seed bank. Germination in the large-seeded plants is not affected by the r:fr ratio and have a shorter lifespan in the seed bank.

- Dispersal drives variable recruitment through space. The diversity of dispersal syndromes in plants is testament to the selection pressures that plants face in their quest for safe sites for establishment.

- The standard approach for estimating establishment probability is to estimate seed production on the plot across all the individuals in the plot in year t, count the number of recruits in the plot in year $t + 1$, then take the ratio of recruits in year $t + 1$ to seeds produced in year t to estimate an establishment probability. These values can range from 0.00 to 0.99, with an empirical median establishment probability of 0.0001076. The peak of this distribution will be a spike very close to zero with a thin tail that extends all the way to one.

- Plant recruitment is a spatially and temporally dynamic process that poses serious challenges for linking traits to demography. However, if we continue to focus most of our research effort into understanding the traits that govern growth and survival at the expense of recruitment, we are missing half of the story. We need to get the fecundity function right.

12.9 Questions

1. Consider your favorite plant. What conditions must it overcome to successfully produce seed, disperse, germinate, emerge, and establish as a seedling? If it is a clonal plant, what conditions favor clonal propagation?

2. How could you incorporate dispersal syndrome into a fecundity kernel for multiple species in an integral projection model?

3. What traits, beyond seed mass, are critical for understanding plant recruitment?

Epilogue

If you have been lucky enough to travel around the world, you have had the opportunity to acquaint yourself with unrelated plant species from different continents that share similar traits because they evolved under similar environmental selection pressures. Such observations of convergent evolution have inspired multiple generations of plant ecologists to explain the processes that produced the diversity of plant form and function that we observe today.

Plant strategy models that were developed early in the search for plant strategies emphasized either demographic life history traits to explain temporal dynamics in response to disturbances, or they emphasized functional traits to explain responses to gradients in resource availability. Many plant strategy models were successfully developed in specific ecosystems, but the problem became much more difficult when the goal broadened to explain global variation in plant form and function. The objective of this book is to articulate a framework, applicable to all plants globally, to describe how functional traits determine plant demographic responses to environmental gradients and to their neighbors.

The jury may still be out over the power of traits to predict plant population performance, but we know that traits matter. Their importance has been confirmed for centuries and the examples of phenotypic adaptations described in this book demonstrate that. Yet, we often remain humbled by the vast natural variation observed in this world, much of it remaining unexplained by our most sophisticated statistical models.

If you have read this book from cover to cover, it is my hope that you are convinced that plant strategies cannot easily be reduced to simple geometries. It is wishful thinking for a triangle to explain a tesseract. It may be disappointing that we cannot draw plant strategies on a whiteboard, but it is unsurprising, given the riotous functional diversity of plants. Plant strategies are multidimensional phenotypes that yield demographic persistence in the environments to which they are most adapted. Testing the explanatory power of traits will require diligence, patience, and perseverance to keep measuring important traits, to keep measuring the environmental conditions that matter to plants, and to link these traits and environments to population performance. A robust test of plant strategy theory is a two-step process.

First, we must determine what is phenotypically possible by determining the dimensionality of plant traits. Many traits are correlated and redundant, so this step simplifies our search for the leading dimensions of plant form and function. Tremendous progress has been made in the last few decades regarding our understanding of plant trait dimensionality. However, we must not become complacent and rely on the biased sample of traits that are available in current global databases, especially regarding the difficult-to-measure traits, such as traits of fine roots, clonal growth organs, and bud banks. We must continue to do the hard work of measuring traits on local plants and share our data globally. We must also continue to determine what is demographically possible and advance a life history theory that adequately describes the boundaries of demographic behavior in plants.

Second, we must test the power of traits for predicting population growth rates across gradients of resource limitation, disturbance regimes, and temperature, and across gradients of densities and frequencies of other plant strategies. Determining which, where, and when traits matter for explaining population performance is the critical step for advancing the explanatory power of plant strategy theory. This approach holds great promise for making general predictions of phenotypic responses to global change. Observational demographic datasets often lack measurements of population declines outside their natural range of environmental conditions precisely because the species cannot live in those conditions; however, observing failed introductions into unsuitable habitats is necessary for mapping the contours of fitness landscapes. Experimental common gardens beyond species ranges may provide the gold standard because they provide experimental control over environmental conditions, competitive interactions, and can force the observations of demographic failures. That said, observational datasets that are sufficiently large should also capture demographic failures. I believe that a synthesis of empirical modeling with game theoretical approaches will provide the most compelling path toward the maturation of plant strategy theory.

This book began with a description of the classic duels that occurred between population ecologists and comparative functional ecologists. Any attempt to merge these two disciplines runs the risk that plant demographers may think the depths of demographic mechanisms have not adequately been plunged, and the functional ecologists might think the discussion of physiological processes has barely scratched the surface. Nevertheless, this book provides a clear path toward synthesizing these disciplines and highlights recent progress, but a new schism is already emerging among the very scientists that are forging this path. The theoreticians' assumptions and model complexities raise suspicions among the empiricists, and the empiricists' confounding factors and inability to control for competitive interactions raise doubts among the theoreticians. Continued collaboration between these two groups—long known for their feuds in many areas of science—is needed for one very important reason: the youngest generations have inherited a planet under threat.

Every effort must be made to conserve declining species and restore degraded ecosystems or else humankind and all other species with which we share this planet may be the culprit of planetary upheaval. Plant strategies are directly relevant to conservation biology and restoration ecology because strategies can be applied to generalize conservation actions across thousands of species in danger of extinction. Indeed, the strongest test of plant strategy theory may be the application of traits to restore biodiversity in ecosystems that have been disturbed by human activities and where climate change is causing unprecedented plant mortality events. Plant strategy theory will mature at pace when we test our predictions with new observations, successfully apply the theory in the field to yield positive conservation outcomes, and adapt our management plans on the ground. Let us not confine plant strategies to pages in books, but rather let us apply this knowledge to restore the remarkable diversity of plants on our precious planet Earth.

References

Abbey, E. (1977). *The journey home: Some words in defense of the American West*. New York: Dutton Adult.

Abella, S.R. and Fornwalt, P.J. (2015). "Ten years of vegetation assembly after a North American mega fire," *Global Change Biology*, 21, 789–802.

Abrahamson, W.G. and Gadgil, M. (1973). "Growth form and reproductive effort in goldenrods (*Solidago*, Compositae)," *The American Naturalist*, 107, 651–661.

Ackerly, D.D. (2003). "Community assembly, niche conservatism, and adaptive evolution in changing environments," *International Journal of Plant Sciences*, 164, S165–S184.

Ackerly, D.D. (2009). "Conservatism and diversification of plant functional traits: Evolutionary rates versus phylogenetic signal," *Proceedings of the National Academy of Sciences of the United States of America*, 106, 19699–19706.

Ackerly, D.D. and Donoghue, M.J. (1995). "Phylogeny and ecology reconsidered," *Journal of Ecology*, 83, 730–733.

Ackerly, D.D. and Monson, R.K. (2003). "Waking the sleeping giant: The evolutionary foundations of plant function," *International Journal of Plant Sciences*, 164, S1–S6.

Ackerly, D.D., et al. (2000). "The evolution of plant ecophysiological traits: Recent advances and future directions," *Bioscience*, 50, 979–995.

Adams, H.D., et al. (2017). "A multi-species synthesis of physiological mechanisms in drought-induced tree mortality," *Nature Ecology & Evolution*, 1, 1285–1291.

Adams, R.P., et al. (1981). "The south-western USA and northern Mexico one-seeded junipers: Their volatile oils and evolution," *Biochemical Systematics and Ecology*, 9, 93–96.

Adams, T.P., Purves, D.W., and Pacala, S.W. (2007). "Understanding height-structured competition in forests: is there an R* for light?" *Proceedings of the Royal Society B: Biological Sciences*, 274, 3039–3047.

Adams, V.M., Marsh, D.M., and Knox, J.S. (2005). "Importance of the seed bank for population viability and population monitoring in a threatened wetland herb," *Biological Conservation*, 124, 425–436.

Adler, P.B., Ellner, S.P., and Levine, J.M. (2010). "Coexistence of perennial plants: an embarrassment of niches," *Ecology Letters*, 13, 1019–1029.

Adler, P.B., Tyburczy, W.R., and Laurenroth, W.K. (2007). "Long-term mapped quadrats from Kansas prairie: A unique source of demographic information for herbaceous plants," *Ecology*, 88, 2673.

Adler, P.B., et al. (2004). "Functional traits of graminoids in semi-arid steppes: a test of grazing histories," *Journal of Applied Ecology*, 41, 653–663.

Adler, P.B., et al. (2006). "Climate variability has a stabilizing effect on the coexistence of prairie grasses," *Proceedings of the National Academy of Sciences of the United States of America*, 103, 12793–12798.

Adler, P.B., et al. (2014). "Functional traits explain variation in plant life history strategies," *Proceedings of the National Academy of Sciences of the United States of America*, 111, 740–745.

Adler, P.B., et al. (2018a). "Weak interspecific interactions in a sagebrush steppe? Conflicting evidence from observations and experiments," *Ecology*, 99, 1621–1632.

Adler, P.B., et al. (2018b). "Competition and coexistence in plant communities: intraspecific competition is stronger than interspecific competition," *Ecology Letters*, 21, 1319–1329.

Aerts, R. (1995). "The advantages of being evergreen," *Trends in Ecology & Evolution*, 10, 402–407.

Aerts, R. and Chapin, F.S. (1999). "The mineral nutrition of wild plants revisited: A re-evaluation of processes and patterns," *Advances in Ecological Research*, 30, 1–67.

Aerts, R. and van der Peijl, M.J. (1993). "A simple model to explain the dominance of low-productive perennials in nutrient-poor habitats," *Oikos*, 66, 144–147.

Aerts, R., Boot, R.G.A., and van der Aart, P.J.M. (1991). "The relation between above- and belowground biomass allocation patterns and competitive ability," *Oecologia*, 87, 551–559.

Agee, J.K. (1996). *Fire ecology of Pacific Northwest forests*. Washington, DC: Island Press.

Agrawal, A.A. (2007). "Macroevolution of plant defense strategies," *Trends in Ecology & Evolution*, 22, 103–109.

Agrawal, A.A. (2020). "A scale-dependent framework for trade-offs, syndromes, and specialization in organismal biology," *Ecology*, 101, e02924.

Agrawal, A.A. and Fishbein, M. (2006). "Plant defense syndromes," *Ecology*, 87, S132–S149.

Agrawal, A.A., Conner, J.K., and Rasmann, S. (2010). "Tradeoffs and negative correlations in evolutionary ecology," in Bell, M.A., et al. (eds.) *Evolution since Darwin: The first 150 years*. Sunderland, MA: Sinauer Associates, Inc., pp. 243–268.

Agrawal, A.A., et al. (2012). "Toxic cardenolides: chemical ecology and coevolution of specialized plant–herbivore interactions," *New Phytologist*, 194, 28–45.

Aitken, S.N., et al. (2008). "Adaptation, migration or extirpation: climate change outcomes for tree populations," *Evolutionary Applications*, 1, 95–111.

Akin, W.E. (1991). *Global patterns: climate, vegetation and soils*. Norman: University of Oklahoma Press.

Akman, M., et al. (2012). "Wait or escape? Contrasting submergence tolerance strategies of *Rorippa amphibia*, *Rorippa sylvestris* and their hybrid," *Annals of Botany*, 109, 1263–1276.

Alexander, J.M., et al. (2016). "When climate reshuffles competitors: A call for experimental macroecology," *Trends in Ecology & Evolution*, 31, 831–841.

Algeo, T.J. and Scheckler, S.E. (1998). "Terrestrial-marine teleconnections in the Devonian: links between the evolution of land plants, weathering processes, and marine anoxic events," *Philosophical Transactions of the Royal Society of London. Series B: Biological Sciences*, 353, 113–130.

Allen, C.D., et al. (2010). "A global overview of drought and heat-induced tree mortality reveals emerging climate change risks for forests," *Forest Ecology and Management*, 259, 660–684.

Allsopp, N. and Stock, W. (1995). "Relationships between seed reserves, seedling growth and mycorrhizal responses in 14 related shrubs (Rosidae) from a low-nutrient environment," *Functional Ecology*, 9, 248–254.

Ally, D., Ritland, K., and Otto, S.P. (2008). "Can clone size serve as a proxy for clone age? An exploration using microsatellite divergence in *Populus tremuloides*," *Molecular Ecology*, 17, 4897–4911.

Alvarez-Clare, S. and Kitajima, K. (2007). "Physical defence traits enhance seedling survival of neotropical tree species," *Functional Ecology*, 21, 1044–1054.

Alves, R.J.V., et al. (2013). "Longevity of the Brazilian underground tree *Jacaranda decurrens* Cham," *Anais da Academia Brasileira de Ciências*, 85, 671–678.

Anderegg, L.D.L., et al. (2022). "Representing plant diversity in land models: An evolutionary approach to make 'functional types' more functional," *Global Change Biology*, 28, 2541–2554.

Anderegg, W.R.L., Berry, J.A., and Field, C.B. (2012). "Linking definitions, mechanisms, and modeling of drought-induced tree death," *Trends in Plant Science*, 17, 693–700.

Anderegg, W.R.L., Kane, J.M., and Anderegg, L.D.L. (2013). "Consequences of widespread tree mortality triggered by drought and temperature stress," *Nature Climate Change*, 3, 30–36.

Anderson, J., McClaran, M.P., and Adler, P.B. (2012). "Cover and density of semi-desert grassland plants in permanent quadrats mapped from 1915 to 1947," *Ecology*, 93, 1492-1492.

Anderson, J.T. (2016). "Plant fitness in a rapidly changing world," *New Phytologist*, 210, 81–87.

Andrew, S.C., et al. (2022). "Assessing the vulnerability of plant functional trait strategies to climate change," *Global Ecology and Biogeography*, 31, 1194–1206.

Andrewartha, H.G. and Birch, L.C. (1954). *The distribution and abundance of animals*. Chicago, IL: University of Chicago Press.

Andrews, H.N. (1961). *Studies in paleobotany*. New York: Wiley.

Angert, A.L. (2006). "Demography of central and marginal population of monkeyflowers (*Mimulus cardinalis* and *M. lewisii*)," *Ecology*, 87, 2014–2025.

Angert, A.L. (2009). "The niche, limits to species' distributions, and spatiotemporal variation in demography across the elevation ranges of two monkeyflowers," *Proceedings of the National Academy of Sciences of the United States of America*, 106, 19693.

Angert, A.L. and Schemske, D.W. (2005). "The evolution of species' distributions: Reciprocal transplants across the elevation ranges of *Mimulus cardinalis* and *M. lewisii*," *Evolution*, 59, 1671–1684.

Angert, A.L., et al. (2009). "Functional tradeoffs determine species coexistence via the storage effect," *Proceedings of the National Academy of Sciences of the United States of America*, 106, 11641–11645.

Anten, N.P.R. (2016). "Optimization and game theory in canopy models," in Hikosaka, K., Niinemets, Ü., and Anten, N.P.R. (eds.) *Canopy photosynthesis: From basics to applications*. Dordrecht: Springer, pp. 355–377.

Arber, A. (1950). *The natural philosophy of plant form*. Cambridge: Cambridge University Press.

Archibald, S. and Bond, W.J. (2003). "Growing tall vs growing wide: Tree architecture and allometry of *Acacia karroo* in forest, savanna, and arid environments," *Oikos*, 102, 3–14.

Archibald, S., Hempson, G.P., and Lehmann, C. (2019). "A unified framework for plant life-history strategies shaped by fire and herbivory," *New Phytologist*, 224, 1490–1503.

Archibald, S., et al. (2013). "Defining pyromes and global syndromes of fire regimes," *Proceedings of the National Academy of Sciences of the United States of America*, 110, 6442.

Arcus, V.L. and Mulholland, A.J. (2020). "Temperature, dynamics, and enzyme-catalyzed reaction rates," *Annual Review of Biophysics*, 49, 163–180.

Armstrong, D.P. and Westoby, M. (1993). "Seedlings from large seeds tolerated defoliation better: A test using phylogenetically independent contrasts," *Ecology*, 74, 1092–1100.

Arnold, S.J. (1983). "Morphology, performance and fitness," *American Zoologist*, 23, 347–361.

Arnold, S.J. (2003). "Performance surfaces and adaptive landscapes," *Integrative and Comparative Biology*, 43, 367–375.

Arroyo, J.I., et al. (2022). "A general theory for temperature dependence in biology," *Proceedings of the National Academy of Sciences of the United States of America*, 119, e2119872119.

Attiwill, P.M. and Adams, M.A. (2013). "Mega-fires, inquiries and politics in the eucalypt forests of Victoria, south-eastern Australia," *Forest Ecology and Management*, 294, 45–53.

Augé, R.M., Toler, H.D., and Saxton, A.M. (2015). "Arbuscular mycorrhizal symbiosis alters stomatal conductance of host plants more under drought than under amply watered conditions: A meta-analysis," *Mycorrhiza*, 25, 13–24.

Augustine, S.M. (2016). "Function of heat-shock proteins in drought tolerance regulation of plants," in Hossain, M.A., et al. (eds.) *Drought stress tolerance in plants*, Vol 1: *Physiology and biochemistry*. Cham: Springer International Publishing, pp. 163–185.

Austin, M.P. (1990). Community theory and competition in vegetation. in Grace, J.B. and Tilman, D. (eds.) *Perspectives on plant competition*. San Diego: Academic Press, Inc., pp. 215–239.

Austin, M.P. (1999). "The potential contribution of vegetation ecology to biodiversity research," *Ecography*, 22, 465–484.

Austin, M.P. and Smith, T.M. (1989). "A new model for the continuum concept," *Vegetatio*, 83, 35–47.

Baas, P., et al. (2004). "Evolution of xylem physiology," in Hemsley, A. and Poole, I. (eds.) *The evolution of plant physiology*. London: Elsevier Academic Press, pp. 273–295.

Bagousse-Pinguet, Y.L., et al. (2021). "Functional rarity and evenness are key facets of biodiversity to boost multifunctionality," *Proceedings of the National Academy of Sciences of the United States of America*, 118, e2019355118.

Bahr, L.M. (1982). "Functional taxonomy: An immodest proposal," *Ecological Modelling*, 15, 211–233.

Bailey-Serres, J. and Voesenek, L. (2008). "Flooding stress: Acclimations and genetic diversity," *Annual Review of Plant Biology*, 59, 313–339.

Baker, H.G. (1972). "Seed weight in relation to environmental conditions in California," *Ecology*, 53, 997–1010.

Balachowski, J.A. and Volaire, F.A. (2018). "Implications of plant functional traits and drought survival strategies for ecological restoration," *Journal of Applied Ecology*, 55, 631–640.

Balazs, K.R., et al. (2022). "Directional selection shifts trait distributions of planted species in dryland restoration," *Journal of Ecology*, 110, 540–552.

Balazs, K.R., et al. (2020). "The right trait in the right place at the right time: Matching traits to environment improves restoration outcomes," *Ecological Applications*, 30, e02110.

Balding, M. and Williams, K.J.H. (2016). "Plant blindness and the implications for plant conservation," *Conservation Biology*, 30, 1192–1199.

Ballhorn, D.J., et al. (2016). "Friend or foe—light availability determines the relationship between mycorrhizal fungi, rhizobia and lima bean (*Phaseolus lunatus* L.)," *PLoS One*, 11, e0154116.

Baltzer, J.L. and Thomas, S.C. (2007b). "Determinants of whole-plant light requirements in Bornean rain forest tree saplings," *Journal of Ecology*, 95, 1208–1221.

Baltzer, J.L. and Thomas, S.C. (2007a). "Physiological and morphological correlates of whole-plant light compensation point in temperate deciduous tree seedlings," *Oecologia*, 153, 209–223.

Baquedano, F. and Castillo, F. (2006). "Comparative ecophysiological effects of drought on seedlings of the Mediterranean water-saver *Pinus halepensis* and water-spenders *Quercus coccifera* and *Quercus ilex*," *Trees*, 20, 689.

Barabás, G., D'Andrea, R., and Stump, S.M. (2018). "Chesson's coexistence theory," *Ecological Monographs*, 88, 277–303.

Barkman, J. (1988). "New systems of plant growth forms and phenological plant types," in Werger, M. (ed.) *Plant form and vegetation structure*. The Hague: SPB Academic Publishing, pp. 9–44.

Barrett, S.C.H., Harder, L.D., and Worley, A.C. (1997). "The comparative biology of pollination and mating in flowering plants," in Silvertown, J., Franco, M., and Harper, J. (eds.) *Plant life histories: Ecology, phylogeny, and evolution*. Cambridge: Cambridge University Press, pp. 57–76.

Bartlett, M.K., et al. (2012a). "Rapid determination of comparative drought tolerance traits: using an osmometer to predict turgor loss point," *Methods in Ecology and Evolution*, 3, 880–888.

Bartlett, M.K., Scoffoni, C., and Sack, L. (2012b). "The determinants of leaf turgor loss point and prediction of drought tolerance of species and biomes: A global meta-analysis," *Ecology Letters*, 15, 393–405.

Baskin, C.C. and Baskin, J.M. (1998). *Seeds: Ecology, biogeography, and evolution of dormancy and germination*. San Diego: Academic Press.

Baudisch, A., et al. (2013). "The pace and shape of senescence in angiosperms," *Journal of Ecology*, 101, 596–606.

Bäurle, I. and Laux, T. (2003). "Apical meristems: The plant's fountain of youth," *Bioessays*, 25, 961–970.

Baylis, G.T.S. (1975). "The magnolioid mycorrhiza and mycotrophy in root systems derived from it," in Sanders, F.E., Mosse, B., and Tinker, P.B. (eds.) *Endomycorrhizas: Proceedings of a symposium*. London: Academic Press, pp. 373–389.

Bazzaz, F.A. (1979). "The physiological ecology of plant succession," *Annual Review of Ecology and Systematics*, 10, 351–371.

Bazzaz, F.A. (1996). *Plants in changing environments: Linking physiological, population, and community ecology*. Cambridge: Cambridge University Press.

Bazzaz, F.A, et al. (1987). "Allocating resources to reproduction and defense," *Bioscience*, 37, 58–67.

Beck, J.J., et al. (2020). "Asymmetrical pegs in square holes? Functional and phylogenetic determinants of plant community assembly in temperate forest understories," *bioRxiv* [online], 2020.2010.2014.312512.

Bell, T.J., Bowles, M.L., and McEachern, A.K. (2003). "Projecting the success of plant population restoration with viability analysis," in Brigham, C.A. and Schwartz, M.W. (eds.) *Population viability in plants*. Berlin: Springer, pp. 313–348.

Bellingham, P.J. and Sparrow, A.D. (2000). "Resprouting as a life history strategy in woody plant communities," *Oikos*, 89, 409–416.

Bellman, R.E. (1961). *Adaptive control processes: A guided tour*. Princeton: Princeton University Press.

Berendse, F. and Elberse, W.T. (1990). "Competition and nutrient availability in heathland and grassland ecosystems," in Grace, J.B. and D. Tilman, (eds.) *Perspectives on plant competition*. San Diego: Academic Press, Inc., pp. 93–116.

Berendse, F., Elberse, W.T., and Geerts, R. (1992). "Competition and nitrogen loss from plants in grassland ecosystems," *Ecology*, 73, 46–53.

Bergmann, J., et al. (2020). "The fungal collaboration gradient dominates the root economics space in plants," *Science Advances*, 6, eaba3756.

Beyer, M., et al. (2018). "Examination of deep root water uptake using anomalies of soil water stable isotopes, depth-controlled isotopic labeling and mixing models," *Journal of Hydrology*, 566, 122–136.

Bidlack, J.E. and Jansky, S.H. (2011). *Stern's introductory plant biology*. 12th edn. New York: McGraw-Hill.

Bierzychudek, P. (1982). "Life histories and demography of shade-tolerant temperate forest herbs: A review," *New Phytologist*, 90, 757–776.

Binkley, D. (2021). *Forest ecology: An evidence-based approach.* Oxford: John Wiley & Sons, Ltd.

Birch, L.C. (1960). "The genetic factor in population ecology," *The American Naturalist,* 94, 5–24.

Blackman, C.J., Brodribb, T.J., and Jordan, G.J. (2010). "Leaf hydraulic vulnerability is related to conduit dimensions and drought resistance across a diverse range of woody angiosperms," *New Phytologist,* 188, 1113–1123.

Blackman, V.H. (1919). "The compound interest law and plant growth," *Annals of Botany,* 33, 353–360.

Blomberg, S.P., et al. (2012). "Independent contrasts and PGLS regression estimators are equivalent," *Systematic Biology,* 61, 382–391.

Blonder, B., et al. (2018). "Microenvironment and functional-trait context dependence predict alpine plant community dynamics," *Journal of Ecology,* 106, 1323–1337.

Blonder, B., et al. (2014). "Plant ecological strategies shift across the Cretaceous–Paleogene boundary," *PLoS Biology,* 12, e1001949.

Blonder, B., et al. (2011). "Venation networks and the origin of the leaf economics spectrum," *Ecology Letters,* 14, 91–100.

Bloom, A.J., Chapin, F.S., and Mooney, H.A. (1985). "Resource limitation in plants—an economic analogy," *Annual Review of Ecology and Systematics,* 16, 363–392.

Blumenthal, D.M., et al. (2020). "Traits link drought resistance with herbivore defence and plant economics in semi-arid grasslands: The central roles of phenology and leaf dry matter content," *Journal of Ecology,* 108, 2336–2351.

Bochet, E. and García-Fayos, P. (2015). "Identifying plant traits: A key aspect for species selection in restoration of eroded roadsides in semiarid environments," *Ecological Engineering,* 83, 444–451.

Bocsi, T., et al. (2016). "Plants' native distributions do not reflect climatic tolerance," *Diversity and Distributions,* 22, 615–624.

Bolker, B.M. and Pacala, S.W. (1999). "Spatial moment equations for plant competition: understanding spatial strategies and the advantages of short dispersal," *The American Naturalist,* 153, 575–602.

Bombo, A.B., Siebert, F., and Fidelis, A. (2022). "Fire and herbivory shape belowground bud banks in a semi-arid African savanna," *African Journal of Range & Forage Science,* 39, 16–26.

Bond, W.J. (1997). "Functional types for predicting changes in biodiversity: A case study of Cape fynbos," in Smith, T.M., Shugart, H.H., and Woodward, F.I. (eds.) *Plant functional types: their relevance to ecosystem properties and global change.* Cambridge: Cambridge University Press, pp. 174–194.

Bond, W.J. (2005). "Large parts of the world are brown or black: A different view on the 'Green World' hypothesis," *Journal of Vegetation Science,* 16, 261–266.

Bond, W.J. (2019). *Open ecosystems: ecology and evolution beyond the forest edge.* Oxford: Oxford University Press.

Bond, W.J. (2021). "Out of the shadows: ecology of open ecosystems," *Plant Ecology & Diversity,* 14, 205–222.

Bond, W.J. and Keeley, J.E. (2005). "Fire as a global 'herbivore': The ecology and evolution of flammable ecosystems," *Trends in Ecology & Evolution,* 20, 387–394.

Bond, W.J. and van Wilgen, B.W. (1996). *Fire and plants.* London: Chapman and Hall.

Bond, W.J., Woodward, F.I., and Midgley, G.F. (2005). "The global distribution of ecosystems in a world without fire," *New Phytologist,* 165, 525–538.

Bonser, S.P. and Aarssen, L.W. (1996). "Meristem allocation: A new classification theory for adaptive strategies in herbaceous plants," *Oikos*, 77, 347–352.

Borer, E.T., et al. (2014). "Herbivores and nutrients control grassland plant diversity via light limitation," *Nature*, 508, 517–520.

Bormann, F.H. and Likens, G.E. (1994). *Pattern and process in a forested ecosystem: Disturbance, development and the steady state based on the Hubbard Brook ecosystem study.* New York: Springer Science & Business Media.

Both, A.J., et al. (2015). "Guidelines for measuring and reporting environmental parameters for experiments in greenhouses," *Plant Methods*, 11, 43.

Bowles, A.M.C., Paps, J., and Bechtold, U. (2022). "Water-related innovations in land plants evolved by different patterns of gene cooption and novelty," *New Phytologist*, 235, 732–742.

Box, E.O. (1981). *Macroclimate and plant forms: An introduction to predictive modeling in phytogeography.* London: Springer Science & Business Media.

Box, G.E.P. (1976). "Science and statistics," *Journal of the American Statistical Association*, 71, 791–799.

Boyce, M.S. (1984). "Restitution of r-and K-selection as a model of density-dependent natural selection," *Annual Review of Ecology and Systematics*, 15, 427–447.

Bradford, K.J. (2002). "Applications of hydrothermal time to quantifying and modeling seed germination and dormancy," *Weed Science*, 50, 248–260.

Bradshaw, A.D. (1987a). "Comparison—its scope and limits," *New Phytologist*, 106, 3–21.

Bradshaw, A.D. (1987b). "Restoration: An acid test for ecology," in Jordan, W.R., Gilpin, M.E., and Aber, J.D. (eds.) *Restoration ecology: A synthetic approach to ecological research.* Cambridge: Cambridge University Press, pp. 23–30.

Braga, J., et al. (2018). "Integrating spatial and phylogenetic information in the fourth-corner analysis to test trait–environment relationships," *Ecology*, 99, 2667–2674.

Braun-Blanquet, J. (1964). *Pflanzensoziologie.* New York: Aufl. Springer Wein.

Breckle, S-W. (2002). *Walter's vegetation of the Earth.* 4th edn. New York: Springer.

Breshears, D.D., et al. (2005). "Regional vegetation die-off in response to global-change-type drought," *Proceedings of the National Academy of Sciences of the United States of America*, 102, 15144–15148.

Brigham, C.A. and Schwartz, M.W. editors. (2003). *Population viability in plants: Conservation, management, and modeling of rare plants.* Berlin: Springer-Verlag.

Brondizio, E.S., Settele, J., and Díaz, S., (eds.) (2019). *Global assessment report on biodiversity and ecosystem services of the Intergovernmental Science-Policy Platform on Biodiversity and Ecosystem Services* (Version 1). Bonn: IPBES secretariat.

Brooks, A. and Farquhar, G.D. (1985). "Effect of temperature on the CO^2/O^2 specificity of ribulose-1,5-bisphosphate carboxylase/oxygenase and the rate of respiration in the light," *Planta*, 165, 397–406.

Brown, J.S. and Vincent, T.L. (1987). "Coevolution as an evolutionary game," *Evolution*, 41, 66–79.

Brudvig, L.A., et al. (2017). "Interpreting variation to advance predictive restoration science," *Journal of Applied Ecology*, 54, 1018–1027.

Bruelheide, H., et al. (2018). "Global trait–environment relationships of plant communities," *Nature Ecology & Evolution*, 2, 1906–1917.

Bruna, E.M., Fiske, I.J., and Trager, M.D. (2009). "Habitat fragmentation and plant populations: is what we know demographically irrelevant?" *Journal of Vegetation Science*, 20, 569–576.

Bucharova, A. (2017). "Assisted migration within species range ignores biotic interactions and lacks evidence," *Restoration Ecology*, 25, 14–18.

Bueno, A., et al. (2019). "Effects of temperature on the cuticular transpiration barrier of two desert plants with water-spender and water-saver strategies," *Journal of Experimental Botany*, 70, 1613–1625.

Bullock, J.M., et al. (1995). "Gap colonization as a source of grassland community change: Effects of gap size and grazing on the rate and mode of colonization by different species," *Oikos*, 72, 273–282.

Burghardt, L.T. and Metcalf, C.J.E. (2017). "The evolution of senescence in annual plants: the importance of phenology and the potential for plasticity," in Shefferson, R., Jones, O., and Salguero-Gómez, R. (eds.) *The evolution of senescence in the tree of life*. Cambridge: Cambridge University Press, pp. 284–302.

Burke, M.J.W. and Grime, J.P. (1996). "An experimental study of plant community invasibility," *Ecology*, 77, 776–790.

Burrows, G.E. (2013). "Buds, bushfires and resprouting in the eucalypts," *Australian Journal of Botany*, 61, 331–349.

Burrows, G.E. (2002). "Epicormic strand structure in *Angophora, Eucalyptus* and *Lophostemon* (Myrtaceae): Implications for fire resistance and recovery," *New Phytologist*, 153, 111–131.

Burton, A. (2016). "Staving off extinction," *Frontiers in Ecology and the Environment*, 14, 580–580.

Burton, P.J., Jentsch, A., and Walker, L.R. (2020). "The ecology of disturbance interactions," *Bioscience*, 70, 854–870.

Butler, E.E., et al. (2017). "Mapping local and global variability in plant trait distributions," *Proceedings of the National Academy of Sciences of the United States of America*, 114, E10937–E10946.

Butrim, M.J. and Royer, D.L. (2020). "Leaf-economic strategies across the Eocene–Oligocene transition correlate with dry season precipitation and paleoelevation," *American Journal of Botany*, 107, 1772–1785.

Butt, N. and Gallagher, R. (2018). "Using species traits to guide conservation actions under climate change," *Climatic Change*, 151, 317–332.

Butterfield, B.J. (2015). "Environmental filtering increases in intensity at both ends of climatic gradients, though driven by different factors, across woody vegetation types of the southwest USA," *Oikos*, 124, 1374–1382.

Butterfield, B.J. and Briggs, J.M. (2011). "Regeneration niche differentiates functional strategies of desert woody plant species," *Oecologia*, 165, 477–487.

Cadotte, M.W. and Davies, T.J. (2016). *Phylogenies in ecology: A guide to concepts and methods*. Princeton: Princeton University Press.

Cadotte, M.W., et al. (2006). "On testing the competition-colonization trade-off in a multispecies assemblage," *The American Naturalist*, 168, 704–709.

Cadotte, M.W., et al. (2015). "Predicting communities from functional traits," *Trends in Ecology & Evolution*, 30, 510–511.

Cain, S.A. (1950). "Life-forms and phytoclimate," *Botanical Review*, 16, 1–32.

Campbell, B.D. and Grime, J.P. (1992). "An experimental test of plant strategy theory," *Ecology*, 73, 15–29.

Campbell, B.D., et al. (1991). "The quest for a mechanistic understanding of resource competition in plant communities: The role of experiments," *Functional Ecology*, 5, 241–253.

Canadell, J., et al. (1996). "Maximum rooting depth of vegetation types at the global scale," *Oecologia*, 108, 583–595.

Carlquist, S. (1975). *Ecological strategies of xylem evolution*. Berkeley: University of California Press.

Carlquist, S., Baldwin, B.G., and Carr, G.D. (2003). *Tarweeds & silverswords: Evolution of the Madiinae (Asteraceae)*. St. Louis: Missouri Botanical Garden Press.

Carlucci, M.B., et al. (2020). Functional traits and ecosystem services in ecological restoration. *Restoration Ecology*, 28, 1372–1383.

Carmona, C.P., Azcárate, F.M., and Peco, B. (2013). "Does cattle dung cause differences between grazing increaser and decreaser germination response?" *Acta Oecologica*, 47, 1–7.

Carmona, C.P., et al. (2021). "Fine-root traits in the global spectrum of plant form and function," *Nature*, 597, 683–687.

Carter, T.A., et al. (2022). "Understory plant community responses to widespread spruce mortality in a subalpine forest," *Journal of Vegetation Science*, 33, e13109.

Caswell, H. (1989). "Analysis of life table response experiments I. Decomposition of effects on population growth rate," *Ecological Modelling*, 46, 221–237.

Caswell, H. (2001). *Matrix population models*. 2nd edn. Sunderland, MA: Sinauer Associates, Inc.

Caswell, H. and Salguero-Gómez, R. (2013). "Age, stage and senescence in plants," *Journal of Ecology*, 101, 585–595.

Caswell, H. and Shyu, E. (2017). "Senescence, selection gradients and mortality," *in* Shefferson, R., Jones, O., and Salguero-Gómez, R. (eds.) *The evolution of senescence in the tree of life*. Cambridge: Cambridge University Press, pp. 56–82.

Caswell, H. and Takada, T. (2004). "Elasticity analysis of density-dependent matrix population models: The invasion exponent and its substitutes," *Theoretical Population Biology*, 65, 401–411.

Catford, J.A., Jansson, R., and Nilsson, C. (2009). "Reducing redundancy in invasion ecology by integrating hypotheses into a single theoretical framework," *Diversity and Distributions*, 15, 22–40.

Catford, J.A., et al. (2020). "Community diversity outweighs effect of warming on plant colonization," *Global Change Biology*, 26, 3079–3090.

Catford, J.A., et al. (2016). "Disentangling the four demographic dimensions of species invasiveness," *Journal of Ecology*, 104, 1745–1758.

Caughlin, T.T., et al. (2019). "Landscape heterogeneity is key to forecasting outcomes of plant reintroduction," *Ecological Applications*, 29, e01850.

Caughlin, T.T., et al. (2015). "Loss of animal seed dispersal increases extinction risk in a tropical tree species due to pervasive negative density dependence across life stages," *Proceedings of the Royal Society B: Biological Sciences*, 282, 20142095.

Certini, G. (2005). "Effects of fire on properties of forest soils: A review," *Oecologia*, 143, 1–10.

Chalmandrier, L., et al. (2021). "Linking functional traits and demography to model species-rich communities," *Nature communications*, 12, 2724.

Chalmandrier, L., et al. (2022). "Predictions of biodiversity are improved by integrating trait-based competition with abiotic filtering," *Ecology Letters*, 25, 1277–1289.

Chang-Yang, C.H., et al. (2021). "Closing the life cycle of forest trees: The difficult dynamics of seedling-to-sapling transitions in a subtropical rainforest," *Journal of Ecology*, 109, 2705–2716.

Chapin, F.S. (1988). "Ecological aspects of plant mineral nutrition," *Advances in Plant Nutrition (USA)*, 3, 161–191.

Chapin, F.S. (1980). "The mineral nutrition of wild plants," *Annual Review of Ecology and Systematics*, 11, 233–260.

Chapin, F.S., Autumn, K., and Pugnaire, F. (1993). "Evolution of suites of traits in response to environmental stress," *The American Naturalist*, 142, S78–S92.

Charles-Dominique, T., et al. (2015a). "Bud protection: A key trait for species sorting in a forest-savanna mosaic," *New Phytologist*, 207, 1052–1060.

Charles-Dominique, T., et al. (2015b). "Functional differentiation of biomes in an African savanna/forest mosaic," *South African Journal of Botany*, 101, 82–90.

Charles-Dominique, T., et al. (2018). "Steal the light: Shade vs fire adapted vegetation in forest-savanna mosaics," *New Phytologist*, 218, 1419–1429.

Charlesworth, B. (1980). *Evolution in age-structured populations*. Cambridge: Cambridge University Press.

Charlesworth, B. (2000). "Fisher, Medawar, Hamilton and the evolution of aging," *Genetics*, 156, 927–931.

Charnov, E.L. (1993). *Life history invariants: Some explorations of symmetry in evolutionary ecology*. Oxford: Oxford University Press.

Charnov, E.L. and Schaffer, W.M. (1973). "Life-history consequences of natural selection: Cole's result revisited," *The American Naturalist*, 107, 791–793.

Chave, J., et al. (2009). "Towards a worldwide wood economics spectrum," *Ecology Letters*, 12, 351–366.

Chen, G. and Sun, W. (2018). "The role of botanical gardens in scientific research, conservation, and citizen science," *Plant Diversity*, 40, 181–188.

Chesson, P. (2000). "Mechanisms of maintenance of species diversity," *Annual Review of Ecology and Systematics*, 31, 343–366.

Chiba, N. and Hirose, T. (1993). "Nitrogen acquisition and use in three perennials in the early stage of primary succession," *Functional Ecology*, 7, 287–292.

Chisholm, R.A. and Fung, T. (2020). "Janzen–Connell effects are a weak impediment to competitive exclusion," *The American Naturalist*, 196, 649–661.

Choat, B. (2013). "Predicting thresholds of drought-induced mortality in woody plant species," *Tree Physiology*, 33, 669–671.

Choat, B., et al. (2012). "Global convergence in the vulnerability of forests to drought," *Nature*, 491, 752–755.

Choi, Y.D., et al. (2008). "Ecological restoration for future sustainability in a changing environment," *Ecoscience*, 15, 53–64.

Chu, C. and Adler, P.B. (2015). "Large niche differences emerge at the recruitment stage to stabilize grassland coexistence," *Ecological Monographs*, 85, 373–392.

Chu, C., et al. (2014). "Life form influences survivorship patterns for 109 herbaceous perennials from six semi-arid ecosystems," *Journal of Vegetation Science*, 25, 947–954.

Clapham, A. (1956). "Autecological studies and the biological flora of the British Isles," *Journal of Ecology*, 44, 1–11.

Clark, D.L., et al. (2012). "Plant traits—a tool for restoration?" *Applied Vegetation Science*, 15, 449–458.

Clarke, P.J., et al. (2005). "Landscape patterns of woody plant response to crown fire: Disturbance and productivity influence sprouting ability," *Journal of Ecology*, 93, 544–555.

Clarke, P.J., et al. (2013). "Resprouting as a key functional trait: How buds, protection and resources drive persistence after fire," *New Phytologist*, 197, 19–35.

Clarke, P.J., et al. (2015). "A synthesis of postfire recovery traits of woody plants in Australian ecosystems," *Science of the Total Environment*, 534, 31–42.

Clarkson, B.R. and Clarkson, B.D. (1995). "Recent vegetation changes on Mount Tarawera, Rotorua, New Zealand," *New Zealand Journal of Botany*, 33, 339–354.

Clements, F.E. (1920). *Plant indicators: The relation of plant communities to process and practice.* Washington, DC: Carnegie Institution of Washington.

Clements, F.E. (1905). *Research methods in ecology.* Lincoln, NE: University Publishing Co.

Cody, M. and Mooney, H. (1978). "Convergence versus nonconvergence in Mediterranean-climate ecosystems," *Annual Review of Ecology and Systematics*, 9, 265–321.

Cole, L.C. (1954). "The population consequences of life history phenomena," *The Quarterly Review of Biology*, 29, 103–137.

Coley, P.D. (1983). "Herbivory and defensive characteristics of tree species in a lowland tropical forest," *Ecological Monographs*, 53, 209–229.

Coley, P.D., Bryant, J.P., and Chapin, F.S. (1985). "Resource availability and plant antiherbivore defense," *Science*, 230, 895–899.

Collins, L. (2020). "Eucalypt forests dominated by epicormic resprouters are resilient to repeated canopy fires," *Journal of Ecology*, 108, 310–324.

Collins, S.L., et al. (2011). "An integrated conceptual framework for long-term social–ecological research," *Frontiers in Ecology and the Environment*, 9, 351–357.

Collinson, A.S. (1988). *Introduction to world vegetation.* 2nd edn. London: Unwin Hyman Ltd.

Comas, L.H., Callahan, H.S., and Midford, P.E. (2014). "Patterns in root traits of woody species hosting arbuscular and ectomycorrhizas: Implications for the evolution of belowground strategies," *Ecology and Evolution*, 4, 2979–2990.

Comas, L.H., et al. (2012). "Evolutionary patterns and biogeochemical significance of angiosperm root traits," *International Journal of Plant Sciences*, 173, 584–595.

Condit, R., Hubbell, S.P., and Foster, R.B. (1996). "Assessing the response of plant functional types to climatic change in tropical forests," *Journal of Vegetation Science*, 7, 405–416.

Connell, J.H. (1978). "Diversity in tropical rain forests and coral reefs: High diversity of trees and corals is maintained only in a nonequilibrium state," *Science*, 199, 1302–1310.

Connell, J.H., Den Boer, P.J., and Gradwell, G.R. (1971). "On the role of natural enemies in preventing competitive exclusion in some marine animals and in rain forest trees," in den Boer, P.J. and Gradwell, G.R. (eds.) *Dynamics of populations.* Wageningen: Centre for Agricultural Publishing and Documentation, pp. 298–312.

Cook, R. (1980). "The biology of seeds in the soil," in Solbrig, O.T. (ed.) *Demography and evolution in plant populations.* Botanical Monograph No. 15. Berkeley, CA: University of California Press, pp. 107–129.

Coomes, D.A. and Grubb, P.J. (1996). "Amazonian caatinga and related communities at La Esmeralda, Venezuela: Forest structure, physiognomy and floristics, and control by soil factors," *Vegetatio*, 122, 167–191.

Coomes, D.A. and Grubb, P.J. (2003). "Colonization, tolerance, competition and seed-size variation within functional groups," *Trends in Ecology & Evolution*, 18, 283–291.

Cornelissen, J.H.C., Castro-Diez, P., and Carnelli, A.L. (1998). "Variation in relative growth rate among woody species," in Lambers, H., Poorter, H., and Van Vuuren, M.M.I. (eds.) *Inherent variation in plant growth: Physiological mechanisms and ecological consequences.* Leiden: Backhuys Publishers, pp. 363–392.

Cornelissen, J.H.C., Diez, P.C., and Hunt, R. (1996). "Seedling growth, allocation and leaf attributes in a wide range of woody plant species and types," *Journal of Ecology*, 84, 755–765.

Corrêa Scalon, M., et al. (2020). "Diversity of functional trade-offs enhances survival after fire in Neotropical savanna species," *Journal of Vegetation Science*, 31, 139–150.

Craine, J.M. (2006). "Competition for nutrients and optimal root allocation," *Plant and Soil*, 285, 171–185.

Craine, J.M. (2005). "Reconciling plant strategy theories of Grime and Tilman," *Journal of Ecology*, 93, 1041–1052.

Craine, J.M. (2009). *Resource strategies of wild plants*. Princeton: Princeton University Press.

Craine, J.M. and Dybzinski, R. (2013). "Mechanisms of plant competition for nutrients, water and light," *Functional Ecology*, 27, 833–840.

Craine, J.M. and Reich, P.B. (2005). "Leaf-level light compensation points in shade-tolerant woody seedlings," *New Phytologist*, 166, 710–713.

Craine, J.M., et al. (2012b). "Flowering phenology as a functional trait in a tallgrass prairie," *New Phytologist*, 193, 673–682.

Craine, J.M., et al. (2002). "Functional traits, productivity and effects on nitrogen cycling of 33 grassland species," *Functional Ecology*, 16, 563–574.

Craine, J.M., et al. (2009). "Global patterns of foliar nitrogen isotopes and their relationships with climate, mycorrhizal fungi, foliar nutrient concentrations, and nitrogen availability," *New Phytologist*, 183, 980–992.

Craine, J.M., et al. (1999). "Measurement of leaf longevity of 14 species of grasses and forbs using a novel approach," *New Phytologist*, 142, 475–481.

Craine, J.M., et al. (2001). "The relationships among root and leaf traits of 76 grassland species and relative abundance along fertility and disturbance gradients," *Oikos*, 93, 274–285.

Craine, J.M., et al. (2012a). "Resource limitation, tolerance, and the future of ecological plant classification," *Frontiers in Plant Science*, 3, 246.

Crone, E.E., et al. (2013). "Ability of matrix models to explain the past and predict the future of plant populations," *Conservation Biology*, 27, 968–978.

Crone, E.E., et al. (2011). "How do plant ecologists use matrix population models?" *Ecology Letters*, 14, 1–8.

Cunha, H.F.V., et al. (2022). "Direct evidence for phosphorus limitation on Amazon forest productivity," *Nature*, 608, 558–562.

Currano, E.D., et al. (2021). "Scars on fossil leaves: An exploration of ecological patterns in plant–insect herbivore associations during the Age of Angiosperms," *Palaeogeography, Palaeoclimatology, Palaeoecology*, 582, 110636.

Currano, E.D., et al. (2008). "Sharply increased insect herbivory during the Paleocene–Eocene Thermal Maximum," *Proceedings of the National Academy of Sciences of the United States of America*, 105, 1960–1964.

Currey, D.R. (1965). "An ancient bristlecone pine stand in eastern Nevada," *Ecology*, 46, 564–566.

da Silva, G.S., et al. (2021). "Resprouting strategies of three native shrub Cerrado species from a morphoanatomical and chemical perspective," *Australian Journal of Botany*, 69, 527–542.

Dahlgren, J.P. and Roach, D.A. (2017). "Demographic senescence in herbaceous plants," in Shefferson, R., Jones, O., and Salguero-Gómez, R. (eds.) *The evolution of senescence in the tree of life*. Cambridge: Cambridge University Press, pp. 303–319.

Dahlgren, J.P., Bengtsson, K., and Ehrlén, J. (2016). "The demography of climate-driven and density-regulated population dynamics in a perennial plant," *Ecology*, 97, 899–907.

Daily, G.C. (ed). (1997). *Nature's services: Societal dependence on natural ecosystems*. Washington, DC: Island Press.

Dalgleish, H.J. and Hartnett, D.C. (2006). "Below-ground bud banks increase along a precipitation gradient of the North American Great Plains: A test of the meristem limitation hypothesis" *New Phytologist*, 171, 81–89.

Dalgleish, H.J., Koons, D.N., and Adler, P.B. (2010). "Can life-history traits predict the response of forb populations to changes in climate variability?" *Journal of Ecology*, 98, 209–217.

Dansereau, P. (1951). "Description and recording of vegetation upon a structural basis," *Ecology*, 32, 172–229.

Dantas, V.L. and Pausas, J.G. (2013). "The lanky and the corky: Fire-escape strategies in savanna woody species," *Journal of Ecology*, 101, 1265–1272.

Dantas, V.L. and Pausas, J.G. (2022). "The legacy of the extinct Neotropical megafauna on plants and biomes," *Nature Communications*, 13, 129.

Dantas, V.L. and Pausas, J.G. (2020). "Megafauna biogeography explains plant functional trait variability in the tropics," *Global Ecology and Biogeography*, 29, 1288–1298.

Daou, L. and Shipley, B. (2020). "Simplifying the protocol for the quantification of generalized soil fertility gradients in grassland community ecology," *Plant and Soil*, 457, 457–468.

Darwin, C. (1859). *On the origin of species by means of natural selection*. London: John Murray.

Darwin, C. and Wallace, A.R. (1858). "On the tendency of species to form varieties; and on the perpetuation of varieties and species by natural means of selection," *Zoological Journal of the Linnean Society*, 3, 46–62.

Davis, M.B. and Shaw, R.G. (2001). "Range shifts and adaptive responses to Quaternary climate change," *Science*, 292, 673–679.

de Bello, F., et al. (2021). *Handbook of trait-based ecology: From theory to R tools*. Cambridge: Cambridge University Press.

de Bello, F., et al. (2015). "On the need for phylogenetic 'corrections' in functional trait-based approaches," *Folia Geobotanica*, 50, 349–357.

de Groot, R.S., Wilson, M.A., and Boumans, R.M.J. (2002). "A typology for the classification, description and valuation of ecosystem functions, goods and services," *Ecological Economics*, 41, 393–408.

de Jong, T. and Klinkhamer, P. (2005). *Evolutionary ecology of plant reproductive strategies*. Cambridge: Cambridge University Press.

de Kroon, H., et al. (1986). "Elasticity: The relative contribution of demographic parameters to population growth rate," *Ecology*, 67, 1427–1431.

de Witte, L.C. and Stöcklin, J. (2010). "Longevity of clonal plants: Why it matters and how to measure it," *Annals of Botany*, 106, 859–870.

Del Grosso, S., et al. (2008). "Global potential net primary production predicted from vegetation class, precipitation, and temperature," *Ecology*, 89, 2117–2126.

DeLucia, E.H., et al. (1996). "Contribution of intercellular reflectance to photosynthesis in shade leaves," *Plant, Cell & Environment*, 19, 159–170.

Del-Val, E. and Crawley, M.J. (2005). "Are grazing increaser species better tolerators than decreasers? An experimental assessment of defoliation tolerance in eight British grassland species," *Journal of Ecology*, 93, 1005–1016.

Deno, N.C. (1993). *Seed germination, theory and practice*. State College, PA: Self published.

Detto, M., et al. (2019). "Bias in the detection of negative density dependence in plant communities," *Ecology Letters*, 22, 1923–1939.

Díaz, S. (2022). "A fabric of life view of the world," *Science*, 375, 1204.

Díaz, S., Noy-Meir, I., and Cabido, M. (2001). "Can grazing response of herbaceous plants be predicted from simple vegetative traits?" *Journal of Applied Ecology*, 38, 497–508.

Díaz, S., et al. (2016). "The global spectrum of plant form and function," *Nature*, 529, 167–171.

Díaz, S., et al. (2007). "Plant trait responses to grazing—a global synthesis," *Global Change Biology*, 13, 313–341.

Diekmann, O. (2004). "Beginners guide to adaptive dynamics," *Banach Center Publications*, 63, 47–86.

Dixit, A. and Kolb, T. (2020). "Variation in seedling budburst phenology and structural traits among southwestern ponderosa pine provenances," *Canadian Journal of Forest Research*, 50, 872–879.

Dixon, K.W., Roche, S., and Pate, J.S. (1995). "The promotive effect of smoke derived from burnt native vegetation on seed germination of Western Australian plants," *Oecologia*, 101, 185–192.

Doak, D.F. and Morris, W.F. (2010). "Demographic compensation and tipping points in climate-induced range shifts," *Nature*, 467, 959–962.

Doak, D.F., Thomson, D., and Jules, E.S. (2002). "Population viability analysis for plants: Understanding the demographic consequences of seed banks for population health," in Beissinger, S.R. and McCullough, D.R. (eds.) *Population viability analysis*. Chicago: University of Chicago Press, pp. 312–337.

Doak, D.F., et al. (2021). "A critical comparison of integral projection and matrix projection models for demographic analysis," *Ecological Monographs*, 91, e01447.

Donoho, D.L. (2000). "High-dimensional data analysis: The curses and blessings of dimensionality," *AMS Math Challenges Lecture*, 1, 1–32.

Dornelas, M., et al. (2014). "Assemblage time series reveal biodiversity change but not systematic loss," *Science*, 344, 296–299.

Drayton, B. and Primack, R.B. (2012). "Success rates for reintroductions of eight perennial plant species after 15 years," *Restoration Ecology*, 20, 299–303.

Drenovsky, R.E., et al. (2012). "A functional trait perspective on plant invasion," *Annals of Botany*, 110, 141–153.

Drude, O. (1886). "Deutschlands Pflanzengeographie," in Handbuch Deutsch Landes- und Volksk. Stuttgart: Verlag von J. Enelhorn.

Drude, O. (1913). *Die Ökologie der Pflanzen*. Braunschweig: Die Wissenschaft. F. Vieweg.

Du Rietz, G.E. (1931). "Life-forms of terrestrial flowering plants, 1," *Acta Phytogeographica Suecica*, 3, 1–95.

Duarte, C.M., et al. (1995). "Comparative functional plant ecology: Rationale and potentials," *Trends in Ecology & Evolution*, 10, 418–421.

Dubrovsky, J.G., North, G.B., and Nobel, P.S. (1998). "Root growth, developmental changes in the apex, and hydraulic conductivity for *Opuntia ficus-indica* during drought," *New Phytologist*, 138, 75–82.

Duckworth, J.C., Kent, M., and Ramsay, P.M. (2000). "Plant functional types: An alternative to taxonomic plant community description in biogeography?" *Progress in Physical Geography*, 24, 515–542.

Duffy, K.A., et al. (2021). "How close are we to the temperature tipping point of the terrestrial biosphere?" *Science Advances*, 7, eaay1052.

Duncan, D.J. (2002). *My story as told by water: Confessions, druidic rants, reflections, bird-watchings, fish-stalkings, visions, songs and prayers refracting light, from living rivers, in the Age of the Industrial Dark*. Berkeley: University of California Press.

During, H.J., et al. (1985). "Dynamics of plant coenotic populations," in White, J. (ed.) *The population structure of vegetation*. Dordrecht: Dr W. Junk Publishers, pp. 341–370.

Dybzinski, R., Farrior, C.E., and Pacala, S.W. (2015). "Increased forest carbon storage with increased atmospheric CO2 despite nitrogen limitation: A game-theoretic allocation model for trees in competition for nitrogen and light," *Global Change Biology*, 21, 1182–1196.

Dybzinski, R., et al. (2011). "Evolutionarily stable strategy carbon allocation to foliage, wood, and fine roots in trees competing for light and nitrogen: An analytically tractable, individual-based model and quantitative comparisons to data," *The American Naturalist*, 177, 153–166.

Dyksterhuis, E.J. (1949). "Condition and management of range land based on quantitative ecology," *Journal of Range Management*, 2, 104–115.

Ebert, T.A. (1999). *Plant and animal populations: Methods in demography*. San Diego: Academic Press.

Edwards, E.J. and Ogburn, R.M. (2012). "Angiosperm responses to a low-CO2 world: CAM and C4 photosynthesis as parallel evolutionary trajectories," *International Journal of Plant Sciences*, 173, 724–733.

Egerton, F.N. (2018). "History of ecological sciences, Part 61A: Terrestrial biogeography and paleobiogeography, 1700–1830s," *The Bulletin of the Ecological Society of America*, 99, 192–241.

Ehrlén, J. (1999). "Modelling and measuring plant life histories," in Vuorisalo, T.O. and Mutikainen, P.K. (eds.) *Life history evolution in plants*. Dordrecht: Kluwer Academic Publishers, pp. 26–61.

Eiseley, L. (1957). *The immense journey*. New York: Vintage Books.

Elberse, W.T. and Berendse, F. (1993). "A comparative study of the growth and morphology of eight grass species from habitats with different nutrient availabilities," *Functional Ecology*, 7, 223–229.

Elderd, B.D. and Miller, T.E.X. (2016). "Quantifying demographic uncertainty: Bayesian methods for integral projection models," *Ecological Monographs*, 86, 125–144.

El-Kassaby, Y.A. and Barclay, H.J. (1992). "Cost of reproduction in Douglas-fir," *Canadian Journal of Botany*, 70, 1429–1432.

Ellenberg, H. (1988). *Vegetation of Central Europe*. 4th edn. Cambridge: Cambridge University Press.

Ellenberg, H. and Mueller-Dombois, D. (1967). "A key to Raunkiaer plant life forms with revised subdivisions," *Berlin Geobotanical Institute Rübel*, 37, 21–55.

Ellner, S.P. (1985a). "ESS germination strategies in randomly varying environments. I. Logistic-type models," *Theoretical Population Biology*, 28, 50–79.

Ellner, S.P. (1985b). "ESS germination strategies in randomly varying environments. II. Reciprocal yield-law models," *Theoretical Population Biology*, 28, 80–116.

Ellner, S.P. (1986). "Germination dimorphisms and parent-offspring conflict in seed germination," *Journal of Theoretical Biology*, 123, 173–185.

Ellner, S.P. and Rees, M. (2006). "Integral projection models for species with complex demography," *The American Naturalist*, 167, 410–428.

Ellner, S.P., Childs, D.Z., and Rees, M. (2016). *Data-driven modelling of structured populations: A practical guide to the integral projection model*. Cham: Springer.

Ellner, S.P., et al. (2022). "A critical comparison of integral projection and matrix projection models for demographic analysis: Comment," *Ecology*, 103, e3605.

Elser, J.J., et al. (2007). "Global analysis of nitrogen and phosphorus limitation of primary producers in freshwater, marine and terrestrial ecosystems," *Ecology Letters*, 10, 1135–1142.

Endara, M.J. and Coley, P.D. (2011). "The resource availability hypothesis revisited: A meta-analysis," *Functional Ecology*, 25, 389–398.

Enquist, B.J. (2010). "Wanted: A general and predictive theory for trait-based plant ecology," *Bioscience*, 60, 854–855.

Enquist, B.J., Brown, J.H., and West, G.B. (1998). "Allometric scaling of plant energetics and population density," *Nature*, 395, 163–165.

Enquist, B.J., et al. (2007). "A general integrative model for scaling plant growth, carbon flux, and functional trait spectra," *Nature*, 449, 218–222.

Enquist, B.J., et al. (1999). "Allometric scaling of production and life-history variation in vascular plants," *Nature*, 401, 907–911.

Epling, C., Lewis, H., and Ball, F.M. (1960). "The breeding group and seed storage: A study in population dynamics," *Evolution*, 14, 238–255.

Epstein, E. and Bloom, A.J. (2005). *Mineral nutrition of plants: Principles and perspectives*. 2nd edn. Sunderland, MA: Sinauer Associates, Inc.

Eriksson, O. and Kiviniemi, K. (1999). "Evolution of plant dispersal," in Vuorisalo, T.O. and Mutikainen, P.K. (eds.) *Life history evolution in plants*. Dordrecht: Kluwer Academic Publishers, pp. 215–238.

Eshel, A. and Beeckman, T. (2013). *Plant roots: The hidden half*. New York: CRC Press.

Etterson, J.R., et al. (2020). "Assisted migration across fixed seed zones detects adaptation lags in two major North American tree species," *Ecological Applications*, 30, e02092.

Evans, G.C. (1972). *The quantitative analysis of plant growth*. Berkeley: University of California Press.

Evert, R.F. (2006). *Esau's plant anatomy: Meristems, cells, and tissues of the plant body: Their structure, function, and development*. 3rd edn. Hoboken: John Wiley & Sons.

Falster, D.S. and Westoby, M. (2003). "Plant height and evolutionary games," *Trends in Ecology and Evolution*, 18, 337–343.

Falster, D.S., Duursma, R.A., and FitzJohn, R.G. (2018). "How functional traits influence plant growth and shade tolerance across the life cycle," *Proceedings of the National Academy of Sciences of the United States of America*, 115, E6789–E6798.

Falster, D.S., et al. (2021a). "AusTraits, a curated plant trait database for the Australian flora," *Scientific Data*, 8, 254.

Falster, D.S., et al. (2021b). "Emergent shapes of trait-based competition functions from resource-based models: A Gaussian is not normal in plant communities," *The American Naturalist*, 198, 253–267.

Falster, D.S., et al. (2011). "Influence of four major plant traits on average height, leaf-area cover, net primary productivity, and biomass density in single-species forests: A theoretical investigation," *Journal of Ecology*, 99, 148–164.

Falster, D.S., et al. (2017). "Multitrait successional forest dynamics enable diverse competitive coexistence," *Proceedings of the National Academy of Sciences of the United States of America*, 114, E2719–E2728.

Falster, D.S., et al. (2016). "plant: A package for modelling forest trait ecology and evolution," *Methods in Ecology and Evolution*, 7, 136–146.

Fan, Y., et al. (2017). "Hydrologic regulation of plant rooting depth," *Proceedings of the National Academy of Sciences of the United States of America*, 114, 10572.

Fan, Z-X., et al. (2012). "Hydraulic conductivity traits predict growth rates and adult stature of 40 Asian tropical tree species better than wood density," *Journal of Ecology*, 100, 732–741.

Farrior, C.E., et al. (2013a). "Competition for water and light in closed-canopy forests: A tractable model of carbon allocation with implications for carbon sinks," *The American Naturalist*, 181, 314–330.

Farrior, C.E., et al. (2013b). "Resource limitation in a competitive context determines complex plant responses to experimental resource additions," *Ecology*, 94, 2505–2517.

Felsenstein, J. (1985). "Phylogenies and the comparative method," *The American Naturalist*, 125, 1–15.

Fenner, M. (1985). *Seed ecology*. London: Springer Science & Business Media.

Fidelis, A. and Zirondi, HL. (2021). "And after fire, the Cerrado flowers: A review of post-fire flowering in a tropical savanna," *Flora*, 280, 151849.

Fidelis, A., et al. (2014). "Does disturbance affect bud bank size and belowground structures diversity in Brazilian subtropical grasslands?" *Flora-Morphology, Distribution, Functional Ecology of Plants*, 209, 110–116.

Fiedler, S., et al. (2021). "Global change shifts trade-offs among ecosystem functions in woodlands restored for multifunctionality," *Journal of Applied Ecology*, 58, 1705–1717.

Fierer, N., et al. (2005). "Litter quality and the temperature sensitivity of decomposition," *Ecology*, 86, 320–326.

Filartiga, A.L., Klimešová, J., and Appezzato-da-Glória, B. (2017). "Underground organs of Brazilian Asteraceae: Testing the CLO-PLA database traits," *Folia Geobotanica*, 52, 367–385.

Firbank, L.G. and Watkinson, A.R. (1990). "On the effects of competition: From monocultures to mixtures," in Grace, J.B. and Tilman, D. (eds.) *Perspectives on plant competition*. San Diego: Academic Press, pp. 165–192.

Fischer, L.K., von der Lippe, M., and Kowarik, I. (2013). "Urban grassland restoration: Which plant traits make desired species successful colonizers?" *Applied Vegetation Science*, 16, 272–285.

Fischer, R.A. and Maurer, R. (1978). "Drought resistance in spring wheat cultivars. I. Grain yield responses," *Australian Journal of Agricultural Research*, 29, 897–912.

Fisher, R.A. (1930). *The genetical theory of natural selection*. Oxford: Clarendon Press.

Flores, O., et al. (2014). "An evolutionary perspective on leaf economics: Phylogenetics of leaf mass per area in vascular plants," *Ecology and Evolution*, 4, 2799–2811.

Floyd, S.K. and Ranker, T.A. (1998). "Analysis of a transition matrix model for *Gaura neomexicana* ssp. *coloradensis* (Onagraceae) reveals spatial and temporal demographic variability," *International Journal of Plant Sciences*, 159, 853–863.

Foden, W.B., et al. (2019). "Climate change vulnerability assessment of species," *WIREs Climate Change*, 10, e551.

Fonseca, C.R., et al. (2000). "Shifts in trait-combinations along rainfall and phosphorus gradients," *Journal of Ecology*, 88, 964–977.

Fowler, N.L., Overath, R.D., and Pease, C.M. (2006). "Detection of density dependence requires density manipulations and calculation of λ," *Ecology*, 87, 655–664.

Fox, G.A. and Gurevitch, J. (2000). "Population numbers count: Tools for near-term demographic analysis," *The American Naturalist*, 156, 242–256.

Franco, M. and Silvertown, J. (1996). "Life history variation in plants: An exploration of the fast-slow continuum hypothesis," *Philosophical Transactions of the Royal Society of London Series B: Biological Sciences*, 351, 1341–1348.

Franco, M. and Silvertown, J. (2004). "A comparative demography of plants based upon elasticities of vital rates," *Ecology*, 85, 531–538.

Franklin, O., et al. (2020). "Organizing principles for vegetation dynamics," *Nature Plants*, 6, 444–453.

Freckleton, R.P., et al. (2006). "Census error and the detection of density dependence," *Journal of Animal Ecology*, 75, 837–851.

Freschet, G.T., et al. (2010). "Evidence of the 'plant economics spectrum' in a subarctic flora," *Journal of Ecology*, 98, 362–373.

Freschet, G.T., et al. (2021). "A starting guide to root ecology: Strengthening ecological concepts and standardising root classification, sampling, processing and trait measurements," *New Phytologist*, 232, 973–1122.

Friedman, J. (2020). "The evolution of annual and perennial plant life histories: Ecological correlates and genetic mechanisms," *Annual Review of Ecology, Evolution, and Systematics*, 51, 461–481.

Fristoe, T.S., et al. (2021). "Dimensions of invasiveness: Links between local abundance, geographic range size, and habitat breadth in Europe's alien and native floras," *Proceedings of the National Academy of Sciences of the United States of America*, 118, e2021173118.

Fry, J.D. (2003). "Detecting ecological trade-offs using selection experiments," *Ecology*, 84, 1672–1678.

Fulé, P.Z., Covington, W.W., and Moore, M.M. (1997). "Determining reference conditions for ecosystem management of southwestern ponderosa pine forests," *Ecological Applications*, 7, 895–908.

Fuller, R.N. and Del Moral, R. (2003). "The role of refugia and dispersal in primary succession on Mount St. Helens, Washington," *Journal of Vegetation Science*, 14, 637–644.

Funk, J.L. and Cornwell, W.K. (2013). "Leaf traits within communities: Context may affect the mapping of traits to function," *Ecology*, 94, 1893–1897.

Funk, J.L. and McDaniel, S. (2010). "Altering light availability to restore invaded forest: The predictive role of plant traits," *Restoration Ecology*, 18, 865–872.

Funk, J.L. and Wolf, A.A. (2016). "Testing the trait-based community framework: Do functional traits predict competitive outcomes?" *Ecology*, 97, 2206–2211.

Funk, J.L., Hoffacker, M.K., and Matzek, V. (2015). "Summer irrigation, grazing and seed addition differentially influence community composition in an invaded serpentine grassland," *Restoration Ecology*, 23, 122–130.

Funk, J.L., et al. (2008). "Restoration through reassembly: Plant traits and invasion resistance," *Trends in Ecology and Evolution*, 23, 695–703.

Funk, J.L., et al. (2017). "Revisiting the Holy Grail: Using plant functional traits to understand ecological processes," *Biological Reviews* 92, 1156–1173.

Gallagher, R.V., et al. (2015). "Assisted colonization as a climate change adaptation tool," *Austral Ecology*, 40, 12–20.

Futuyma, D.J. and Moreno, G. (1988). "The evolution of ecological specialization." *Annual Review of Ecology and Systematics*, 19, 207–233.

Gallagher, R.V., et al. (2021). "A guide to using species trait data in conservation," *One Earth*, 4, 927–936.

Garbowski, M., Johnston, D.B., and Brown, C.S. (2021). "Leaf and root traits, but not relationships among traits, vary with ontogeny in seedlings," *Plant and Soil*, 460, 247–261.

Garcia, L.C., et al. (2015). "Flower functional trait responses to restoration time," *Applied Vegetation Science*, 18, 402–412.

García, M.B. and Antor, R.J. (1995). "Age and size structure in populations of a long-lived dioecious geophyte: *Borderea pyrenaica* (Dioscoreaceae)," *International Journal of Plant Sciences*, 156, 236–243.

García, M.B., Dahlgren, J.P., and Ehrlen, J. (2011). "No evidence of senescence in a 300-year-old mountain herb," *Journal of Ecology*, 99, 1424–1430.

García-Callejas, D., Godoy, O., and Bartomeus, I. (2020). "cxr: A toolbox for modelling species coexistence in R," *Methods in Ecology and Evolution*, 11, 1221–1226.

Gardiner, R., Shoo, L.P., and Dwyer, J.M. (2019). "Look to seedling heights, rather than functional traits, to explain survival during extreme heat stress in the early stages of subtropical rainforest restoration," *Journal of Applied Ecology*, 56, 2687–2697.

Garnier, E. (1992). "Growth analysis of congeneric annual and perennial grass species," *Journal of Ecology*, 80, 665–675.

Garnier, E., Navas, M-L., and Grigulis, K. (2016). *Plant functional diversity: Organism traits, community structure, and ecosystem properties*. Oxford: Oxford University Press.

Garnier, E., et al. (2007). "Assessing the effects of land-use change on plant traits, communities and ecosystem functioning in grasslands: A standardized methodology and lessons from an application to 11 European sites," *Annals of Botany*, 99, 967–985.

Garnier, E., et al. (2001). "A standardized protocol for the determination of specific leaf area and leaf dry matter content," *Functional Ecology*, 15, 688–695.

Gaston, K.J. (2009). "Geographic range limits: Achieving synthesis," *Proceedings of the Royal Society B: Biological Sciences*, 276, 1395–1406.

Geritz, S.A. (1995). "Evolutionarily stable seed polymorphism and small-scale spatial variation in seedling density," *The American Naturalist*, 146, 685–707.

Geritz, S.A., et al. (1997). "Dynamics of adaptation and evolutionary branching," *Physical Review Letters*, 78, 2024–2027.

Giannini, T.C., et al. (2017). "Selecting plant species for practical restoration of degraded lands using a multiple-trait approach," *Austral Ecology*, 42, 510–521.

Gibert, A., Tozer, W., and Westoby, M. (2019). "Plant performance response to eight different types of symbiosis," *New Phytologist*, 222, 526–542.

Gibert, A., et al. (2016). "On the link between functional traits and growth rate: Meta-analysis shows effects change with plant size, as predicted," *Journal of Ecology*, 104, 1488–1503.

Gibson, D.J. (2015). *Methods in comparative plant population ecology*. Oxford: Oxford University Press.

Gilardelli, F., et al. (2015). "Ecological filtering and plant traits variation across quarry geomorphological surfaces: Implication for restoration," *Environmental Management*, 55, 1147–1159.

Gill, A.M. (1975). "Fire and the Australian flora: A review," *Australian Forestry*, 38, 4–25.

Gill, R.A. and Jackson, R.B. (2000). "Global patterns of root turnover for terrestrial ecosystems," *New Phytologist*, 147, 13–31.

Gillespie, R.G., et al. (2012). "Long-distance dispersal: A framework for hypothesis testing," *Trends in Ecology & Evolution*, 27, 47–56.

Gillison, A.N. (1981). "Towards a functional vegetation classification," in Gillison, A. and Anderson, D.J. (eds.) *Vegetation classification in Australia*. Canberra: ANU Press, pp. 30–41.

Gitay, H. and Noble, I.R. (1997). "What are functional types and how should we seek them?" in Smith, T.M., Shugart, H.H., and Woodward, F.I. (eds.) *Plant functional types: Their relevance to ecosystem properties and global change*. Cambridge: Cambridge University Press, pp. 3–19.

Givnish, T.J. (1982). "On the adaptive significance of leaf height in forest herbs," *The American Naturalist*, 120, 353–381.

Givnish, T.J. (ed). (1986). *On the economy of plant form and function: Proceedings of the Sixth Maria Moors Cabot Symposium*. Cambridge: Cambridge University Press.

Givnish, T.J. (1995). "Plant stems: Biomechanical adaptation for energy capture and influence on species distributions," *in* Gartner, B. (ed.) *Plant stems: Physiology and functional morphology*. San Diego: Academic Press, pp. 3–49.

Gleason, S.M., et al. (2016). "Weak tradeoff between xylem safety and xylem-specific hydraulic efficiency across the world's woody plant species," *New Phytologist*, 209, 123–136.

Goldberg, D.E. (1996). "Competitive ability: Definitions, contingency and correlated traits," *Philosophical Transactions of the Royal Society of London. Series B: Biological Sciences*, 351, 1377–1385.

Goldberg, D.E. (1990). "Components of resource competition in plant communities," in Grace, J.B. and Tilman, D. (eds.) *Perspectives on plant competition*. San Diego: Academic Press, Inc, pp. 27–49.

Goldberg, D.E. and Landa, K. (1991). "Competitive effect and response: Hierarchies and correlated traits in the early stages of competition," *Journal of Ecology*, 79, 1013–1030.

Goldsmith, F. and Harrison, C. (1976). "Description and analysis of vegetation," in Chapman, S.B. (ed.) *Methods in plant ecology*. Oxford: Blackwell Scientific Publications, pp. 85–155.

Gommers, C.M.M., et al. (2013). "Shade tolerance: When growing tall is not an option," *Trends in Plant Science*, 18, 65–71.

Gould, S.J. and Lewontin, R.C. (1979). "The spandrels of San Marco and the Panglossian paradigm: A critique of the adaptationist programme," *Proceedings of the Royal Society of London Series B: Biological Sciences*, 205, 581–598.

Grace, J.B. (1990). "On the relationship between plant traits and competitive ability," in Grace, J.B. and Tilman, D. (eds.) *Perspectives on plant competition*. San Diego: Academic Press, Inc, pp. 51–65/

Grady, K.C., et al. (2015). "A bridge too far: Cold and pathogen constraints to assisted migration of riparian forests," *Restoration Ecology*, 23, 811–820.

Grady, K.C., et al. (2017). "Local biotic adaptation of trees and shrubs to plant neighbors," *Oikos*, 126, 583–593.

Grass, G. (1959). *Die Blechtrommel (The tin drum)*. Darmstadt: Hermann Luchterhand Verlag.

Greene, D.F. and Johnson, E.A. (1989). "A model of wind dispersal of winged or plumed seeds," *Ecology*, 70, 339–347.

Greenslade, P.J.M. (1983). "Adversity selection and the habitat templet," *The American Naturalist*, 122, 352–365.

Gremer, J., et al. (2017). "Complex life histories and senescence in plants," in Shefferson, R., Jones, O., and Salguero-Gómez, R. (eds.) *The evolution of senescence in the tree of life*. Cambridge: Cambridge University Press, p. 320.

Griffin-Nolan, R.J., et al. (2019). "Extending the osmometer method for assessing drought tolerance in herbaceous species," *Oecologia*, 189, 353–363.

Griffin-Nolan, R.J., et al. (2021). "Friend or foe? The role of biotic agents in drought-induced plant mortality," *Plant Ecology*, 222, 537–548.

Griffiths, M. and York, L.M. (2020). "Targeting root ion uptake kinetics to increase plant productivity and nutrient use efficiency," *Plant Physiology*, 182, 1854–1868.

Grime, J.P. (1994). "Defining the scope and testing the validity of CSR theory: A response to Midgley, Laurie and Le Maitre," *Bulletin of the South African Institute of Ecologists*, 13, 4–7.

Grime, J.P. (1984). "The ecology of species, families and communities of the contemporary British flora," *New Phytologist*, 98, 15–33.

Grime, J.P. (1977). "Evidence for the existence of three primary strategies in plants and its relevance to ecological and evolutionary theory," *The American Naturalist*, 111, 1169–1194.

Grime, J.P. (1979). *Plant strategies and vegetation processes*. Chichester: Wiley.

Grime, J.P. (2001). *Plant strategies, vegetation processes, and ecosystem properties*. 2nd edn. Oxford: John Wiley & Sons.

Grime, J.P. (1983). "Prediction of weed and crop response to climate based upon measurements of nuclear DNA content," *Aspects of Applied Biology*, 4, 87–98.

Grime, J.P. (1989). "The stress debate: Symptom of impending synthesis?" *Biological Journal of the Linnean Society*, 37, 3–17.

Grime, J.P. (1974). "Vegetation classification by reference to strategies," *Nature*, 250, 26–31.

Grime, J.P. and Hunt, R. (1975). "Relative growth-rate: Its range and adaptive significance in a local flora," *Journal of Ecology*, 63, 393–422.

Grime, J.P. and Jeffrey, D.W. (1965). "Seedling establishment in vertical gradients of sunlight," *Journal of Ecology*, 53, 621–642.

Grime, J.P. and Pierce, S. (2012). *The evolutionary strategies that shape ecosystems.* Oxford: John Wiley & Sons.

Grime, J.P., Hodgson, J.G., and Hunt, R. (1988). *Comparative plant ecology: A functional approach to common British species.* Dordrecht: Springer.

Grime, J.P., Shacklock, J.M.L., and Brand, S.R. (1985). "Nuclear DNA contents, shoot phenology and species co-existence in a limestone grassland community," *The New Phytologist*, 100, 435–445.

Grime, J.P., Hunt, R., and Krzanowski, W.J. (1987). "Evolutionary physiological ecology of plants," in Calow, P. (ed.) *Evolutionary physiological ecology.* Cambridge: Cambridge University Press, pp. 105–125.

Grime, J.P., et al. (1997a). "Functional types: Testing the concept in Northern England," in Smith, T.M., Shugart, H.H., and Woodward, F.I. (eds.) *Plant functional types: Their relevance to ecosystem properties and global change.* Cambridge: Cambridge University Press, pp. 122–150.

Grime, J.P., et al. (1997b). "Integrated screening validates primary axes of specialisation in plants," *Oikos*, 79, 259–281.

Grisebach, A. (1872). *Die Vegetation der Erde nach ihrer klimatischen Anordnung.* Leipzig: W. Engelmann.

Grishin, S.Y., et al. (1996). "Succession following the catastrophic eruption of Ksudach volcano (Kamchatka, 1907)," *Vegetatio*, 127, 129–153.

Gross, H. (1972). "Crown deterioration and reduced growth associated with excessive seed production by birch," *Canadian Journal of Botany*, 50, 2431–2437.

Grubb, P.J. (1977). "The maintenance of species-richness in plant communities: The importance of the regeneration niche," *Biological Reviews*, 52, 107–145.

Grubb, P.J. (1985). "Plant populations and vegetation in relation to habitat, disturbance and competition: Problems of generalization," in White, J. (ed.) *The population structure of vegetation.* Dordrecht: Dr. W. Junk Publishers, pp. 595–621.

Grubb, P.J. (1992). "Presidential address: A positive distrust in simplicity—lessons from plant defences and from competition among plants and among animals," *Journal of Ecology*, 80, 585–610.

Grubb, P.J. (1980). "Review: *Plant strategies and vegetation processes* by J. P. Grime," *The New Phytologist*, 86, 123–124.

Grubb, P.J. (1998a). "A reassessment of the strategies of plants which cope with shortages of resources," *Perspectives in Plant Ecology, Evolution and Systematics*, 1, 3–31.

Grubb, P.J. (1987). "Some generalizing ideas about colonization and succession in green plants and fungi," in Gray, A.J. and Crawley, M.J. (eds.) *Colonization, succession, and stability.* Oxford: Blackwell, pp. 81–102

Grubb, P.J. (1998b). "Seeds and fruits of tropical rainforest plants: Interpretation of the range in seed size, degree of defence and flesh/seed quotients," in Newbery, D.M., Prins, H.H.T., and Brown, N.D. (eds.) *Dynamics of tropical communities: The 37th symposium of the British Ecological Society, Cambridge University, 1996.* Oxford: Blackwell Science, Ltd, pp. 1–24.

Grubb, P.J. (1976). "A theoretical background to the conservation of ecologically distinct groups of annuals and biennials in the chalk grassland ecosystem," *Biological Conservation*, 10, 53–76.

Grubb, P.J. (2016). "Trade-offs in interspecific comparisons in plant ecology and how plants overcome proposed constraints," *Plant Ecology & Diversity*, 9, 3–33.

Grubb, P.J., Coomes, D.A., and Metcalfe, D.J. (2005). "Comment on 'A brief history of seed size'," *Science*, 310, 783-783.

Guerrero-Ramírez, N.R., et al. (2021). "Global root traits (GRooT) database," *Global Ecology and Biogeography*, 30, 25–37.

Guimarães, P.R., Galetti, M., and Jordano, P. (2008). "Seed dispersal anachronisms: Rethinking the fruits extinct megafauna ate," *PLoS One*, 3, e1745.

Hacke, U.G., et al. (2001). "Trends in wood density and structure are linked to prevention of xylem implosion by negative pressure," *Oecologia*, 126, 457–461.

Hairston, N.G., Smith, F.E., and Slobodkin, L.B. (1960). "Community structure, population control, and competition," *The American Naturalist*, 94, 421–425.

Haldane, J. (1937). "The effect of variation of fitness," *The American Naturalist*, 71, 337–349.

Hallé, F., Oldeman, R.A., and Tomlinson, P.B. (1978). *Tropical trees and forests: An architectural analysis.* New York: Springer.

Hallett, L.M., et al. (2019). "Rainfall variability maintains grass–forb species coexistence," *Ecology Letters*, 22, 1658–1667.

Hammond, W.M., et al. (2019). "Dead or dying? Quantifying the point of no return from hydraulic failure in drought-induced tree mortality," *New Phytologist*, 223, 1834–1843.

Hansen, T.F. (2014). "Use and misuse of comparative methods in the study of adaptation," in Garamszegi, L.Z. (ed.) *Modern phylogenetic comparative methods and their application in evolutionary biology.* Berlin: Springer-Verlag, pp. 351–379.

Hargreaves, A.L., Samis, K.E., and Eckert, C.G. (2014). "Are species' range limits simply niche limits writ large? A review of transplant experiments beyond the range," *The American Naturalist*, 183, 157–173.

Harper, J.L. (1982). "After description," in Newman, E.I. (ed.) The plant community as a working mechanism. Oxford: Blackwell Scientific Publications, pp. 11–25.

Harper, J.L. (1967). "A Darwinian approach to plant ecology," *Journal of Applied Ecology*, 4, 267–290.

Harper, J.L. (1977). *Population biology of plants.* San Diego: Academic Press.

Harper, J.L. (1981). "The population biology of modular organisms," in May, R.M. (ed.) *Theoretical ecology.* Sunderland, MA: Sinauer Associates, pp. 53–77.

Harper, J.L. and Ogden, J. (1970). "The reproductive strategy of higher plants: I. The concept of strategy with special reference to *Senecio vulgaris* L.," *The Journal of Ecology*, 58, 681–698.

Harper, J.L. and White, J. (1974). "The demography of plants," *Annual Review of Ecology and Systematics*, 5, 419–463.

Harper, J.L., Lovell, P., and Moore, K. (1970). "The shapes and sizes of seeds," *Annual Review of Ecology and Systematics*, 1, 327–356.

Harper, J.R.M., et al. (2022). "Application of a trait-based climate change vulnerability assessment to determine management priorities at protected area scale," *Conservation Science and Practice*, 4, e12756.

Harpole, W.S., et al. (2011). "Nutrient co-limitation of primary producer communities," *Ecology Letters*, 14, 852–862.

Harris, J.A., et al. (2006). "Ecological restoration and global climate change," *Restoration Ecology*, 14, 170–176.

Harrison, S., et al. (2015). "Plant communities on infertile soils are less sensitive to climate change," *Annals of Botany*, 116, 1017–1022.

Harrison, S.P., et al. (2021). "Eco-evolutionary optimality as a means to improve vegetation and land-surface models," *New Phytologist*, 231, 2125–2141.

Hart, R. (1977). "Why are biennials so few?" *The American Naturalist*, 111, 792–799.

Hartmann, H., et al. (2022). "Climate change risks to global forest health: Emergence of unexpected events of elevated tree mortality worldwide," *Annual Review of Plant Biology*, 73, 673–702.

Hartshorn, G.S. (1978). "Tree falls and tropical forest dynamics," in Tomlinson, P.B. and Zimmermann, M.H. (eds.) *Tropical trees as living systems*. Cambridge: Cambridge University Press, pp. 617–665.

Harvey, H.T., Shellhammer, H.S., and Stecker, R.E. (1980). *Giant sequoia ecology: Fire and reproduction*. Washington, DC: US Department of the Interior, National Park Service.

Harvey, P.H. and Pagel, M.D. (1991). *The comparative method in evolutionary biology*. Oxford: Oxford University Press.

Harvey, P.H., Read, A.F., and Nee, S. (1995). "Why ecologists need to be phylogenetically challenged," *Journal of Ecology*, 83, 535–536.

Hawkins, C.P. and MacMahon, J.A. (1989). "Guilds: The multiple meanings of a concept," *Annual Review of Entomology*, 34, 423–451.

Heal, O.W. and Grime, J.P. (1991). "Comparative analysis of ecosystems: Past lessons and future directions," in Cole, J., Lovett, G., and Findlay, S. (eds.) *Comparative analyses of ecosystems: Patterns, mechanisms, and theories*. New York: Springer, pp. 7–23.

Hedberg, P., et al. (2013). "A functional trait approach to fen restoration analysis," *Applied Vegetation Science*, 16, 658–666.

Heidel, B., Cox, S., and Blomquist, F. (2014). "Dune habitat trends of an endangered species, *Penstemon haydenii* (blowout penstemon) in Wyoming," Prepared for Bureau of Land Management (BLM). Laramie, WY: Wyoming Natural Diversity Database.

Heidel, B., Handley, J., and Tuthill, D. (2019). "31-year population trends of Colorado butterfly plant (*Oenothera coloradensis*; Onagraceae), a short-lived riparian species on F. E. Warren Air Force Base, Laramie County, Wyoming," Prepared for U.S. Fish and Wildlife Service and F. E. Warren Air Force Base. Laramie, WY: Wyoming Natural Diversity Database.

Hempson, G.P., Archibald, S., and Bond, W.J. (2015). "A continent-wide assessment of the form and intensity of large mammal herbivory in Africa," *Science*, 350, 1056–1061.

Henery, M.L. and Westoby, M. (2001). "Seed mass and seed nutrient content as predictors of seed output variation between species," *Oikos*, 92, 479–490.

Herben, T. and Klimešová, J. (2020). "Evolution of clonal growth forms in angiosperms," *New Phytologist*, 225, 999–1010.

Herben, T., Klimešová, J., and Chytrý, M. (2018). "Philip Grime's fourth corner: Are there plant species adapted to high disturbance and low productivity?" *Oikos*, 127, 1125–1131.

Herben, T., Šerá, B., and Klimešová, J. (2015). "Clonal growth and sexual reproduction: Tradeoffs and environmental constraints," *Oikos*, 124, 469–476.

Herms, D.A. and Mattson, W.J. (1992). "The dilemma of plants: To grow or defend," *Quarterly Review of Biology*, 67, 283–335.

Hesse, E., Rees, M., and Müller-Schärer, H. (2008). "Life-history variation in contrasting habitats: Flowering decisions in a clonal perennial herb (*Veratrum album*)," *The American Naturalist*, 172, E196–E213.

Hetherington, A.J. and Dolan, L. (2018). "Stepwise and independent origins of roots among land plants," *Nature*, 561, 235–238.

Hewitt, N., et al. (2011). "Taking stock of the assisted migration debate," *Biological Conservation*, 144, 2560–2572.

Heyduk, K., et al. (2019). "The genetics of convergent evolution: Insights from plant photosynthesis," *Nature Reviews Genetics*, 20, 485–493.

Higgs, E., et al. (2014). "The changing role of history in restoration ecology," *Frontiers in Ecology and the Environment*, 12, 499–506.

Hilde, C.H., et al. (2020). "The demographic buffering hypothesis: Evidence and challenges," *Trends in Ecology & Evolution*, 35, 523–538.

Hobbs, R.J. and Cramer, V.A. (2008). "Restoration ecology: Interventionist approaches for restoring and maintaining ecosystem function in the face of rapid environmental change," *Annual Review of Environment and Resources*, 33, 39–61.

Hobbs, R.J., Higgs, E., and Harris, J.A. (2009). "Novel ecosystems: Implications for conservation and restoration," *Trends in Ecology & Evolution*, 24, 599–605.

Hodgson, J.G., et al. (1999). "Allocating CSR plant functional types: A soft approach to a hard problem," *Oikos*, 85, 282–294.

Hodson, M.J., et al. (2005). "Phylogenetic variation in the silicon composition of plants," *Annals of Botany*, 96, 1027–1046.

Hooper, D.U. and Johnson, L. (1999). "Nitrogen limitation in dryland ecosystems: Responses to geographical and temporal variation in precipitation," *Biogeochemistry*, 46, 247–293.

Hooper, D.U., et al. (2005). "Effects of biodiversity on ecosystem functioning: A consensus of current knowledge," *Ecological Monographs*, 75, 3–35.

Hunt, R. and Cornelissen, J. (1997a). "Components of relative growth rate and their interrelations in 59 temperate plant species," *New Phytologist*, 135, 395–417.

Hunt, R. and Cornelissen, J. (1997b). "Physiology, allocation, and growth rate: A reexamination of the Tilman model," *The American Naturalist*, 150, 122–130.

Hurlburt, D.P. (1999). *Population ecology and economic botany of* Echinacea angustifolia, *a native prairie medicinal plant*. PhD Dissertation, University of Kansas, Lawrence.

Huston, M. (1979). "A general hypothesis of species diversity," *The American Naturalist*, 113, 81–101.

Huston, M.A. (1994). *Biological diversity: The coexistence of species on changing landscapes*. Cambridge: Cambridge University Press.

Hutchings, J.A. (2021). *A primer of life histories: Ecology, evolution, and application*. Oxford: Oxford University Press.

Hutchinson, G.E. (1965). *The ecological theater and the evolutionary play*. New Haven: Yale University Press.

Huxman, T.E., Winkler, D.E., and Mooney, K.A. (2022). "A common garden super-experiment: An impossible dream to inspire possible synthesis," *Journal of Ecology*, 110, 997–1004.

Ingestad, T. (1979). "Nitrogen stress in birch seedlings," *Physiologia Plantarum*, 45, 149–157.

Iversen, C.M., et al. (2017). "A global Fine-Root Ecology Database to address below-ground challenges in plant ecology," *New Phytologist*, 215, 15–26.

Jackson, R.B., et al. (1996). "A global analysis of root distributions for terrestrial biomes," *Oecologia*, 108, 389–411.

Jackson, S.T. and Hobbs, R.J. (2009). "Ecological restoration in the light of ecological history," *Science*, 325, 567–569.

Jackson, W.D. (1968). "Fire, air, water and earth: An elemental ecology of Tasmania," *Proceedings of the Ecological Society of Australia*, 8, 9–16

Jager, M.M., et al. (2015). "Soil fertility induces coordinated responses of multiple independent functional traits," *Journal of Ecology*, 103, 374–385.

Jakobsson, A. and Eriksson, O. (2000). "A comparative study of seed number, seed size, seedling size and recruitment in grassland plants," *Oikos*, 88, 494–502.

James, J.J., Svejcar, T.J., and Rinella, M.J. (2011). "Demographic processes limiting seedling recruitment in arid grassland restoration," *Journal of Applied Ecology*, 48, 961–969.

Jamil, T., et al. (2013). "Selecting traits that explain species–environment relationships: A generalized linear mixed model approach," *Journal of Vegetation Science*, 24, 988–1000.

Jankowska-Blaszczuk, M. and Daws, M. (2007). "Impact of red:far red ratios on germination of temperate forest herbs in relation to shade tolerance, seed mass and persistence in the soil," *Functional Ecology*, 21, 1055–1062.

Janovský, Z., Herben, T., and Klimešová, J. (2017). "Accounting for clonality in comparative plant demography – growth or reproduction?" *Folia Geobotanica*, 52, 433–442.

Janzen, D.H. (1970). "Herbivores and the number of tree species in tropical forests," *The American Naturalist*, 104, 501–528.

Janzen, D.H. (1971). "Seed predation by animals," *Annual Review of Ecology and Systematics*, 2, 465–492.

Jentsch, A. and White, P. (2019). "A theory of pulse dynamics and disturbance in ecology," *Ecology*, 100, e02734.

Ji, W., et al. (2020). "Functional ecology of congeneric variation in the leaf economics spectrum," *New Phytologist*, 225, 196–208.

Jiménez-Sierra, C., Mandujano, M.C., and Eguiarte, L.E. (2007). "Are populations of the candy barrel cactus (*Echinocactus platyacanthus*) in the desert of Tehuacán, Mexico at risk? Population projection matrix and life table response analysis," *Biological Conservation*, 135, 278–292.

Johnson, L.C., et al. (2022). "Reciprocal transplant gardens as gold standard to detect local adaptation in grassland species: New opportunities moving into the 21st century," *Journal of Ecology*, 110, 1054–1071.

Johnson, N.C., Graham, J., and Smith, F. (1997). "Functioning of mycorrhizal associations along the mutualism–parasitism continuum," *The New Phytologist*, 135, 575–585.

Johnson, N.C., et al. (2015). "Mycorrhizal phenotypes and the Law of the Minimum," *New Phytologist*, 205, 1473–1484.

Johnson, R.A. and Wichern, D.W. (1992). *Applied multivariate statistical analysis*. Englewood Cliffs: Prentice Hall.

Johnson, W.C., Werner, B., and Guntenspergen, GR. (2016). "Non-linear responses of glaciated prairie wetlands to climate warming," *Climatic Change*, 134, 209–223.

Jones, O.R., et al. (2022). "Rcompadre and Rage—two R packages to facilitate the use of the COMPADRE and COMADRE databases and calculation of life history traits from matrix population models," *Methods in Ecology and Evolution*, 13, 770–781.

Jones, O.R. and Vaupel, J.W. (2017). "Senescence is not inevitable," *Biogerontology*, 18, 965–971.

Jones, O.R., et al. (2014). "Diversity of ageing across the Tree of Life," *Nature*, 505, 169–173.

Jordan, W.R. (2003). *The sunflower forest*. Berkeley: University of California Press.

Joswig, J.S., et al. (2022). "Climatic and soil factors explain the two-dimensional spectrum of global plant trait variation," *Nature Ecology & Evolution*, 6, 36–50.

Jump, A.S. and Penuelas, J. (2005). "Running to stand still: Adaptation and the response of plants to rapid climate change," *Ecology Letters*, 8, 1010–1020.

Jurado, E. and Westoby, M. (1992). "Seedling growth in relation to seed size among species of arid Australia," *Journal of Ecology*, 80, 407–416.

Kahmen, S. and Poschlod, P. (2008). "Effects of grassland management on plant functional trait composition," *Agriculture, Ecosystems & Environment*, 128, 137–145.

Kahmen, S., Poschlod, P., and Schreiber, K-F. (2002). "Conservation management of calcareous grasslands. Changes in plant species composition and response of functional traits during 25 years," *Biological Conservation*, 104, 319–328.

Kambach, S., et al. (2022). "Consistency of demographic trade-offs across 13 (sub)tropical forests," *Journal of Ecology*, 110, 1485–1496.

Kattge, J., et al. (2020). "TRY plant trait database—enhanced coverage and open access," *Global Change Biology*, 26, 119–188.

Kattge, J., et al. (2011). "TRY—a global database of plant traits," *Global Change Biology*, 17, 2905–2935.

Keddy, P.A. (1992). "Assembly and response rules: Two goals for predictive community ecology," *Journal of Vegetation Science*, 3, 157–164.

Keddy, P.A. and Laughlin, DC. (2022). *A framework for community ecology: Species pools, filters and traits.* Cambridge: Cambridge University Press.

Keddy, P.A. and MacLellan, P. (1990). "Centrifugal organization in forests," *Oikos*, 59, 75–84.

Keddy, P.A. and Shipley, B. (1989). "Competitive hierarchies in herbaceous plant communities," *Oikos*, 54, 234–241.

Keeley, J.E., et al. (2011). "Fire as an evolutionary pressure shaping plant traits," *Trends in Plant Science*, 16, 406–411.

Kelly, D., et al. (2000). "Predator satiation and extreme mast seeding in 11 species of *Chionochloa* (Poaceae)," *Oikos*, 90, 477–488.

Kelly, D. and Sullivan, J.J. (1997). "Quantifying the benefits of mast seeding on predator satiation and wind pollination in *Chionochloa pallens* (Poaceae)," *Oikos*, 78, 143–150.

Kelly, R., et al. (2021). "Climatic and evolutionary contexts are required to infer plant life history strategies from functional traits at a global scale," *Ecology Letters*, 24, 970–983.

Kenrick, P. and Strullu-Derrien, C. (2014). "The origin and early evolution of roots," *Plant Physiology*, 166, 570–580.

Kikuzawa, K. and Ackerly, D.D. (1999). "Significance of leaf longevity in plants," *Plant Species Biology*, 14, 39–45.

Kimball, S., et al. (2012). "Fitness and physiology in a variable environment," *Oecologia*, 169, 319–329.

Kimberley, A., et al. (2013). "Identifying the trait syndromes of conservation indicator species: How distinct are British ancient woodland indicator plants from other woodland species?" *Applied Vegetation Science*, 16, 667–675.

Kimmerer, R. (2013). Braiding sweetgrass: Indigenous wisdom, scientific knowledge, and the teachings of plants. London: Milkweed Editions.

Kindscher, K. (1992). *Medicinal wild plants of the prairie: An ethnobotanical guide.* Lawrence: University Press of Kansas.

King, D.A., et al. (2006). "The role of wood density and stem support costs in the growth and mortality of tropical trees," *Journal of Ecology*, 94, 670–680.

Kirkwood, T.B.L. (1977). "Evolution of ageing," *Nature*, 270, 301–304.

Kirkwood, T.B.L. (2017). "The disposable soma theory: Origins and evolution," in Shefferson, R., Jones, O., and Salguero-Gómez, R. (eds.) *The evolution of senescence in the tree of life.* Cambridge: Cambridge University Press, pp. 23–39.

Kitajima, K. (1994). "Relative importance of photosynthetic traits and allocation patterns as correlates of seedling shade tolerance of 13 tropical trees," *Oecologia*, 98, 419–428.

Klein, T. (2014). "The variability of stomatal sensitivity to leaf water potential across tree species indicates a continuum between isohydric and anisohydric behaviours," *Functional Ecology*, 28, 1313–1320.

Klimešová, J. and Herben, T. (2015). "Clonal and bud bank traits: Patterns across temperate plant communities," *Journal of Vegetation Science*, 26, 243–253.

Klimešová, J. and Klimeš, L. (2007). "Bud banks and their role in vegetative regeneration—a literature review and proposal for simple classification and assessment," *Perspectives in Plant Ecology, Evolution and Systematics*, 8, 115–129.

Klimešová, J., Martínková, J., and Ottaviani, G. (2018). "Belowground plant functional ecology: Towards an integrated perspective," *Functional Ecology*, 32, 2115–2126.

Klimešová, J., Tackenberg, O., and Herben, T. (2016). "Herbs are different: Clonal and bud bank traits can matter more than leaf–height–seed traits," *New Phytologist*, 210, 13–17.

Klimešová, J., et al. (2019). "Handbook of standardized protocols for collecting plant modularity traits," *Perspectives in Plant Ecology, Evolution and Systematics*, 40, 125485.

Kneitel, J.M. and Chase, J.M. (2004). "Trade-offs in community ecology: Linking spatial scales and species coexistence," *Ecology Letters*, 7, 69–80.

Knight, D.H. (1965). "A gradient analysis of Wisconsin prairie vegetation on the basis of plant structure and function," *Ecology*, 46, 744–747.

Knight, D.H. and Loucks, O.L. (1969). "A quantitative analysis of Wisconsin forest vegetation on the basis of plant function and gross morphology," *Ecology*, 50, 219–234.

Kokko, H. (2007). *Modelling for field biologists and other interesting people.* Cambridge: Cambridge University Press.

Kokko, H. (2021). "The stagnation paradox: The ever-improving but (more or less) stationary population fitness," *Proceedings of the Royal Society B: Biological Sciences*, 288, 20212145.

Komatsu, K.J., et al. (2019). "Global change effects on plant communities are magnified by time and the number of global change factors imposed," *Proceedings of the National Academy of Sciences of the United States of America*, 116, 17867–17873.

Kong, D., et al. (2014). "Leading dimensions in absorptive root trait variation across 96 subtropical forest species," *New Phytologist*, 203, 863–872.

König, P., et al. (2018). "Advances in flowering phenology across the Northern Hemisphere are explained by functional traits," *Global Ecology and Biogeography*, 27, 310–321.

Kooyers, N.J. (2015). "The evolution of drought escape and avoidance in natural herbaceous populations," *Plant Science*, 234, 155–162.

Körner, C. (2016). "Plant adaptation to cold climates," F1000Research [online: version 1; peer review: 2 approved], 5(F1000 Faculty Rev): 2769. https://doi.org/10.12688/f1000research.9107.1

Kozłowski, J. and Wiegert, R.G. (1987). "Optimal age and size at maturity in annuals and perennials with determinate growth," *Evolutionary Ecology*, 1, 231–244.

Kraft, N.J.B., Godoy, O., and Levine, J.M. (2015b). "Plant functional traits and the multidimensional nature of species coexistence," *Proceedings of the National Academy of Sciences of the United States of America*, 112, 797–802.

Kraft, N.J.B., et al. (2015a). "Community assembly, coexistence and the environmental filtering metaphor," *Functional Ecology*, 29, 592–599.

Kramer-Walter, K.R., et al. (2016). "Root traits are multidimensional: Specific root length is independent from root tissue density and the plant economic spectrum," *Journal of Ecology*, 104, 1299–1310.

Kress, W.J. and Krupnick, G.A. (2022). "Lords of the biosphere: Plant winners and losers in the Anthropocene," *Plants, People, Planet*, 4, 350–366.

Kuhn, T.S. (1970). *The structure of scientific revolutions*. Chicago: University of Chicago Press.

Kulpa, S.M. and Leger, E.A. (2013). "Strong natural selection during plant restoration favors an unexpected suite of plant traits," *Evolutionary Applications*, 6, 510–523.

Kunstler, G., et al. (2016). "Plant functional traits have globally consistent effects on competition," *Nature*, 529, 204–207.

Kuswandi, R. and Murdjoko, A. (2015). "Population structures of four tree species in logged-over tropical forest in South Papua, Indonesia: An integral projection model approach," *Indonesian Journal of Forestry Research*, 2, 93–101.

Kutschera, V.L. and Lichtenegger, E. (2002). *Wurzelatlas mitteleuropäischer Waldbäume und Sträucher*. Stuttgart: Leopold Stocker Verlag.

Lack, D. (1947). "The significance of clutch-size," *Ibis*, 89, 302–352.

Ladouceur, E., et al. (2022). "An objective-based prioritization approach to support trophic complexity through ecological restoration species mixes," *Journal of Applied Ecology*, 59, 394–407.

Laliberté, E. (2017). "Below-ground frontiers in trait-based plant ecology," *New Phytologist*, 213, 1597–1603.

Laliberté, E., et al. (2015). "Phosphorus limitation, soil-borne pathogens and the coexistence of plant species in hyperdiverse forests and shrublands," *New Phytologist*, 206, 507–521.

Laliberté, E., et al. (2012). "Which plant traits determine abundance under long-term shifts in soil resource availability and grazing intensity?" *Journal of Ecology*, 100, 662–677.

Lambers, H. and Oliveira, R.S. (2019). *Plant physiological ecology*. 3rd edn. New York: Springer.

Lamontagne, X. and Shipley, B. (2022). "A measure of generalized soil fertility that is largely independent of species identity," *Annals of Botany*, 129, 29–36.

Landres, P.B., Morgan, P., and Swanson, F.J. (1999). "Overview of the use of natural variability concepts in managing ecological systems," *Ecological Applications*, 9, 1179–1188.

Larcher, W. (2003). *Physiological plant ecology: Ecophysiology and stress physiology of functional groups*. 4th edn. Berlin: Springer.

Larson, J.E. and Funk, J.L. (2016). "Regeneration: An overlooked aspect of trait-based plant community assembly models," *Journal of Ecology*, 104, 1284–1298.

Larson, J.E., Ebinger, K.R., and Suding, K.N. (2021). "Water the odds? Spring rainfall and emergence-related seed traits drive plant recruitment," *Oikos*, 130, 1665–1678.

Larson, J.E., et al. (2016). "Do key dimensions of seed and seedling functional trait variation capture variation in recruitment probability?" *Oecologia*, 181, 39–53.

Larson, J.E., et al. (2020). "Ecological strategies begin at germination: Traits, plasticity and survival in the first 4 days of plant life," *Functional Ecology*, 34, 968–979.

Larson, J.E., et al. (2015). "Seed and seedling traits affecting critical life stage transitions and recruitment outcomes in dryland grasses," *Journal of Applied Ecology*, 52, 199–209.

Lauenroth, W.K. and Adler, P.B. (2008). "Demography of perennial grassland plants: Survival, life expectancy and life span," *Journal of Ecology*, 96, 1023–1032.

Lauenroth, W.K., et al. (1994). "The importance of soil water in the recruitment of *Bouteloua gracilis* in the shortgrass steppe," *Ecological Applications*, 4, 741–749.

Lauenroth, W.K., et al. (1997). "Interactions between demographic and ecosystem processes in a semi-arid and an arid grassland: A challenge for plant functional types," in Smith, T.M., Shugart, H.H., and Woodward, F.I. (eds.) *Plant functional types: Their relevance to ecosystem properties and global change*. Cambridge: Cambridge University Press, pp. 234–254.

Laughlin, D.C. (2014b). "Applying trait-based models to achieve functional targets for theory-driven ecological restoration," *Ecology Letters*, 17, 771–784.

Laughlin, D.C. (2014a). "The intrinsic dimensionality of plant traits and its relevance to community assembly," *Journal of Ecology*, 102, 186–193.

Laughlin, D.C. (2018). "Rugged fitness landscapes and Darwinian demons in trait-based ecology," *New Phytologist*, 217, 501–503.

Laughlin, D.C. and Chalmandrier, L. 2018. "Select: Estimates species probabilities based on functional traits," R Package version 1.3.

Laughlin, D.C. and Fulé, P.Z. (2008). "Wildland fire effects on understory plant communities in two fire-prone forests," *Canadian Journal of Forest Research*, 38, 133–142.

Laughlin, D.C. and Messier, J. (2015). "Fitness of multidimensional phenotypes in dynamic adaptive landscapes," *Trends in Ecology & Evolution*, 30, 487–496.

Laughlin, D.C., Moore, M.M., and Fulé, P.Z. (2011b). "A century of increasing pine density and associated shifts in understory plant strategies," *Ecology*, 92, 556–561.

Laughlin, D.C., et al. (2011a). "Climatic constraints on trait-based forest assembly," *Journal of Ecology*, 99, 1489–1499.

Laughlin, D.C., et al. (2020a). "Climatic limits of temperate rainforest tree species are explained by xylem embolism resistance among angiosperms but not among conifers," *New Phytologist*, 226, 727–740.

Laughlin, D.C., et al. (2015b). "Environmental filtering and positive plant litter feedback simultaneously explain correlations between leaf traits and soil fertility," *Ecosystems*, 18, 1269–1280.

Laughlin, D.C., et al. (2018a). "Generating species assemblages for restoration and experimentation: A new method that can simultaneously converge on average trait values and maximize functional diversity," *Methods in Ecology and Evolution*, 9, 1764–1771.

Laughlin, D.C., et al. (2017c). "The hierarchy of predictability in ecological restoration: Are vegetation structure and functional diversity more predictable than community composition?" *Journal of Applied Ecology*, 54, 1058–1069.

Laughlin, D.C., et al. (2017a). "Intraspecific trait variation can weaken interspecific trait correlations when assessing the whole-plant economic spectrum," *Ecology and Evolution*, 7, 8936–8949.

Laughlin, D.C., et al. (2010). "A multi-trait test of the leaf-height-seed plant strategy scheme with 133 species from a pine forest flora," *Functional Ecology*, 24, 493–501.

Laughlin, D.C., et al. (2020b). "The net effect of functional traits on fitness," *Trends in Ecology & Evolution*, 35, 1037–1047.

Laughlin, D.C., et al. (2015a). "Quantifying multimodal trait distributions improves trait-based predictions of species abundances and functional diversity," *Journal of Vegetation Science*, 26, 46–57.

Laughlin, D.C., et al. (2021). "Root traits explain plant species distributions along climatic gradients yet challenge the nature of ecological trade-offs," *Nature Ecology & Evolution*, 5, 1123–1134.

Laughlin, D.C., et al. (2018b). "Survival rates indicate that correlations between community-weighted mean traits and environments can be unreliable estimates of the adaptive value of traits," *Ecology Letters*, 21, 411–421.

Laughlin, D.C., et al. (2017b). "Using trait-based ecology to restore resilient ecosystems: Historical conditions and the future of montane forests in western North America," *Restoration Ecology*, 25, S135–S146.

Lavorel, S. (2013). "Plant functional effects on ecosystem services," *Journal of Ecology*, 101, 4–8.

Lavorel, S. and Garnier, E. (2002). "Predicting changes in community composition and ecosystem functioning from plant traits: Revisiting the Holy Grail," *Functional Ecology*, 16, 545–556.

Lavorel, S., et al. (2007). "Plant functional types: Are we getting any closer to the Holy Grail?" in Canadell, J.G., Pataki, D.E., and Pitelka, L.F. (eds.) *Terrestrial ecosystems in a changing world*. Berlin: Springer, pp. 149–164.

Law, R., Bradshaw, A.D., and Putwain, P.D. (1977). "Life-history variation in *Poa annua*," *Evolution*, 31, 233–246.

Lawes, M.J., Midgley, J.J., and Clarke, P.J. (2013). "Costs and benefits of relative bark thickness in relation to fire damage: A savanna/forest contrast," *Journal of Ecology*, 101, 517–524.

Lay, D.C. (2006). *Linear algebra and its applications*. 3rd edn. Boston: Pearson Education.

Le Coeur, C., et al. (2022). "Life history adaptations to fluctuating environments: Combined effects of demographic buffering and lability," *Ecology Letters*, 25, 2107–2119.

Lee, D.W., Lowry, J.B., and Stone, B. (1979). "Abaxial anthocyanin layer in leaves of tropical rain forest plants: Enhancer of light capture in deep shade," *Biotropica*, 11, 70–77.

Lee, J.A. and Verleyson, M. (2007). *Nonlinear dimensionality reduction*. New York: Springer.

Lefkovitch, L.P. (1965). "The study of population growth in organisms grouped by stages," *Biometrics*, 21, 1–18.

Leger, E.A., et al. (2021). "Selecting native plants for restoration using rapid screening for adaptive traits: Methods and outcomes in a Great Basin case study," *Restoration Ecology*, 29, e13260.

Leimar, O., et al. (2013). "Limiting similarity, species packing, and the shape of competition kernels," *Journal of Theoretical Biology*, 339, 3–13.

Leishman, M.R. and Westoby, M. (1992). "Classifying plants into groups on the basis of associations of individual traits—evidence from Australian semi-arid woodlands," *Journal of Ecology*, 80, 417–424.

Leishman, M.R. and Westoby, M. (1994a). "The role of large seed size in shaded conditions: Experimental evidence," *Functional Ecology*, 8, 205–214.

Leishman, M.R. and Westoby, M. (1994b). "The role of seed size in seedling establishment in dry soil conditions—experimental evidence from semi-arid species," *Journal of Ecology*, 82, 249–258.

Leishman, M.R., Westoby, M., and Jurado, E. (1995). "Correlates of seed size variation: A comparison among five temperate floras," *Journal of Ecology*, 83, 517–529.

Leishman, M.R., et al. (2000). "The evolutionary ecology of seed size," in Fenner, M. (ed.) *Seeds: The ecology of regeneration in plant communities*. Wallingford: CAB International, pp. 31–57.

Lembrechts, J.J., et al. (2020). "SoilTemp: A global database of near-surface temperature," *Global Change Biology*, 26, 6616–6629.

Lenoir, J., et al. (2008). "A significant upward shift in plant species optimum elevation during the 20th Century," *Science*, 320, 1768–1771.

Leopold, A. (1949). *A Sand County almanac with essays on conservation from Round River*. Oxford: Oxford University Press.

Leslie, P.H. (1945). "On the use of matrices in certain population mathematics," *Biometrika*, 33, 183–212.

Lessmann, J.M., et al. (2001). "Effect of climatic gradients on the photosynthetic responses of four *Phragmites australis* populations," *Aquatic Botany*, 69, 109–126.

Levin, S.C., Crandall, R.M., and Knight, T.M. (2019). "Population projection models for 14 alien plant species in the presence and absence of aboveground competition," *Ecology*, 100, e02681.

Levin, S.C., et al. (2021). "ipmr: Flexible implementation of Integral Projection Models in R," *Methods in Ecology and Evolution*, 12, 1826–1834.

Levin, S.C., et al. (2022). "Rpadrino: An R package to access and use PADRINO, an open access database of Integral Projection Models," *Methods in Ecology and Evolution*, 13, 1923–1929.

Levine, J.I., et al. (2022). "Competition for water and species coexistence in phenologically structured annual plant communities," *Ecology Letters*, 25, 1110–1125.

Levine, J.M. (2016). "A trail map for trait-based studies," *Nature*, 529, 163–164.

Levine, J.M. and HilleRisLambers, J. (2009). "The importance of niches for the maintenance of species diversity," *Nature*, 461, 254–257.

Levine, J.M. and Rees, M. (2002). "Coexistence and relative abundance in annual plant assemblages: The roles of competition and colonization," *The American Naturalist*, 160, 452–467.

Levins, R. (1968). *Evolution in changing environments: Some theoretical explorations.* Princeton: Princeton University Press.

Levitt, J. (1980). *Responses of plants to environmental stress*, Volume 2: *Water, radiation, salt and other stresses.* San Diego: Academic Press.

Li, D. and Ives, A.R. (2017). "The statistical need to include phylogeny in trait-based analyses of community composition," *Methods in Ecology and Evolution*, 8, 1192–1199.

Liang, L.L., et al. (2018). "Macromolecular rate theory (MMRT) provides a thermodynamics rationale to underpin the convergent temperature response in plant leaf respiration," *Global Change Biology*, 24, 1538–1547.

Lieberman, D., et al. (1985). "Growth rates and age-size relationships of tropical wet forest trees in Costa Rica," *Journal of Tropical Ecology*, 1, 97–109.

Linkies, A., et al. (2010). "The evolution of seeds," *New Phytologist*, 186, 817–831.

Liu, H., et al. (2012). "Overcoming extreme weather challenges: Successful but variable assisted colonization of wild orchids in southwestern China," *Biological Conservation*, 150, 68–75.

Loehle, C. (1988b). "Problems with the triangular model for representing plant strategies," *Ecology*, 69, 284–286.

Loehle, C. (1988a). "Tree life history strategies: The role of defenses," *Canadian Journal of Forest Research*, 18, 209–222.

Logofet, D.O., Kazantseva, E.S., and Onipchenko, V.G. (2020). "Seed bank as a persistent problem in matrix population models: From uncertainty to certain bounds," *Ecological Modelling*, 438, 109284.

Lortie, C.J. and Hierro, J.L. (2022). "A synthesis of local adaptation to climate through reciprocal common gardens," *Journal of Ecology*, 110, 1015–1021.

Loss, S.R., Terwilliger, L.A., and Peterson, A.C. (2011). "Assisted colonization: Integrating conservation strategies in the face of climate change," *Biological Conservation*, 144, 92–100.

Louthan, A.M., Doak, D.F., and Angert, A.L. (2015). "Where and when do species interactions set range limits?" *Trends in Ecology & Evolution*, 30, 780–792.

Louault, F., et al. (2005). "Plant traits and functional types in response to reduced disturbance in a semi-natural grassland," *Journal of Vegetation Science*, 16, 151–160.

Lundgren, M.R. and Des Marais, D.L. (2020). "Life history variation as a model for understanding trade-offs in plant–environment interactions," *Current Biology*, 30, R180–R189.

Lusk, C.H. and Jorgensen, M.A. (2013). "The whole-plant compensation point as a measure of juvenile tree light requirements," *Functional Ecology*, 27, 1286–1294.

Lusk, C.H. and Laughlin, D.C. (2016). "Regeneration patterns, environmental filtering and tree species coexistence in a temperate forest," *New Phytologist*, 213, 657–668.

Lusk, C.H., Wiser, S.K., and Laughlin, D.C. (2020). "Macroclimate and topography interact to influence the abundance of divaricate plants in New Zealand," *Frontiers in Plant Science*, 11, 507.

Lusk, C.H., et al. (2018). "Frost and leaf-size gradients in forests: Global patterns and experimental evidence," *New Phytologist*, 219, 565–573.

Lynch, A., et al. (1998). "Genetic evidence that *Lomatia tasmanica* (Proteaceae) is an ancient clone," *Australian Journal of Botany*, 46, 25–33.

M'Gonigle, L.K., et al. (2016). "A tool for selecting plants when restoring habitat for pollinators," *Conservation Letters*, 10, 105–111.

Ma, L., et al. (2022). "Global evaluation of the Ecosystem Demography model (ED v3.0)," *Geoscientific Model Development*, 15, 1971–1994.

Ma, Z., et al. (2018). "Evolutionary history resolves global organization of root functional traits," *Nature*, 555, 94–97.

MacArthur, R.H. (1972). *Geographical ecology: Patterns in the distribution of species*. Princeton: Princeton University Press.

MacArthur, R.H. (1968). "The theory of the niche," in Lewontin, R.C. (ed.) *Population biology and evolution*. Syracuse: Syracuse University Press, pp. 159–176.

MacArthur, R.H. and Wilson, E.O. (1967). *The theory of island biogeography*. Princeton: Princeton University Press.

MacColl, A.D. (2011). "The ecological causes of evolution," *Trends in Ecology & Evolution*, 26, 514–522.

Mack, A.L. (1993). "The sizes of vertebrate-dispersed fruits: A neotropical-paleotropical comparison," *The American Naturalist*, 142, 840–856.

Madon, O. and Médail, F. (1997). "The ecological significance of annuals on a Mediterranean grassland (Mt Ventoux, France)," *Plant Ecology*, 129, 189–199.

Maglianesi, M.A., et al. (2014). "Morphological traits determine specialization and resource use in plant–hummingbird networks in the neotropics," *Ecology*, 95, 3325–3334.

Mäkelä, A., et al. (2002). "Challenges and opportunities of the optimality approach in plant ecology," *Silva Fennica*, 36, 605–614.

Mallet, J. (2012). "The struggle for existence. How the notion of carrying capacity, K, obscures the links between demography, Darwinian evolution and speciation," *Evolutionary Ecology Research*, 14, 627–665.

Mandujano, M.C., et al. (2015). "The population dynamics of an endemic collectible cactus," *Acta Oecologica*, 63, 1–7.

Mangel, M. and Stamps, J. (2001). "Trade-offs between growth and mortality and the maintenance of individual variation in growth," *Evolutionary Ecology Research*, 3, 611–632.

Manion, P.D. (1991). *Tree disease concepts*. 2nd edn. Hoboken: Prentice-Hall.

Maracahipes, L., et al. (2018). "How to live in contrasting habitats? Acquisitive and conservative strategies emerge at inter- and intraspecific levels in savanna and forest woody plants," *Perspectives in Plant Ecology, Evolution and Systematics*, 34, 17–25.

Marder, M. (2013). *Plant-thinking: A philosophy of vegetal life.* New York: Columbia University Press.

Marks, C.O. and Lechowicz, M.J. (2006). "Alternative designs and the evolution of functional diversity," *The American Naturalist*, 167, 55–66.

Martínez-Berdeja, A., et al. (2019). "Evidence for population differentiation among Jeffrey and Ponderosa pines in survival, growth and phenology," *Forest Ecology and Management*, 434, 40–48.

Martínez-Garza, C., Bongers, F., and Poorter, L. (2013). "Are functional traits good predictors of species performance in restoration plantings in tropical abandoned pastures?" *Forest Ecology and Management*, 303, 35–45.

Martínez-Vilalta, J., et al. (2010). "Interspecific variation in functional traits, not climatic differences among species ranges, determines demographic rates across 44 temperate and Mediterranean tree species," *Journal of Ecology*, 98, 1462–1475.

Mason, C.M. and Donovan, L.A. (2015). "Does investment in leaf defenses drive changes in leaf economic strategy? A focus on whole-plant ontogeny," *Oecologia*, 177, 1053–1066.

Mason, C.M., et al. (2016). "Phylogenetic structural equation modelling reveals no need for an 'origin' of the leaf economics spectrum," *Ecology Letters*, 19, 54–61.

Mason, N.W.H., et al. (2012). "Changes in coexistence mechanisms along a long-term soil chronosequence revealed by functional trait diversity," *Journal of Ecology*, 100, 678–689.

Matthews, J.W., Spyreas, G., and Long, C.M. (2015). "A null model test of Floristic Quality Assessment: Are plant species' Coefficients of Conservatism valid?" *Ecological Indicators*, 52, 1–7.

Mauseth, J.D. (2009). *Botany: An introduction to plant biology.* 4th edn. Sudbury, MA: Jones and Bartlett Publishers.

May, L.H. and Milthorpe, F.L. (1962). "Drought resistance of crop plants," *Field Crop Abstracts*, 15, 171–179.

Maynard Smith, J. (1982). *Evolution and the theory of games.* Cambridge: Cambridge University Press.

Maynard Smith, J. and Price, G.R. (1973). "The logic of animal conflict," *Nature*, 246, 15–18.

McArthur, C., Hagerman, A.E., and Robbins, C.T. (1991). "Physiological strategies of mammalian herbivores against plant defenses," in Palo, R.T. and Robbins, C.T. (eds.) *Plant defenses against mammalian herbivory.* Boca Raton: CRC Press, Inc, pp. 103–114.

McCarthy, J.K. (2018). *Predicting the diversity and functional composition of woody plant communities under climate change.* PhD Thesis, Brisbane: University of Queensland.

McCormack, M.L., et al. (2015). "Redefining fine roots improves understanding of below-ground contributions to terrestrial biosphere processes," *New Phytologist*, 207, 505–518.

McCrae, R.R. and Costa, P.T. (2003). *Personality in adulthood: A five-factor theory perspective.* 2nd edn. New York: Guilford Press.

McDowell, N., et al. (2008). "Mechanisms of plant survival and mortality during drought: Why do some plants survive while others succumb to drought?" *New Phytologist*, 178, 719–739.

McDowell, N.G., et al. (2022). "Mechanisms of woody-plant mortality under rising drought, CO2 and vapour pressure deficit," *Nature Reviews Earth & Environment*, 3, 294–308.

McGhee, G.R. (2011). *Convergent evolution: Limited forms most beautiful.* Cambridge, MA: MIT Press.

McGill, B.J. (2012). "Trees are rarely most abundant where they grow best," *Journal of Plant Ecology*, 5, 46–51.

McGill, B.J. and Brown, J.S. (2007). "Evolutionary game theory and adaptive dynamics of continuous traits," *Annual Review of Ecology, Evolution, and Systematics*, 38, 403–435.

McGill, B.J., et al. (2006). "Rebuilding community ecology from functional traits," *Trends in Ecology & Evolution*, 21, 178–185.

McGraw, J.B. and Caswell, H. (1996). "Estimation of individual fitness from life-history data," *The American Naturalist*, 147, 47–64.

McNickle, G.G. and Dybzinski, R. (2013). "Game theory and plant ecology," *Ecology Letters*, 16, 545–555.

McNickle, G.G. and Evans, W.D. (2018). "Toleration games: Compensatory growth by plants in response to enemy attack is an evolutionarily stable strategy," *AoB PLANTS*, 10, ply035.

McNickle, G.G., et al. (2016). "The world's biomes and primary production as a triple tragedy of the commons foraging game played among plants," Proceedings of the Royal Society B: Biological Sciences, 283, 20161993.

Meller, P., et al. (2022). "Correlates of geoxyle diversity in Afrotropical grasslands," *Journal of Biogeography*, 49, 339–352.

Mencuccini, M. and Munné-Bosch, S. (2017). "Physiological and biochemical processes related to ageing and senescence in plants," in Shefferson, R., Jones, O., and Salguero-Gómez, R. (eds.) *The evolution of senescence in the tree of life*. Cambridge: Cambridge University Press, pp. 257–283.

Meng, T.T., et al. (2015). "Responses of leaf traits to climatic gradients: Adaptive variation versus compositional shifts," *Biogeosciences*, 12, 5339–5352.

Merchant, T.K., et al. (2022). "Four reasons why functional traits are not being used in restoration practice," *Restoration Ecology*, 30, e13788.

Merow, C., et al. (2017). "Climate change both facilitates and inhibits invasive plant ranges in New England," *Proceedings of the National Academy of Sciences of the United States of America*, 114, E3276.

Merow, C., et al. (2014a). "Advancing population ecology with integral projection models: A practical guide," *Methods in Ecology and Evolution*, 5, 99–110.

Merow, C., et al. (2014b). "On using integral projection models to generate demographically driven predictions of species' distributions: Development and validation using sparse data," *Ecography*, 37, 1167–1183.

Metcalf, C.J.E., et al. (2009). "A time to grow and a time to die: A new way to analyze the dynamics of size, light, age, and death of tropical trees," *Ecology*, 90, 2766–2778.

Metcalf, C.J.E., et al. (2013). "IPMpack: An R package for integral projection models," *Methods in Ecology and Evolution*, 4, 195–200.

Metcalf, C.J.E. and Pavard, S. (2007). "Why evolutionary biologists should be demographers," *Trends in Ecology & Evolution*, 22, 205–212.

Metcalf, C.J.E., et al. (2006). "Growth–survival trade-offs and allometries in rosette-forming perennials," *Functional Ecology*, 20, 217–225.

Metcalf, C.J.E., et al. (2008). "Evolution of flowering decisions in a stochastic, density-dependent environment," *Proceedings of the National Academy of Sciences of the United States of America*, 105, 10466–10470.

Metcalf, C.J.E., Rose, K.E., and Rees, M. (2003). "Evolutionary demography of monocarpic perennials," *Trends in Ecology & Evolution*, 18, 471–480.

Metz, J.A.J. (2012). "Adaptive dynamics," in Alan, H. and Louis, G. (eds.) *Encyclopedia of theoretical ecology*. Berkeley: University of California Press, pp. 7–17.

Metz, J.A.J, Nisbet, R.M., and Geritz, S.A.H. (1992). "How should we define 'fitness' for general ecological scenarios?" *Trends in Ecology & Evolution*, 7, 198–202.

Milchunas, D.G., Sala, O.E., and Lauenroth, W.K. (1988). "A generalized model of the effects of grazing by large herbivores on grassland community structure," *The American Naturalist*, 132, 87–106.

Miller, J.E.D., Damschen, E.I., and Ives, A.R. (2018). "Functional traits and community composition: A comparison among community-weighted means, weighted correlations, and multilevel models," *Methods in Ecology and Evolution*, 10, 415–425.

Moles, A.T. (2018). "Being John Harper: Using evolutionary ideas to improve understanding of global patterns in plant traits," *Journal of Ecology*, 106, 1–18.

Moles, A.T. and Leishman, M.R. (2008). "The seedling as part of a plant's life history strategy," in Leck, M.A., Parker, V.T., and Simpson, R.L. (eds.) *Seedling ecology and evolution*. Cambridge: Cambridge University Press, pp. 217–238.

Moles, A.T. and Westoby, M. (2006). "Seed size and plant strategy across the whole life cycle," *Oikos*, 113, 91–105.

Moles, A.T., Hodson, D.W., and Webb, C.J. (2000). "Seed size and shape and persistence in the soil in the New Zealand flora," *Oikos*, 89, 541–545.

Moles, A.T., et al. (2005b). "A brief history of seed size," *Science*, 307, 576–580.

Moles, A.T., et al. (2005a). "Factors that shape seed mass evolution," Proceedings of the National Academy of Sciences of the United States of America, 102, 10540–10544.

Moles, A.T., et al. (2007). "Global patterns in seed size," *Global Ecology and Biogeography*, 16, 109–116.

Moles, A.T., et al. (2009). "Global patterns in plant height," *Journal of Ecology*, 97, 923–932.

Monro, A.K., et al. (2018). "Discovery of a diverse cave flora in China," *PLoS One*, 13, e0190801.

Mooney, H.A. and Dunn, E.L. (1970). "Convergent evolution of Mediterranean-climate evergreen sclerophyll shrubs," *Evolution*, 24, 292–303.

Mooney, H.A. and Gulmon, S.L. (1979). "Topics in plant population biology," in Otto, T.S., et al. (eds.) *Environmental and evolutionary constraints on the photosynthetic characteristics of higher plants*. New York: Columbia University Press, pp. 316–337.

Moore, M.M., Wallace Covington, W., and Fulé, P.Z. (1999). "Reference conditions and ecological restoration: A southwestern ponderosa pine perspective," *Ecological Applications*, 9, 1266–1277.

Moore, M.M., et al. (2022). "Cover and density of southwestern ponderosa pine understory plants in permanent chart quadrats (2002–2020)," *Ecology*, 103, e3661.

Morales, M., et al. (2013). "Photo-oxidative stress markers reveal absence of physiological deterioration with ageing in *Borderea pyrenaica*, an extraordinarily long-lived herb," *Journal of Ecology*, 101, 555–565.

Morris, C.D. (2016). "Is the grazing tolerance of mesic decreaser and increaser grasses altered by soil nutrients and competition?" *African Journal of Range & Forage Science*, 33, 235–245.

Morris, W.F. and Doak, D.F. (1998). "Life history of the long-lived gynodioecious cushion plant *Silene acaulis* (Caryophyllaceae), inferred from size-based population projection matrices," *American Journal of Botany*, 85, 784–793.

Morris, W.F. and Doak, D.F. (2002). *Quantitative conservation biology: Theory and practice of population viability analysis*. Sunderland, MA: Sinauer.

Morton, A.G. (1981). *History of botanical science: An account of the development of botany from ancient times to the present day*. Academic Press.

Mouillot, D., et al. (2021). "The dimensionality and structure of species trait spaces," *Ecology Letters*, 24, 1988–2009.

Mrad, A., et al. (2021). "The roles of conduit redundancy and connectivity in xylem hydraulic functions," *New Phytologist*, 231, 996–1007.

Mueller, J.M. and Hellmann, J.J. (2008). "An assessment of invasion risk from assisted migration," *Conservation Biology*, 22, 562–567.

Mueller-Dombois, D. and Ellenberg, H. (1974). *Aims and methods of vegetation ecology*. Hoboken: John Wiley and Sons, Inc.

Muller-Landau, H.C. (2004). "Interspecific and inter-site variation in wood specific gravity of tropical trees," *Biotropica*, 36, 20–32.

Munné-Bosch, S. (2007). "Aging in perennials," *Critical Reviews in Plant Sciences*, 26, 123–138.

Munné-Bosch, S. (2014). "Perennial roots to immortality," *Plant Physiology*, 166, 720–725.

Murcia, C., et al. (2014). "A critique of the 'novel ecosystem' concept," *Trends in Ecology & Evolution*, 29, 548–553.

Mutch, R.W. (1970). "Wildland fires and ecosystems—a hypothesis," *Ecology*, 51, 1046–1051.

Myers, I.B. and Myers, P.B. (1980). *Gifts differing: Understanding personality types*. Boston: Davies-Black Publishing.

Nardini, A. and Luglio, J. (2014). "Leaf hydraulic capacity and drought vulnerability: Possible trade-offs and correlations with climate across three major biomes," *Functional Ecology*, 28, 810–818.

Nascimento, H.E., et al. (2005). "Demographic and life-history correlates for Amazonian trees," *Journal of Vegetation Science*, 16, 625–634.

Nathan, M.J. (2021). "Does anybody really know what time it is? From biological age to biological time," *History and philosophy of the life sciences*, 43, 26-26.

Nathan, R. (2006). "Long-distance dispersal of plants," *Science*, 313, 786–788.

Navarro-Cano, J.A., Goberna, M., and Verdú, M. (2019). "Using plant functional distances to select species for restoration of mining sites," *Journal of Applied Ecology*, 56, 2353–2362.

Needham, J., et al. (2016). "Forest community response to invasive pathogens: The case of ash dieback in a British woodland," *Journal of Ecology*, 104, 315–330.

Needham, J., et al. (2018). "Inferring forest fate from demographic data: From vital rates to population dynamic models," Proceedings of the Royal Society B: Biological Sciences, 285, 20172050.

Nelson, R.A. (1979). *Handbook of Rocky Mountain plants*. 3rd edn. Ringwood: Skyland Publishers.

Nguyen, V., et al. (2019). "Consequences of neglecting cryptic life stages from demographic models," *Ecological Modelling*, 408, 108723.

Niinemets, Ü. and Tenhunen, J. (1997). "A model separating leaf structural and physiological effects on carbon gain along light gradients for the shade-tolerant species *Acer saccharum*," *Plant, Cell & Environment*, 20, 845–866.

Niinemets, Ü. and Valladares, F. (2006). "Tolerance to shade, drought, and waterlogging of temperate northern hemisphere trees and shrubs," *Ecological Monographs*, 76, 521–547.

Niklas, K.J. (1994). "Predicting the height of fossil plant remains: An allometric approach to an old problem," *American Journal of Botany*, 81, 1235–1242.

Niklas, K.J. (1997). *The evolutionary biology of plants*. Chicago: University of Chicago Press.

Nobel, P.S. (1991). "Achievable productivities of certain CAM plants: Basis for high values compared with C3 and C4 plants," *New Phytologist*, 119, 183–205.

Noble, I.R. and Slatyer, R. (1980). "The use of vital attributes to predict successional changes in plant communities subject to recurrent disturbances," *Vegetatio*, 43, 5–21.

Nolan, R.H., et al. (2021). "Limits to post-fire vegetation recovery under climate change," *Plant, Cell & Environment*, 44, 3471–3489.

Noodén, L.D. (1988a). "Whole plant senescence and aging," in Noodén, L.D. and Leopold, A.C. (eds.) *Senescence and aging in plants*. San Diego: Academic Press, Inc, pp. 392–439.

Noodén, L.D. (1988b). "The phenomena of senescence and aging," in Noodén, L.D. and Leopold, A.C. (eds.) *Senescence and aging in plants*. San Diego: Academic Press, Inc, pp. 2–50.

Norton, J. and Ouyang, Y. (2019). "Controls and adaptive management of nitrification in agricultural soils," *Frontiers in Microbiology*, 10, 1931.

Noy-Meir, I. (1973). "Desert ecosystems: Environment and producers," *Annual Review of Ecology and Systematics*, 4, 25–51.

Ocheltree, T.W., Nippert, J.B., and Prasad, P.V.V. (2016). "A safety vs efficiency trade-off identified in the hydraulic pathway of grass leaves is decoupled from photosynthesis, stomatal conductance and precipitation," *New Phytologist*, 210, 97–107.

Ohnmeiss, T.E. and Baldwin, I.T. (1994). "The allometry of nitrogen to growth and an inducible defense under nitrogen-limited growth," *Ecology*, 75, 995–1002.

Oksanen, J. (1990). "Predation, herbivory, and plant strategies along gradients of primary productivity," in Grace, J.B. and Tilman, D. (eds.) *Perspectives on plant competition*. San Diego: Academic Press, Inc, pp. 445–474.

Oldeman, R.A.A. and Van Dijk, J. (1991). "Diagnosis of the temperament of tropical rain forest trees," in Gomez-Pompa, A., Whitmore, T.C., and Hadley, M. (eds.) *Rain forest regeneration and management*. Man and Biosphere Series 6. Paris: UNESCO, pp. 21–65.

Oliveira, R.S., et al. (2019). "Embolism resistance drives the distribution of Amazonian rainforest tree species along hydro-topographic gradients," *New Phytologist*, 221, 1457–1465.

Oliveira, R.S., et al. (2021). "Linking plant hydraulics and the fast–slow continuum to understand resilience to drought in tropical ecosystems," *New Phytologist*, 230, 904–923.

Ondei, S., et al. (2016). "Post-fire resprouting strategies of rainforest and savanna saplings along the rainforest–savanna boundary in the Australian monsoon tropics," *Plant Ecology*, 217, 711–724.

Ordonez, A., Wright, I.J., and Olff, H. (2010). "Functional differences between native and alien species: A global-scale comparison," *Functional Ecology*, 24, 1353–1361.

Ordoñez, J.C., et al. (2009). "A global study of relationships between leaf traits, climate and soil measures of nutrient fertility," *Global Ecology and Biogeography*, 18, 137–149.

Orians, G.H. and Solbrig, O.T. (1977). "A cost-income model of leaves and roots with special reference to arid and semiarid areas," *The American Naturalist*, 111, 677–690.

Orme, D., et al. (2018). "caper: Comparative analyses of phylogenetics and evolution in R," R package version 1.0.1.

Ostertag, R., et al. (2015). "Using plant functional traits to restore Hawaiian rainforest," *Journal of Applied Ecology*, 52, 805–809.

Ott, J.P., Klimešová, J., and Hartnett, D.C. (2019). "The ecology and significance of below-ground bud banks in plants," *Annals of Botany*, 123, 1099–1118.

Ottaviani, G., et al. (2017). "On plant modularity traits: Functions and challenges," *Trends in Plant Science*, 22, 648–651.

Paine, C.E.T., et al. (2015). "Globally, functional traits are weak predictors of juvenile tree growth, and we do not know why," *Journal of Ecology*, 103, 978–989.

Paine, C.E.T., et al. (2012). "How to fit nonlinear plant growth models and calculate growth rates: An update for ecologists," *Methods in Ecology and Evolution*, 3, 245–256.

Pakeman, R.J. (2004). "Consistency of plant species and trait responses to grazing along a productivity gradient: A multi-site analysis," *Journal of Ecology*, 92, 893–905.

Palma, E., et al. (2021). "Plant functional traits reflect different dimensions of species invasiveness," *Ecology*, 102, e03317.

Palmer, B. (2019). "Restoration ecology: The study of applied optimism," *Restoration Ecology*, 27, 1192–1193.

Pammenter, N.W. and Van der Willigen, C. (1998). "A mathematical and statistical analysis of the curves illustrating vulnerability of xylem to cavitation," *Tree Physiology*, 18, 589–593.

Panchen, Z.A., et al. (2014). "Leaf out times of temperate woody plants are related to phylogeny, deciduousness, growth habit and wood anatomy," *New Phytologist*, 203, 1208–1219.

Paniw, M., de la Riva, E.G., and Lloret, F. (2021). "Demographic traits improve predictions of spatiotemporal changes in community resilience to drought," *Journal of Ecology*, 109, 3233–3245.

Paniw, M., et al. (2017). "Accounting for uncertainty in dormant life stages in stochastic demographic models," *Oikos*, 126, 900–909.

Paquette, A. and Hargreaves, A.L. (2021). "Biotic interactions are more often important at species' warm versus cool range edges," *Ecology Letters*, 24, 2427–2438.

Park, A. and Talbot, C. (2018). "Information underload: Ecological complexity, incomplete knowledge, and data deficits create challenges for the assisted migration of forest trees," *Bioscience*, 68, 251–263.

Parkhurst, D.F. and Loucks, O. (1972). "Optimal leaf size in relation to environment," *Journal of Ecology*, 60, 505–537.

Parmesan, C. and Hanley, M.E. (2015). "Plants and climate change: Complexities and surprises," *Annals of Botany*, 116, 849–864.

Parsley, K.M. (2020). "Plant awareness disparity: A case for renaming plant blindness," *Plants, People, Planet*, 2, 598–601.

Parsons, R.F. (1968). "The significance of growth-rate comparisons for plant ecology," *The American Naturalist*, 102, 595–597.

Paterno, G.B., et al. (2020). "The maleness of larger angiosperm flowers," Proceedings of the National Academy of Sciences of the United States of America, 117, 10921–10926.

Pausas, J.G. (2015). "Bark thickness and fire regime," *Functional Ecology*, 29, 315–327.

Pausas, J.G. and Bradstock, R.A. (2007). "Fire persistence traits of plants along a productivity and disturbance gradient in Mediterranean shrublands of south-east Australia," *Global Ecology and Biogeography*, 16, 330–340.

Pausas, J.G. and Keeley, J.E. (2009). "A burning story: The role of fire in the history of life," *Bioscience*, 59, 593–601.

Pausas, J.G. and Keeley, J.E. (2014). "Evolutionary ecology of resprouting and seeding in fire-prone ecosystems," *New Phytologist*, 204, 55–65.

Pausas, J.G. and Keeley, J.E. (2017). "Epicormic resprouting in fire-prone ecosystems," *Trends in Plant Science*, 22, 1008–1015.

Pausas, J.G., et al. (2004). "Plant functional traits in relation to fire in crown-fire ecosystems," *Ecology*, 85, 1085–1100.

Pavlick, R., et al. (2013). "The Jena Diversity-Dynamic Global Vegetation Model (JeDi-DGVM): A diverse approach to representing terrestrial biogeography and biogeochemistry based on plant functional trade-offs," *Biogeosciences*, 10, 4137–4177.

Pearl, R. and Miner, J.R. (1935). "Experimental studies on the duration of life. XIV. The comparative mortality of certain lower organisms," *The Quarterly Review of Biology*, 10, 60–79.

Pedersen, B. (1999). "Senescence in plants," in Vuorisalo, T.O. and Mutikainen, P.K. (eds.) *Life history evolution in plants*. Dordrecht: Kluwer Academic Publishers, pp. 239–274.

Peltzer, D.A., et al. (2010). "Understanding ecosystem retrogression," *Ecological Monographs*, 80, 509–529.

Peñuelas, J. and Munné-Bosch, S. (2010). "Potentially immortal?" *The New Phytologist*, 187, 564–567.

Peppe, D.J., et al. (2018). "Reconstructing paleoclimate and paleoecology using fossil leaves," in Croft, D.A., Su, D.F., and Simpson, S.W. (eds.) *Methods in paleoecology: Reconstructing Cenozoic terrestrial environments and ecological communities*. Cham: Springer International Publishing, pp. 289–317.

Peppe, D.J., et al. (2011). "Sensitivity of leaf size and shape to climate: Global patterns and paleoclimatic applications," *New Phytologist*, 190, 724–739.

Pérez-Harguindeguy, N., et al. (2013). "New handbook for standardised measurement of plant functional traits worldwide," *Australian Journal of Botany*, 61, 167–234.

Pérez-Ramos, I.M., et al. (2019). "Functional traits and phenotypic plasticity modulate species coexistence across contrasting climatic conditions," *Nature Communications*, 10, 2555.

Perring, M.P., et al. (2015). "Advances in restoration ecology: Rising to the challenges of the coming decades," *Ecosphere*, 6, 1–25.

Petermann, J.S., et al. (2008). "Janzen–Connell effects are widespread and strong enough to maintain diversity in grasslands," *Ecology*, 89, 2399–2406.

Peters, R.H. (1991). *A critique for ecology*. Cambridge: Cambridge University Press.

Pianka, E.R. (1970). "On *r*-and *K*-selection," *The American Naturalist*, 104, 592–597.

Pianka, E.R., et al. (2017). "Toward a periodic table of niches, or exploring the lizard niche hypervolume," *The American Naturalist*, 190, 601–616.

Piao, S., et al. (2019). "Plant phenology and global climate change: Current progresses and challenges," *Global Change Biology*, 25, 1922–1940.

Pickett, S.T.A. and White, P.S. (1985). *The ecology of natural disturbance and patch dynamics*. San Diego: Academic Press.

Pierce, S., et al. (2013). "Allocating CSR plant functional types: The use of leaf economics and size traits to classify woody and herbaceous vascular plants," *Functional Ecology*, 27, 1002–1010.

Pierret, A., et al. (2016). "Understanding deep roots and their functions in ecosystems: An advocacy for more unconventional research," *Annals of Botany*, 118, 621–635.

Pigliucci, M. (2003). "Phenotypic integration: Studying the ecology and evolution of complex phenotypes," *Ecology Letters*, 6, 265–272.

Pilon, N.A.L., et al. (2021). "The diversity of post-fire regeneration strategies in the Cerrado ground layer," *Journal of Ecology*, 109, 154–166.

Pintor, L.M., Brown, J.S., and Vincent, T.L. (2011). "Evolutionary game theory as a framework for studying biological invasions," *The American Naturalist*, 177, 410–423.

Piqueray, J., et al. (2015). "Response of plant functional traits during the restoration of calcareous grasslands from forest stands," *Ecological Indicators*, 48, 408–416.

Pistón, N., et al. (2019). "Multidimensional ecological analyses demonstrate how interactions between functional traits shape fitness and life history strategies," *Journal of Ecology*, 107, 2317–2328.

Plard, F., et al. (2019). "IPM2: Toward better understanding and forecasting of population dynamics," *Ecological Monographs*, 89, e01364.

Polis, G.A. (1999). "Why are parts of the world green? Multiple factors control productivity and the distribution of biomass," *Oikos*, 86, 3–15.

Pollock, L.J., Morris, W.K., and Vesk, P.A. (2012). "The role of functional traits in species distributions revealed through a hierarchical model," *Ecography*, 35, 716–725.

Pons, T.L. (2000). "Seed responses to light," in Fenner, M. (ed.) *Seeds: The ecology of regeneration in plant communities*. Wallingford: CAB International, pp. 237–260.

Poorter, H. (1989). "Interspecific variation in relative growth rate: On ecological causes and physiological consequences," in Lambers, H., et al. (eds.) *Causes and consequences of variation in growth rate and productivity of higher plants*. The Hague: SPB Academic Publishing, pp. 45–68.

Poorter, L. and Bongers, F. (2006). "Leaf traits are good predictors of plant performance across 53 rain forest species," *Ecology*, 87, 1733–1743.

Poorter, H. and Garnier, E. (2007). "Ecological significance of inherent variation in relative growth rate and its components," in Pugnaire, F.I. and Valladares, F. (eds.) *Functional plant ecology*. Boca Raton: CRC Press, pp. 67–100.

Poorter, H. and Remkes, C. (1990). "Leaf area ratio and net assimilation rate of 24 wild species differing in relative growth rate," *Oecologia*, 83, 553–559.

Poorter, H. and Van der Werf, A. (1998). "Is inherent variation in RGR determined by LAR at low irradiance and by NAR at high irradiance? A review of herbaceous species," in Lambers, H., Poorter, H., and Van Vuuren, M.M.I. (eds.) *Inherent variation in plant growth: Physiological mechanisms and ecological consequences*. Leiden: Backhuys Publishers, pp. 309–336.

Poorter, H., et al. (2009). "Causes and consequences of variation in leaf mass per area (LMA): A meta-analysis," *New Phytologist*, 182, 565–588.

Poorter, L., et al. (2010). "The importance of wood traits and hydraulic conductance for the performance and life history strategies of 42 rainforest tree species," *New Phytologist*, 185, 481–492.

Poorter, H., et al. (1991). "Respiratory energy requirements of roots vary with the potential growth rate of a plant species," *Physiologia Plantarum*, 83, 469–475.

Post, E. (2019). *Time in ecology*. Princeton: Princeton University Press.

Prach, K. and Pyšek, P. (1999). "How do species dominating in succession differ from others?" *Journal of Vegetation Science*, 10, 383–392.

Prach, K. and Walker, L.R. (2020). *Comparative plant succession among terrestrial biomes of the world*. Cambridge: Cambridge University Press.

Prentice, I.C., et al. (2014). "Balancing the costs of carbon gain and water transport: Testing a new theoretical framework for plant functional ecology," *Ecology Letters*, 17, 82–91.

Primack, R.B. (1979). "Reproductive effort in annual and perennial species of *Plantago* (Plantaginaceae)," *The American Naturalist*, 114, 51–62.

Puglielli, G., Hutchings, M.J., and Laanisto, L. (2021). "The triangular space of abiotic stress tolerance in woody species: A unified trade-off model," *New Phytologist*, 229, 1354–1362.

Pugnaire, F. and Valladares, F. (eds.) (2007). *Functional plant ecology*. 2nd edn. Boca Raton: CRC Press.

Puhlick, J.J., Laughlin, D.C., and Moore, M.M. (2012). "Factors influencing ponderosa pine regeneration in the southwestern USA," *Forest Ecology and Management*, 264, 10–19.

Puhlick, J.J., et al. (2021). "Soil properties and climate drive ponderosa pine seedling presence in the southwestern USA," *Forest Ecology and Management*, 486, 118972.

Purves, D.W., et al. (2008). "Predicting and understanding forest dynamics using a simple tractable model," Proceedings of the National Academy of Sciences of the United States of America, 105, 17018–17022.

Pywell, R.F., et al. (2003). "Plant traits as predictors of performance in ecological restoration," *Journal of Applied Ecology*, 40, 65–77.

Qiu, T., et al. (2021). "Is there tree senescence? The fecundity evidence," Proceedings of the National Academy of Sciences of the United States of America, 118, e2106130118.

Querejeta, J.I., et al. (2022). "Higher leaf nitrogen content is linked to tighter stomatal regulation of transpiration and more efficient water use across dryland trees," *New Phytologist*, 235, 1351–1364.

Rabinowitz, D. (1981). "Seven forms of rarity," in Synge, H. (ed.) *The biological aspects of rare plant conservation*. Oxford: John Wiley and Sons, pp. 205–217.

Rabotnov, T.A. (1985). "Dynamics of plant coenotic populations," in White, J. (ed.) *The population structure of vegetation*. Dordrecht: Dr W. Junk Publishers, pp. 121–142.

Ramenskii, L.G. (1938). *Introduction to the complex soil-geobotanical investigation of lands* [in Russian]. Moscow: Selkhozgiz.

Raunkiaer, C. (1934). *The life forms of plants and statistical plant geography: Being the collected papers of C. Raunkiaer*. Oxford: Clarendon Press.

Raven, J.A. (2013). "Rubisco: Still the most abundant protein of Earth?" *New Phytologist*, 198, 1–3.

Raven, P.H., Evert, R.F., and Eichhorn, S.E. (2004). *Biology of plants*. New York: W.H. Freeman.

Read, D.J. (1991). "Mycorrhizas in ecosystems," *Experientia*, 47, 376–391.

Rees, M. (1997a). "Seed dormancy," in Crawley, M.J. (ed.) *Plant ecology*. Oxford: Blackwell Scientific, pp. 214–238

Rees, M. (1997b). "Seed dormancy and seed size," in Silvertown, J., Franco, M., and Harper, J. (eds.) *Plant life histories: Ecology, phylogeny, and evolution*. Cambridge: Cambridge University Press, pp. 121–142.

Rees, M. (1993). "Trade-offs among dispersal strategies in British plants," *Nature*, 366, 150–152.

Rees, M. and Ellner, S.P. (2016). "Evolving integral projection models: Evolutionary demography meets eco-evolutionary dynamics," *Methods in Ecology and Evolution*, 7, 157–170.

Rees, M. and Long, M.J. (1993). "The analysis and interpretation of seedling recruitment curves," *The American Naturalist*, 141, 233–262.

Rees, M. and Westoby, M. (1997). "Game-theoretical evolution of seed mass in multi-species ecological models," *Oikos*, 78, 116–126.

Rees, M., Childs, D.Z., and Ellner, S.P. (2014). Building integral projection models: A user's guide. *Journal of Animal Ecology*, 83, 528–545.

Rees, M., Grubb, P.J., and Kelly, D. (1996). "Quantifying the impact of competition and spatial heterogeneity on the structure and dynamics of a four-species guild of winter annuals," *The American Naturalist*, 147, 1–32.

Rees, M., et al. (2010). "Partitioning the components of relative growth rate: How important is plant size variation?" *The American Naturalist*, 176, E152–E161.

Rees, M., et al. (2006). "Seed dormancy and delayed flowering in monocarpic plants: Selective interactions in a stochastic environment," *The American Naturalist*, 168, E53–E71.

Reich, P.B. (2014). "The world-wide 'fast–slow' plant economics spectrum: A traits manifesto," *Journal of Ecology*, 102, 275–301.

Reich, P.B., Walters, M.B., and Ellsworth, D.S. (1997). "From tropics to tundra: Global convergence in plant functioning," *Proceedings of the National Academy of Sciences of the United States of America*, 94, 13730–13734.

Reich, P.B., Walters, M.B., and Ellsworth, D.S. (1992). "Leaf life-span in relation to leaf, plant, and stand characteristics among diverse ecosystems," *Ecological Monographs*, 62, 365–392.

Reich, P.B., et al. (2003). "The evolution of plant functional variation: Traits, spectra, and strategies," *International Journal of Plant Sciences*, 164, S143–S164.

Reich, P.B., et al. (1999). "Generality of leaf trait relationships: A test across six biomes," *Ecology*, 80, 1955–1969.

Revell, L.J. (2012). "phytools: An R package for phylogenetic comparative biology (and other things)," *Methods in Ecology and Evolution*, 3, 217–223.

Revell, L.J. and Harmon, L.J. (2022). *Phylogenetic comparative methods in R*. Princeton: Princeton University Press.

Ricciardi, A. and Simberloff, D. (2009). "Assisted colonization is not a viable conservation strategy," *Trends in Ecology & Evolution*, 24, 248–253.

Richardson, S.J., et al. (2015). "Functional and environmental determinants of bark thickness in fire-free temperate rain forest communities," *American Journal of Botany*, 102, 1590–1598.

Richardson, S.J., et al. (2004). "Rapid development of phosphorus limitation in temperate rainforest along the Franz Josef soil chronosequence," *Oecologia*, 139, 267–276.

Ridley, H.N. (1930). *The dispersal of plants throughout the world*. London: L. Reeve & Company, Limited.

Roach, D.A. (1993). "Evolutionary senescence in plants," *Genetica*, 91, 53–64.

Roddy, A.B., et al. (2021). "Towards the flower economics spectrum," *New Phytologist*, 229, 665–672.

Roderick, M.L. (2000). "On the measurement of growth with applications to the modelling and analysis of plant growth," *Functional Ecology*, 14, 244–251.

Roff, D.A. (1993). *Evolution of life histories: Theory and analysis*. New York: Springer Science & Business Media.

Roff, D.A. (2008). "Defining fitness in evolutionary models," *Journal of Genetics*, 87, 339–348.

Roff, D.A. (1986). "Predicting body size with life history models," *Bioscience*, 36, 316–323.

Rohde, A. and Bhalerao, R.P. (2007). "Plant dormancy in the perennial context," *Trends in Plant Science*, 12, 217–223.

Römer, G., et al. (2021). "Plant demographic knowledge is biased towards short-term studies of temperate-region herbaceous perennials," *bioRxiv* [online], 2021.2004.2025.441327.

Roper, M., Capdevila, P., and Salguero-Gómez, R. (2021). "Senescence: Why and where selection gradients might not decline with age," *Proceedings of the Royal Society B: Biological Sciences*, 288, 20210851.

Rossetto, M., et al. (2017). "From songlines to genomes: Prehistoric assisted migration of a rain forest tree by Australian Aboriginal people," *PLoS One*, 12, e0186663.

Rost, T.L., et al. (1984). *Botany: A brief introduction to plant biology*. 2nd edn. New York: John Wiley & Sons.

Roughgarden, J. (1976). "Resource partitioning among competing species—a coevolutionary approach," *Theoretical Population Biology*, 9, 388–424.

Royer, D.L., et al. (2007). "Fossil leaf economics quantified: Calibration, Eocene case study, and implications," *Paleobiology*, 33, 574–589.

Rüger, N., et al. (2018). "Beyond the fast–slow continuum: Demographic dimensions structuring a tropical tree community," *Ecology Letters*, 21, 1075–1084.

Rüger, N., et al. (2020). "Demographic trade-offs predict tropical forest dynamics," *Science*, 368, 165–168.

Rüger, N., et al. (2012). "Functional traits explain light and size response of growth rates in tropical tree species," *Ecology*, 93, 2626–2636.

Russo, S.E., et al. (2021). "The interspecific growth–mortality trade-off is not a general framework for tropical forest community structure," *Nature Ecology & Evolution*, 5, 174–183.

Saatkamp, A., et al. (2019). "A research agenda for seed-trait functional ecology," *New Phytologist*, 221, 1764–1775.

Sack, L. and Grubb, P.J. (2001). "Why do species of woody seedlings change rank in relative growth rate between low and high irradiance?" *Functional Ecology*, 15, 145–154.

Sack, L., et al. (2013). "How do leaf veins influence the worldwide leaf economic spectrum? Review and synthesis," *Journal of Experimental Botany*, 64, 4053–4080.

Sáenz-Romero, C., et al. (2021). "Assisted migration field tests in Canada and Mexico: Lessons, limitations, and challenges," *Forests*, 12, 9.

Sagan, C. (2006). *The varieties of scientific experience: A personal view of the search for God*. London: Penguin Press.

Sage, R.F. (2004). "The evolution of C4 photosynthesis," *New Phytologist*, 161, 341–370.

Sage, R.F., Christin, P-A., and Edwards, E.J. (2011). "The C4 plant lineages of planet Earth," *Journal of Experimental Botany*, 62, 3155–3169.

Salguero-Gómez, R. (2018). "Implications of clonality for ageing research," *Evolutionary Ecology*, 32, 9–28.

Salguero-Gómez, R. and Casper, B.B. (2010). "Keeping plant shrinkage in the demographic loop," *Journal of Ecology*, 98, 312–323.

Salguero-Gómez, R, and Gamelon, M. (eds.) (2021). *Demographic methods across the Tree of Life*. Oxford: Oxford University Press.

Salguero-Gómez, R., et al. (2015). "The compadre plant matrix database: An open online repository for plant demography," *Journal of Ecology*, 103, 202–218.

Salguero-Gómez, R., et al. (2016). "Fast–slow continuum and reproductive strategies structure plant life-history variation worldwide," Proceedings of the National Academy of Sciences of the United States of America, 113, 230–235.

Salguero-Gómez, R., Shefferson, R.P., and Hutchings, M.J. (2013). "Plants do not count . . . or do they? New perspectives on the universality of senescence," *Journal of Ecology*, 101, 545–554.

Salguero-Gómez, R., et al. (2018). "Delivering the promises of trait-based approaches to the needs of demographic approaches, and vice versa," *Functional Ecology*, 32, 1424–1435.

Salisbury, E.J. (1942). *The reproductive capacity of plants: Studies in quantitative biology*. London: G. Bell and Sons, Ltd.

Sallon, S., et al. (2020). "Origins and insights into the historic Judean date palm based on genetic analysis of germinated ancient seeds and morphometric studies," *Science Advances*, 6, eaax0384.

Sandel, B., Corbin, J., and Krupa, M. (2011). "Using plant functional traits to guide restoration: A case study in California coastal grassland," *Ecosphere*, 2, 1–16.

Santala, K., et al. (2022). "Finding the perfect mix: An applied model that integrates multiple ecosystem functions when designing restoration programs," *Ecological Engineering*, 180, 106646.

Saverimuttu, T. and Westoby, M. (1996). "Components of variation in seedling potential relative growth rate: Phylogenetically independent contrasts," *Oecologia*, 105, 281–285.

Schemske, D.W., et al. (1978). "Flowering ecology of some spring woodland herbs," *Ecology*, 59, 351–366.

Schenk, H.J. and Jackson, R.B. (2002). "Rooting depths, lateral root spreads and below-ground/above-ground allometries of plants in water-limited ecosystems," *Journal of Ecology*, 90, 480–494.

Schimper, A.F.W. (1903). *Plant-geography upon a physiological basis*. Oxford: Clarendon Press.

Schmalzel, R.J., Reichenbacher, F.W., and Rutman, S. (1995). "Demographic study of the rare *Coryphantha robbinsorum* (Cactaceae) in southeastern Arizona," *Madrono*, 42, 332–348.

Scholes, R., et al. (1997). "Plant functional types in African savannas and grasslands," in Smith, T.M., Shugart, H.H., and Woodward, F.I. (eds.) *Plant functional types: Their relevance to ecosystem properties and global change*. Cambridge: Cambridge University Press, pp. 255–268.

Schreiber, S.G., et al. (2013). "Frost hardiness vs. growth performance in trembling aspen: An experimental test of assisted migration," *Journal of Applied Ecology*, 50, 939–949.

Schrodt, F., et al. (2015). "BHPMF—a hierarchical Bayesian approach to gap-filling and trait prediction for macroecology and functional biogeography," *Global Ecology and Biogeography*, 24, 1510–1521.

Schulze, E-D., Beck, E., Müller-Hohenstein, K. (eds.) (2005). *Plant ecology*, tr. G. Lawlor. Heidelberg: Springer.

Schupp, E.W., Jordano, P., and Gómez, J.M. (2010). "Seed dispersal effectiveness revisited: A conceptual review," *New Phytologist*, 188, 333–353.

Schwilk, D.W. and Ackerly, D.D. (2001). "Flammability and serotiny as strategies: Correlated evolution in pines," *Oikos*, 94, 326–336.

Schwinning, S., et al. (2022). "What common-garden experiments tell us about climate responses in plants," *Journal of Ecology*, 110, 986–996.

Scogings, P.F. and Sankaran, M. (2019). *Savanna woody plants and large herbivores*. Oxford: John Wiley & Sons.

Scott, A.C., et al. (2014). *Fire on earth: An introduction*. John Wiley & Sons.

Seastedt, T.R., Hobbs, R.J., and Suding, K.N. (2008). "Management of novel ecosystems: Are novel approaches required?" *Frontiers in Ecology and the Environment*, 6, 547–553.

Segrestin, J., Navas, M-L., and Garnier, E. (2020). "Reproductive phenology as a dimension of the phenotypic space in 139 plant species from the Mediterranean," *New Phytologist*, 225, 740–753.

Selosse, M.A. and Le Tacon, F. (1998). "The land flora: A phototroph-fungus partnership?" *Trends in Ecology & Evolution*, 13, 15–20.

SERI. (2004). *The SER international primer on ecological restoration*. Tucson: Society for Ecological Restoration International.

Seuss, D. (1971). *The Lorax*. New York: Random House Children's Books.

Shackelford, N., et al. (2021). "Drivers of seedling establishment success in dryland restoration efforts," *Nature Ecology & Evolution*, 5, 1283–1290.

Shea, K., Rees, M., and Wood, S.N. (1994). "Trade-offs, elasticities and the comparative method," *Journal of Ecology*, 82, 951–957.

Shefferson, R.P., Jones, O., and Salguero-Gómez, R. (2017). "Introduction: Wilting leaves and rotting branches: Reconciling evolutionary perspectives on senescence," in Shefferson, R., Jones, O., and Salguero-Gómez, R. (eds.) *The evolution of senescence in the tree of life*. Cambridge: Cambridge University Press, pp. 1–20.

Shefferson, R.P., Warren, II, R.J., and Pulliam, H.R. (2014). "Life-history costs make perfect sprouting maladaptive in two herbaceous perennials," *Journal of Ecology*, 102, 1318–1328.

Shen-Miller, J., et al. (1995). "Exceptional seed longevity and robust growth: Ancient Sacred Lotus from China," *American Journal of Botany*, 82, 1367–1380.

Sheth, S.N. and Angert, A.L. (2018). "Demographic compensation does not rescue populations at a trailing range edge," Proceedings of the National Academy of Sciences of the United States of America, 115, 2413–2418.

Shi, G., et al. (2021). "Mesozoic cupules and the origin of the angiosperm second integument," *Nature*, 594, 223–226.

Shigo, A.L. (1986). *A new tree biology: Facts, photos, and philosophies on trees and their problems and proper care.* Durham, NH: Shigo and Trees, Associates.

Shipley, B. (2006). "Net assimilation rate, specific leaf area and leaf mass ratio: Which is most closely correlated with relative growth rate? A meta-analysis," *Functional Ecology*, 20, 565–574.

Shipley, B. (2010). *From plant traits to vegetation structure: Chance and selection in the assembly of ecological communities.* Cambridge: Cambridge University Press.

Shipley, B. and Peters, R.H. (1990). "A test of the Tilman model of plant strategies: Relative growth rate and biomass partitioning," *The American Naturalist*, 136, 139–153.

Shipley, B., et al. (2006). "Fundamental trade-offs generating the worldwide leaf economic spectrum," *Ecology*, 87, 535–541.

Shipley, B., et al. (2016). "Reinforcing loose foundation stones in trait-based plant ecology," *Oecologia*, 180, 923–931.

Shipley, B., et al. (1989). "Regeneration and establishment strategies of emergent macrophytes," *Journal of Ecology*, 77, 1093–1110.

Shoemaker, L.G. and Melbourne, B.A. (2016). "Linking metacommunity paradigms to spatial coexistence mechanisms," *Ecology*, 97, 2436–2446.

Shriver, R.K., et al. (2021). "Quantifying the demographic vulnerabilities of dry woodlands to climate and competition using rangewide monitoring data," *Ecology*, 102, e03425.

Shugart, H.H. (1997). "Plant and ecosystem functional types," in Smith, T.M., Shugart, H.H., and Woodward, F.I. (eds.) *Plant functional types: Their relevance to ecosystem properties and global change.* Cambridge: Cambridge University Press, pp. 20–43.

Shugart, H.H. (1984). *A theory of forest dynamics: The ecological implications of forest succession models.* New York: Springer-Verlag.

Siefert, A. and Laughlin, D.C. (2023). "Estimating the net effect of functional traits on fitness across species and environments," *Methods in Ecology and Evolution*, 14, 1035–1048. https://doi.org/10.1111/2041-210X.14079.

Siefert, A., et al. (2015). "A global meta-analysis of the relative extent of intraspecific trait variation in plant communities," *Ecology Letters*, 18, 1406–1419.

Silva Matos, D.M., Freckleton, R.P., and Watkinson, A.R. (1999). "The role of density dependence in the population dynamics of a tropical palm," *Ecology*, 80, 2635–2650.

Silvertown, J. and Charlesworth, D. (2001). *Introduction to plant population biology.* 4th edn. Oxford: Blackwell Science.

Silvertown, J. and Dodd, M. (1997). "Comparing plants and connecting traits," in Silvertown, J., Franco, M., and Harper, J. (eds.) *Plant life histories: Ecology, phylogeny, and evolution.* Cambridge: Cambridge University Press, pp. 3–16

Silvertown, J. and Franco, M. (1993). "Plant demography and habitat: A comparative approach," *Plant Species Biology*, 8, 67–73.

Silvertown, J., Franco, M., and McConway, K. (1992). "A demographic interpretation of Grime's triangle," *Functional Ecology*, 6, 130–136.

Silvertown, J., Franco, M., and Menges, E. (1996). "Interpretation of elasticity matrices as an aid to the management of plant populations for conservation," *Conservation Biology*, 10, 591–597.

Silvertown, J., Franco, M., and Perez-Ishiwara, R. (2001). "Evolution of senescence in iteroparous perennial plants," *Evolutionary Ecology Research*, 3, 393–412.

Silvertown, J., et al. (1993). "Comparative plant demography—relative importance of life-cycle components to the finite rate of increase in woody and herbaceous perennials," *Journal of Ecology*, 81, 465–476.

Simon, M.F. and Pennington, T. (2012). "Evidence for adaptation to fire regimes in the tropical savannas of the Brazilian Cerrado," *International Journal of Plant Sciences*, 173, 711–723.

Simpson, A.H., Richardson, S.J., and Laughlin, D.C. (2016). "Soil–climate interactions explain variation in foliar, stem, root and reproductive traits across temperate forests," *Global Ecology and Biogeography*, 25, 964–978.

Sitch, S., et al. (2008). "Evaluation of the terrestrial carbon cycle, future plant geography and climate-carbon cycle feedbacks using five Dynamic Global Vegetation Models (DGVMs)," *Global Change Biology*, 14, 2015–2039.

Smart, C.M. (1994). "Gene expression during leaf senescence," *New Phytologist*, 126, 419–448.

Smith, C.C. and Fretwell, S.D. (1974). "The optimal balance between size and number of offspring," *The American Naturalist*, 108, 499–506.

Smith, T. and Huston, M. (1989). "A theory of the spatial and temporal dynamics of plant communities," *Vegetatio*, 83, 49–69.

Smith, T.M., Shugart, H.H., and Woodward, F.I. (eds.) (1997). *Plant functional types: Their relevance to ecosystem properties and global change.* Cambridge: Cambridge University Press.

Snyder, G. (2010). *The practice of the wild: Essays.* Berkeley: Counterpoint Press.

Snyder, R.E. and Ellner, S.P. (2018). "Pluck or luck: Does trait variation or chance drive variation in lifetime reproductive success?" *The American Naturalist*, 191, E90–E107.

Song, X., et al. (2021). "When do Janzen–Connell effects matter? A phylogenetic meta-analysis of conspecific negative distance and density dependence experiments," *Ecology Letters*, 24, 608–620.

Southwood, T.R.E. (1977). "Habitat, the templet for ecological strategies?" *Journal of Animal Ecology*, 46, 337–365.

Sperandii, M.G., et al. (2022). "LOTVS: A global collection of permanent vegetation plots," *Journal of Vegetation Science*, 33, e13115.

Stamp, N. (2003). "Out of the quagmire of plant defense hypotheses," *The Quarterly review of biology*, 78, 23–55.

Stearns, S.C. (1992). *The evolution of life histories.* Oxford: Oxford University Press.

Stears, A.E., et al. (2022a). "plantTracker: An R package to translate maps of plant occurrence into demographic data," *Methods in Ecology and Evolution*, 13, 2129–2137.

Stears, A.E., et al. (2022b). "Water availability dictates how plant traits predict demographic rates," *Ecology*, 103, e3799.

Steffen, W.L. (1996). "A periodic table for ecology? A chemist's view of plant functional types," *Journal of Vegetation Science*, 7, 425–430.

Steiger, T.L. (1930). "Structure of prairie vegetation," *Ecology*, 11, 170–217.

Steiner, U.K. and Tuljapurkar, S. (2012). "Neutral theory for life histories and individual variability in fitness components," *Proceedings of the National Academy of Sciences of the United States of America*, 109, 4684–4689.

Steshenko, A.P. (1976). "On relation between live and dead parts of plants of high mountainous areas of Pamir," *Ekologiia*, 6, 27–34.

Stewart, W.N. and Rothwell, G.W. (1993). *Paleobotany and the evolution of plants.* Cambridge: Cambridge University Press.

Stone, E.L. and Kalisz, P.J. (1991). "On the maximum extent of tree roots," *Forest Ecology and Management*, 46, 59–102.

Strahan, R.T., et al. (2015). "Long-term protection from heavy livestock grazing affects ponderosa pine understory composition and functional traits," *Rangeland Ecology & Management*, 68, 257–265.

Struckman, S., et al. (2019). "The demographic effects of functional traits: An integral projection model approach reveals population-level consequences of reproduction-defence trade-offs," *Ecology Letters*, 22, 1396–1406.

Sumner, E.E., et al. (2022). "Survival and growth of a high-mountain daisy transplanted outside its local range, and implications for climate-induced distribution shifts," *AoB PLANTS*, 14, plac014.

Sussman, R. (2014). *The oldest living things in the world.* Chicago: University of Chicago Press.

Sutherland, S. and Vickery, R. (1988). "Trade-offs between sexual and asexual reproduction in the genus *Mimulus*," *Oecologia*, 76, 330–335.

Swenson, N.G. (2019). *Phylogenetic ecology: A history, critique and remodeling.* Chicago: University of Chicago Press.

Swetnam, T.W., Allen, C.D., and Betancourt, J.L. (1999). "Applied historical ecology: Using the past to manage for the future," *Ecological Applications*, 9, 1189–1206.

Sydes, C. and Grime, J.P. (1981). "Effects of tree leaf litter on herbaceous vegetation in deciduous woodland: I. Field investigations," *The Journal of Ecology*, 69, 237–248.

Taiz, L. and Zeiger, E. (2006). *Plant physiology.* 4th edn. Sunderland, MA: Sinauer Associates, Inc.

Takada, T., Kawai, Y., and Salguero-Gómez, R. (2018). "A cautionary note on elasticity analyses in a ternary plot using randomly generated population matrices," *Population Ecology*, 60, 37–47.

Tamme, R., et al. (2014). "Predicting species' maximum dispersal distances from simple plant traits," *Ecology*, 95, 505–513.

Taylor, D.R., Aarssen, L.W., and Loehle, C. (1990). "On the relationship between r/K selection and environmental carrying capacity: A new habitat templet for plant life history strategies," *Oikos*, 58, 239–250.

Tedersoo, L., Bahram, M., and Zobel, M. (2020). "How mycorrhizal associations drive plant population and community biology," *Science*, 367, eaba1223.

Terborgh, J., et al. (2014). "How many seeds does it take to make a sapling?" *Ecology*, 95, 991–999.

Theobald, E.J., Breckheimer, I., and HilleRisLambers, J. (2017). "Climate drives phenological reassembly of a mountain wildflower meadow community," *Ecology*, 98, 2799–2812.

Theophrastus. (287 BCE/1911). *Historia plantarum (enquiry into plants and minor works on odours and weather signs)*, tr. A. Hort. London: William Heinemann/G.P. Putnam's Sons.

Thomas, H. (2013). "Senescence, ageing and death of the whole plant," *New Phytologist*, 197, 696–711.

Thomas, S.C. (2011). "Age-related changes in tree growth and functional biology: The role of reproduction," in Meinzer, F.C., Lachenbruch, B., and Dawson, T.E. (eds.) *Size-and age-related changes in tree structure and function.* Dordrecht: Springer, pp. 33–64.

Thompson, K., Band, S.R., and Hodgson, J.G. (1993). "Seed size and shape predict persistence in soil," *Functional Ecology*, 7, 236–241.

Thomson, F.J., et al. (2011). "Seed dispersal distance is more strongly correlated with plant height than with seed mass," *Journal of Ecology*, 99, 1299–1307.

Thoreau, H.D. (1913). *Excursions*. New York: Thomas Y. Crowell Company.

Thuiller, W., et al. (2014). "Does probability of occurrence relate to population dynamics?" *Ecography*, 37, 1155–1166.

Thurman, L.L., et al. (2020). "Persist in place or shift in space? Evaluating the adaptive capacity of species to climate change," *Frontiers in Ecology and the Environment*, 18, 520–528.

Tilini, K.L., Meyer, S.E., and Allen, P.S. (2017). "Seed bank dynamics of blowout penstemon in relation to local patterns of sand movement on the Ferris Dunes, south-central Wyoming," *Botany*, 95, 819–828.

Tilman, D. (1994). "Competition and biodiversity in spatially structured habitats," *Ecology*, 75, 2–16.

Tilman, D. (1988). *Plant strategies and the dynamics and structure of plant communities*. Princeton: Princeton University Press.

Tilman, D. (1982). *Resource competition and community structure*. Princeton: Princeton University Press.

Tilman, D. (1985). "The resource-ratio hypothesis of plant succession," *The American Naturalist*, 125, 827–852.

Tíscar, P.A., Lucas-Borja, M.E., and Candel-Pérez, D. (2018). "Lack of local adaptation to the establishment conditions limits assisted migration to adapt drought-prone *Pinus nigra* populations to climate change," *Forest Ecology and Management*, 409, 719–728.

Tjoelker, M.G., et al. (2005). "Linking leaf and root trait syndromes among 39 grassland and savannah species," *New Phytologist*, 167, 493–508.

Tolstoy, L. (2014). *Anna Karenina*, tr. R. Bartlett. Oxford: Oxford University Press.

Tomlinson, K.W., et al. (2012). "Biomass partitioning and root morphology of savanna trees across a water gradient," *Journal of Ecology*, 100, 1113–1121.

Towers, I.R., et al. (2020). "Requirements for the spatial storage effect are weakly evident for common species in natural annual plant assemblages," *Ecology*, 101, e03185.

Tredennick, A.T., Hooten, M.B., and Adler, P.B. (2017). "Do we need demographic data to forecast plant population dynamics?" *Methods in Ecology and Evolution*, 8, 541–551.

Tumber-Dávila, S.J., et al. (2022). "Plant sizes and shapes above and belowground and their interactions with climate," *New Phytologist*, 235, 1032–1056.

Tuomi, J., Augner, M., and Leimar, O. (1999). "Fitness interactions among plants: Optimal defence and evolutionary game theory," in Vuorisalo, T.O. and Mutikainen, P.K. (eds.) *Life history evolution in plants*. Dordrecht: Kluwer Academic Publishers, pp. 63–83.

Turkington, R. (2010). "Obituary: John L. Harper FRS, CBE 1925–2009," *The Bulletin of the Ecological Society of America*, 91, 9–13.

Tyree, M.T. and Zimmerman, M.H. (2013). *Xylem structure and the ascent of sap*. 2nd edn. Berlin: Springer-Verlag.

Uhl, C. and Kauffman, J.B. (1990). "Deforestation, fire susceptibility, and potential tree responses to fire in the eastern Amazon," *Ecology*, 71, 437–449.

Uyeda, J.C., Zenil-Ferguson, R., and Pennell, M.W. (2018). "Rethinking phylogenetic comparative methods," *Systematic Biology*, 67, 1091–1109.

Valladares, F. and Niinemets, Ü. (2008). "Shade tolerance, a key plant feature of complex nature and consequences," *Annual Review of Ecology, Evolution, and Systematics*, 39, 237–257.

Vallejo-Marín, M., Dorken, M.E., and Barrett, S.C. (2010). "The ecological and evolutionary consequences of clonality for plant mating," *Annual Review of Ecology, Evolution, and Systematics*, 41, 193–213.

Valverde-Barrantes, O.J., et al. (2016). "Phylogenetically structured traits in root systems influence arbuscular mycorrhizal colonization in woody angiosperms," *Plant and Soil*, 404, 1–12.

van der Maarel, E. and Sykes, M.T. (1993). "Small-scale plant species turnover in a limestone grassland: The carousel model and some comments on the niche concept," *Journal of Vegetation Science*, 4, 179–188.

Van der Pijl, L. (1982). *Principles of dispersal in higher plants*. 3rd edn. Berlin: Springer-Verlag.

van der Valk, A.G. (1981). "Succession in wetlands: A Gleasonian approach," *Ecology*, 62, 688–696.

van der Valk, A.G. (1985). "Vegetation dynamics of prairie glacial marshes," in White, J. (ed.) *The population structure of vegetation*. Dordrecht: Dr W. Junk Publishers, pp. 293–312.

van der Valk, A.G. and Davis, C. (1978). "The role of seed banks in the vegetation dynamics of prairie glacial marshes," *Ecology*, 59, 322–335.

Van Groenendael, J.M., et al. (1996). "Comparative ecology of clonal plants," *Philosophical Transactions of the Royal Society of London. Series B: Biological Sciences*, 351, 1331–1339.

van Kleunen, M., Bossdorf, O., and Dawson, W. (2018). "The ecology and evolution of alien plants," *Annual Review of Ecology, Evolution, and Systematics*, 49, 25–47.

van Kleunen, M., et al. (2015). "Global exchange and accumulation of non-native plants," *Nature*, 525, 100–103.

van Mantgem, P.J., et al. (2009). "Widespread increase of tree mortality rates in the western United States," *Science*, 323, 521–524.

Van Soest, P.J. (1982). *Nutritional ecology of the ruminant*. Corvallis, OR: O and B Books.

Van Valen, L. (1971). "Group selection and the evolution of dispersal," *Evolution*, 25, 591–598.

Van Valen, L. (1975). "Life, death, and energy of a tree," *Biotropica*, 7, 259–269.

VanderWeide, B.L. and Hartnett, D.C. (2015). "Belowground bud bank response to grazing under severe, short-term drought," *Oecologia*, 178, 795–806.

Vasek, FC. (1980). "Creosote bush: Long-lived clones in the Mojave Desert," *American Journal of Botany*, 67, 246–255.

Vaupel, J.W., et al. (2004). "The case for negative senescence," *Theoretical Population Biology*, 65, 339–351.

Veneklaas, E.J. and Poorter, L. (1998). "Growth and carbon partitioning of tropical tree seedlings in contrasting light environments," in Lambers, H., Poorter, H., and Van Vuuren, M.M.I. (eds.) *Inherent variation in plant growth: Physiological mechanisms and ecological consequences*. Leiden: Backhuys Publishers, pp. 337–362.

Vesk, P.A. and Westoby, M. (2001). "Predicting plant species' responses to grazing," *Journal of Applied Ecology*, 38, 897–909.

Vesk, P.A., Leishman, M.R., and Westoby, M. (2004). "Simple traits do not predict grazing response in Australian dry shrublands and woodlands," *Journal of Applied Ecology*, 41, 22–31.

Vicca, S., et al. (2012). "Urgent need for a common metric to make precipitation manipulation experiments comparable," *New Phytologist*, 195, 518–522.

Vico, G., et al. (2016). "Trade-offs between seed output and life span—a quantitative comparison of traits between annual and perennial congeneric species," *New Phytologist*, 209, 104–114.

Villellas, J., et al. (2015). "Demographic compensation among populations: What is it, how does it arise and what are its implications?" *Ecology Letters*, 18, 1139–1152.

Vincent, T.L. and Brown, J.S. (2005). *Evolutionary game theory, natural selection, and Darwinian dynamics.* Cambridge: Cambridge University Press.

Vincent, T.L., Cohen, Y., and Brown, J.S. (1993). "Evolution via strategy dynamics," *Theoretical Population Biology*, 44, 149–176.

Vines, R.G. (1968). "Heat transfer through bark, and the resistance of trees to fire," *Australian Journal of Botany*, 16, 499–514.

Violle, C., et al. (2017). "Functional rarity: The ecology of outliers," *Trends in Ecology & Evolution*, 32, 356–367.

Violle, C., et al. (2007). "Let the concept of trait be functional!" *Oikos*, 116, 882–892.

Visser, M.D., et al. (2016). "Functional traits as predictors of vital rates across the life cycle of tropical trees," *Functional Ecology*, 30, 168–180.

Vitousek, P.M. (2018). *Nutrient cycling and limitation.* Princeton: Princeton University Press.

Vitt, P., et al. (2010). "Assisted migration of plants: Changes in latitudes, changes in attitudes," *Biological Conservation*, 143, 18–27.

Vleminckx, J., et al. (2021). "Resolving whole-plant economics from leaf, stem and root traits of 1467 Amazonian tree species," *Oikos*, 130, 1193–1208.

Volaire, F. (2018). "A unified framework of plant adaptive strategies to drought: Crossing scales and disciplines," *Global Change Biology*, 24, 2929–2938.

Volaire, F. and Norton, M. (2006). "Summer dormancy in perennial temperate grasses," *Annals of Botany*, 98, 927–933.

von Humboldt, A. (1806). *Ideen zu einer Physiognomik der Gewächse.* Stuttgart: Cotta.

Vuorisalo, T.O. and Mutikainen, P.K. (1999). "Modularity and plant life histories," in Vuorisalo, T.O. and Mutikainen, P.K. (eds.) *Life history evolution in plants.* Dordrecht: Kluwer Academic Publishers, pp. 1–25.

Walker, L.R. and Del Moral, R. (2003). *Primary succession and ecosystem rehabilitation.* Cambridge: Cambridge University Press.

Walker, L.R. and Willig, M.R. (1999). "An introduction to terrestrial disturbances," in Walker, L.R. (ed.) *Ecosystems of disturbed ground.* Ecosystems of the World 16. Amsterdam: Elsevier, pp. 1–16.

Walker, T. and Syers, J. (1976). "The fate of phosphorus during pedogenesis," *Geoderma*, 15, 1–19.

Wandersee, J.H. and Schussler, E.E. (1999). "Preventing plant blindness," *The American Biology Teacher*, 61, 82–86.

Wang, C., et al. (2021). "A web-based software platform for restoration-oriented species selection based on plant functional traits," *Frontiers in Ecology and Evolution*, 9, 122.

Wang, C., et al. (2020). "Application of a trait-based species screening framework for vegetation restoration in a tropical coral island of China," *Functional Ecology*, 34, 1193–1204.

Wang, T., O'Neill, G.A., and Aitken, S.N. (2010). "Integrating environmental and genetic effects to predict responses of tree populations to climate," *Ecological Applications*, 20, 153–163.

Wang, Y., et al. (2019). "Experimental test of assisted migration for conservation of locally range-restricted plants in Alberta, Canada," *Global Ecology and Conservation*, 17, e00572.

Wardle, D.A., Walker, L.R., and Bardgett, R.D. (2004). "Ecosystem properties and forest decline in contrasting long-term chronosequences," *Science*, 305, 509–513.

Waring, R. and Schlesinger, W. (1985). *Forest ecosystems: Concepts and management.* Orlando: Academic Press.

Warming, E. (1909). *Oecology of plants: An introduction to the study of plant communities.* Oxford: Clarendon Press.

Warming, E. (1884). *Om Skudbygning, Overvintring og Foryngelse.* Copenhagen: Festskr. Naturh. Foren.

Warming, E. (1895). *Plantesamfund—Grundtræk af den økologiske Plantegeografi.* Copenhagen: P.G. Philipsens Forlag.

Watkinson, A.R. (1980). "Density-dependence in single-species populations of plants," *Journal of Theoretical Biology*, 83, 345–357.

Watkinson, A.R. and White, J. (1986). "Some life-history consequences of modular construction in plants," *Philosophical Transactions of the Royal Society of London. B, Biological Sciences*, 313, 31–51.

Waxman, D. and Gavrilets, S. (2005). "20 questions on adaptive dynamics," *Journal of Evolutionary Biology*, 18, 1139–1154.

Weaver, J.E. (1919). *The ecological relations of roots.* Washington, DC: Carnegie Institution of Washington.

Weaver, J.E. and Clements, F.E. (1929). *Plant ecology.* New York: McGraw-Hill.

Weber, M.G. and Agrawal, A.A. (2012). "Phylogeny, ecology, and the coupling of comparative and experimental approaches," *Trends in Ecology & Evolution*, 27, 394–403.

Weemstra, M., et al. (2016). "Towards a multidimensional root trait framework: A tree root review," *New Phytologist*, 211, 1159–1169.

Weigelt, A., et al. (2021). "An integrated framework of plant form and function: The belowground perspective," *New Phytologist*, 232, 42–59.

Weiher, E., et al. (1999). "Challenging Theophrastus: A common core list of plant traits for functional ecology," *Journal of Vegetation Science*, 10, 609–620.

Weiher, E., et al. (2004). "Multivariate control of plant species richness and community biomass in blackland prairie," *Oikos*, 106, 151–157.

Weiner, J. (1990). "Asymmetric competition in plant populations," *Trends in Ecology & Evolution*, 5, 360–364.

Wenk, E.H. and Falster, D.S. (2015). "Quantifying and understanding reproductive allocation schedules in plants," *Ecology and evolution*, 5, 5521–5538.

Wenk, E.H., et al. (2018). "Investment in reproduction for 14 iteroparous perennials is large and associated with other life-history and functional traits," *Journal of Ecology*, 106, 1338–1348.

Wepprich, T., et al. (2022.) "Population models and forecasts using long-term monitoring data for a recently delisted riparian forb, *Oenothera coloradensis* (Colorado butterfly plant)," Prepared for U.S. Fish and Wildlife Service and F. E. Warren Air Force Base. Laramie: Wyoming Natural Diversity Database.

Werden, L.K., et al. (2018). "Using soil amendments and plant functional traits to select native tropical dry forest species for the restoration of degraded Vertisols," *Journal of Applied Ecology*, 55, 1019–1028.

Werner, P.A. and Caswell, H. (1977). "Population growth rates and age versus stage-distribution models for teasel (*Dipsacus sylvestris* Huds.)," *Ecology*, 58, 1103–1111.

West, G.B., Brown, J.H., and Enquist, B.J. (1997). "A general model for the origin of allometric scaling laws in biology," *Science*, 276, 122–126.

Westerband, A.C. and Horvitz, C.C. (2017). "Photosynthetic rates influence the population dynamics of understory herbs in stochastic light environments," *Ecology*, 98, 370–381.

Westoby, M. (2022). "Field experiments on mechanisms influencing species boundary movement under climate change," *Plant and Soil*, 476, 527–534.

Westoby, M. (1998). "A leaf–height–seed (LHS) plant ecology strategy scheme," *Plant and Soil*, 199, 213–227.

Westoby, M. (2007). "Generalization in functional plant ecology: The species-sampling problem, plant ecology strategy schemes, and phylogeny," in Pugnaire, F.I. and Valladares, F. (eds.) *Functional plant ecology*. Boca Raton: CRC Press, pp. 685–704.

Westoby, M. and Leishman, M. (1997). "Categorizing plant species into functional types," in Smith, T.M., Shugart, H.H., and Woodward, F.I. (eds.) *Plant functional types: Their relevance to ecosystem properties and global change*. Cambridge: Cambridge University Press, pp. 104–121.

Westoby, M. and Wright, I.J. (2006). "Land-plant ecology on the basis of functional traits," *Trends in Ecology & Evolution*, 21, 261–268.

Westoby, M., Leishman, M.R., and Lord, J. (1997). "Comparative ecology of seed size and dispersal," in Silvertown, J., Franco, M., and Harper, J.L. (eds.) *Plant life histories: Ecology, phylogeny, and evolution*. Cambridge: Cambridge University Press, pp. 143–162.

Westoby, M., Leishman, M.R., and Lord, J.R. (1995b). "Further remarks on phylogenetic correction," *Journal of Ecology*, 83, 727–729.

Westoby, M., Leishman, M.R., and Lord, J.M. (1995a). "On misinterpreting the 'phylogenetic correction'," *Journal of Ecology*, 83, 531–534.

Westoby, M., Schrader, J., and Falster, D. (2022). "Trait ecology of startup plants," *New Phytologist*, 235, 842–847.

Westoby, M., et al. (2017). "How species boundaries are determined: A response to Alexander et al.," *Trends in Ecology & Evolution*, 32, 7–8.

Westoby, M., et al. (2002). "Plant ecological strategies: Some leading dimensions of variation between species," *Annual Review of Ecology and Systematics*, 33, 125–159.

Whelan, R.J. (1995). *The ecology of fire*. Cambridge: Cambridge University Press.

White, J. (1985a). "The census of plants in vegetation," in White, J. (ed.) *The population structure of vegetation*. Dordrecht: Dr W. Junk Publishers, pp. 34–88.

White, J. (1979). "The plant as a metapopulation," *Annual Review of Ecology and Systematics*, 10, 109–145.

White, J. (1985b). "The population structure of vegetation," in White, J. (ed.) *The population structure of vegetation*. Dordrecht: Dr W. Junk Publishers, pp. 1–14.

White, P.S. and Jentsch, A. (2001). "The search for generality in studies of disturbance and ecosystem dynamics," in Esser, K., et al. (eds.) *Progress in botany: Genetics physiology systematics ecology*. Berlin: Springer, pp. 399–450.

White, P.S. and Walker, J.L. (1997). "Approximating nature's variation: Selecting and using reference information in restoration ecology," *Restoration Ecology*, 5, 338–349.

Whitham, T.G. and Slobodchikoff, C. (1981). "Evolution by individuals, plant-herbivore interactions, and mosaics of genetic variability: The adaptive significance of somatic mutations in plants," *Oecologia*, 49, 287–292.

Whitham, T.G., et al. (2020). "Using the Southwest Experimental Garden Array to enhance riparian restoration in response to global environmental change: Identifying and deploying genotypes and populations for current and future environments," in Carothers, S.W., et al. (eds.) *Riparian research and management: Past, present, future*. Volume 2. Gen. Tech. Rep. RMRS-GTR-411. Fort Collins, CO: US Department of Agriculture, pp. 63–79.

Whitmore, T.C. (1987). "'Rain forest' or 'rainforest'?" *Journal of Tropical Ecology*, 3, 24-24.

Whittaker, R. and Goodman, D. (1979). "Classifying species according to their demographic strategy. I. Population fluctuations and environmental heterogeneity," *The American Naturalist*, 113, 185–200.

Wilkinson, D.M. and Sherratt, T.N. (2016). "Why is the world green? The interactions of top–down and bottom–up processes in terrestrial vegetation ecology," *Plant Ecology & Diversity*, 9, 127–140.

Williams, J.L., et al. (2015). "Life history evolution under climate change and its influence on the population dynamics of a long-lived plant," *Journal of Ecology*, 103, 798–808.

Williams, M.I. and Dumroese, R.K. (2013). "Preparing for climate change: Forestry and assisted migration," *Journal of Forestry*, 111, 287–297.

Williams, T. (1945). *The glass menagerie.* New York: Dramatists Play Service, Inc.

Williamson, G.B. and Black, E.M. (1981). "High temperature of forest fires under pines as a selective advantage over oaks," *Nature*, 293, 643–644.

Williamson, M.H. and Fitter, A. (1996). "The characters of successful invaders," *Biological Conservation*, 78, 163–170.

Willis, K.J. and McElwain, J.C. (2014). *The evolution of plants.* 2nd edn. Oxford: Oxford University Press.

Willis, S.G., et al. (2015). "Integrating climate change vulnerability assessments from species distribution models and trait-based approaches," *Biological Conservation*, 190, 167–178.

Wilson, A.M. and Thompson, K. (1989). "A comparative study of reproductive allocation in 40 British grasses," *Functional Ecology*, 3, 297–302.

Wilson, J.B., Agnew, A.D., and Roxburgh, S.H. (2019). *The nature of plant communities.* Cambridge: Cambridge University Press.

Willson, M.F. (1983). *Plant reproductive ecology.* New York: John Wiley & Sons.

Winner, W.E. and Mooney, H.A. (1980). "Responses of Hawaiian plants to volcanic sulfur dioxide: Stomatal behavior and foliar injury," *Science*, 210, 789–791.

Wisheu, I.C. and Keddy, P.A. (1992). "Competition and centrifugal organization of plant communities: Theory and tests," *Journal of Vegetation Science*, 3, 147–156.

Wolkovich, E.M. and Cleland, E.E. (2014). "Phenological niches and the future of invaded ecosystems with climate change," *AoB PLANTS*, 6, plu013.

Wolkovich, E.M. and Ettinger, A.K. (2014). "Back to the future for plant phenology research," *New Phytologist*, 203, 1021–1024.

Woodward, F.I. (1987). *Climate and plant distribution.* Cambridge: Cambridge University Press.

Woodward, F.I. and Kelly, C.K. (1997). "Plant functional types: Towards a definition by environmental constraints," in Smith, T.M., Shugart, H.H., and Woodward, F.I. (eds.) *Plant functional types: Their relevance to ecosystem properties and global change.* Cambridge: Cambridge University Press, pp. 47–65.

Woodward, F.I., Lomas, M.R., and Kelly, C.K. (2004). "Global climate and the distribution of plant biomes," *Philosophical Transactions of the Royal Society of London. Series B: Biological Sciences*, 359, 1465–1476.

Worthy, S.J., et al. (2020). "Alternative designs and tropical tree seedling growth performance landscapes," *Ecology*, 101, e03007.

Wright, I.J., Reich, P.B., and Westoby, M. (2001). "Strategy shifts in leaf physiology, structure and nutrient content between species of high- and low-rainfall and high- and low-nutrient habitats," *Functional Ecology*, 15, 423–434.

Wright, I.J., Reich, P.B., and Westoby, M. (2003). "Least-cost input mixtures of water and nitrogen for photosynthesis," *The American Naturalist*, 161, 98–111.

Wright, I.J., et al. (2017). "Global climatic drivers of leaf size," *Science*, 357, 917–921.

Wright, I.J., et al. (2004). "The worldwide leaf economics spectrum," *Nature*, 428, 821–827.

Wright, S. (1932). "The roles of mutation, inbreeding, crossbreeding, and selection in evolution," Proceedings of the XI International Congress of Genetics, 1, 356–366.

Wright, S.J., et al. (2010). "Functional traits and the growth–mortality trade-off in tropical trees," *Ecology*, 91, 3664–3674.

Xiang, J., et al. (2018). "Overexpressing heat-shock protein OsHSP50.2 improves drought tolerance in rice," *Plant Cell Reports*, 37, 1585–1595.

Yang, J., Cao, M., and Swenson, N.G. (2018). "Why functional traits do not predict tree demographic rates," *Trends in Ecology & Evolution*, 33, 326–336.

Yannelli, F.A., et al. (2018). "Seed density is more effective than multi-trait limiting similarity in controlling grassland resistance against plant invasions in mesocosms," *Applied Vegetation Science*, 21, 411–418.

Yashina, S., et al. (2012). "Regeneration of whole fertile plants from 30,000-y-old fruit tissue buried in Siberian permafrost," Proceedings of the National Academy of Sciences of the United States of America, 109, 4008–4013.

Yirdaw, E. and Leinonen, K. (2002). "Seed germination responses of four Afromontane tree species to red/far-red ratio and temperature," *Forest Ecology and Management*, 168, 53–61.

Yoda, K. (1963). "Intraspecific competition among higher plants. IX. Self-thinning in over-crowded pure stands under cultivation and natural conditions," *Journal of Biology Osaka City University*, 14, 107–109.

Zadworny, M., Comas, L.H., and Eissenstat, D.M. (2018). "Linking fine root morphology, hydraulic functioning and shade tolerance of trees," *Annals of Botany*, 122, 239–250.

Zanne, A.E., et al. (2014). "Three keys to the radiation of angiosperms into freezing environments," *Nature*, 506, 89–92.

Zea-Cabrera, E., et al. (2006). "Tragedy of the commons in plant water use," *Water Resources Research*, 42, 1–12.

Zellweger, F., et al. (2020). "Forest microclimate dynamics drive plant responses to warming," *Science*, 368, 772–775.

Zhu, S-D., et al. (2018). "Leaf turgor loss point is correlated with drought tolerance and leaf carbon economics traits," *Tree Physiology*, 38, 658–663.

Zipkin, E.F. and Saunders, S.P. (2018). "Synthesizing multiple data types for biological conservation using integrated population models," *Biological Conservation*, 217, 240–250.

Zirbel, C.R., et al. (2017). "Plant functional traits and environmental conditions shape community assembly and ecosystem functioning during restoration," *Journal of Applied Ecology*, 54, 1070–1079.

Zirbel, C.R. and Brudvig, L.A. (2020). "Trait–environment interactions affect plant establishment success during restoration," *Ecology*, 101, e02971.

Zucaratto, R., et al. (2021). "Felling the giants: Integral projection models indicate adult management to control an exotic invasive palm," *Plant Ecology*, 222, 93–105.

Zuidema, P.A., et al. (2009). "Do persistently fast-growing juveniles contribute disproportionately to population growth? A new analysis tool for matrix models and its application to rainforest trees," *The American Naturalist*, 174, 709–719.

Zuidema, P.A., et al. (2010). "Integral projection models for trees: A new parameterization method and a validation of model output," *Journal of Ecology*, 98, 345–355.

Zupo, T., et al. (2021). "Post-fire regeneration strategies in a frequently burned Cerrado community," *Journal of Vegetation Science*, 32, e12968.

Index

Functional traits are listed under the organ to which they belong. For example, specific leaf area (SLA) is listed under 'Leaf trait' and specific root length is listed under 'Root trait'. You can also find a list of named plant strategies under 'Plant strategy' and a list of life history and functional trade-offs under 'Trade-off'.